Cases on Models and Methods for STEAM Education

Judith Ann Bazler
Monmouth University, USA

Meta Lee Van Sickle
College of Charleston, USA

A volume in the Advances in Educational Technologies and Instructional Design (AETID) Book Series

IGI Global
DISSEMINATOR OF KNOWLEDGE

Published in the United States of America by
IGI Global
Information Science Reference (an imprint of IGI Global)
701 E. Chocolate Avenue
Hershey PA, USA 17033
Tel: 717-533-8845
Fax: 717-533-8661
E-mail: cust@igi-global.com
Web site: http://www.igi-global.com

Library of Congress Cataloging-in-Publication Data

Names: Bazler, Judith, editor.
Title: Cases on models and methods for STEAM education / Judith Ann Bazler and Meta Lee Van Sickle, editors.
Other titles: Cases on STEAM education in practice | Cases on models and methods for Science, Technology, Engineering, Arts and Mathematics education | Science, technology, engineering, arts and mathematics education
Description: Second Edition. | Hershey, Pennsylvania : Information Science Reference (an imprint of IGI Global), [2019] | Previous edition: 2017.
Identifiers: LCCN 2019005220| ISBN 9781522596318 (Hardcover) | ISBN 9781522596370 (Ebook) | ISBN 9781522596363 (Softcover)
Subjects: LCSH: Science--Study and teaching--Methodology. | Interdisciplinary approach in education. | Science and the arts.
Classification: LCC LB1585 .C33 2019 | DDC 372.35/044--dc23 LC record available at https://lccn.loc.gov/2019005220

This book is published in the IGI Global book series Advances in Educational Technologies and Instructional Design (AETID) (ISSN: 2326-8905; eISSN: 2326-8913)

British Cataloguing in Publication Data
A Cataloguing in Publication record for this book is available from the British Library.

Advances in Educational Technologies and Instructional Design (AETID) Book Series

ISSN:2326-8905
EISSN:2326-8913

Editor-in-Chief: Lawrence A. Tomei, Robert Morris University, USA

MISSION

Education has undergone, and continues to undergo, immense changes in the way it is enacted and distributed to both child and adult learners. In modern education, the traditional classroom learning experience has evolved to include technological resources and to provide online classroom opportunities to students of all ages regardless of their geographical locations. From distance education, Massive-Open-Online-Courses (MOOCs), and electronic tablets in the classroom, technology is now an integral part of learning and is also affecting the way educators communicate information to students.

The **Advances in Educational Technologies & Instructional Design (AETID) Book Series** explores new research and theories for facilitating learning and improving educational performance utilizing technological processes and resources. The series examines technologies that can be integrated into K-12 classrooms to improve skills and learning abilities in all subjects including STEM education and language learning. Additionally, it studies the emergence of fully online classrooms for young and adult learners alike, and the communication and accountability challenges that can arise. Trending topics that are covered include adaptive learning, game-based learning, virtual school environments, and social media effects. School administrators, educators, academicians, researchers, and students will find this series to be an excellent resource for the effective design and implementation of learning technologies in their classes.

COVERAGE

- E-Learning
- Digital Divide in Education
- Web 2.0 and Education
- Instructional Design Models
- K-12 Educational Technologies
- Collaboration Tools
- Online Media in Classrooms
- Bring-Your-Own-Device
- Adaptive Learning
- Curriculum Development

IGI Global is currently accepting manuscripts for publication within this series. To submit a proposal for a volume in this series, please contact our Acquisition Editors at Acquisitions@igi-global.com or visit: http://www.igi-global.com/publish/.

The Advances in Educational Technologies and Instructional Design (AETID) Book Series (ISSN 2326-8905) is published by IGI Global, 701 E. Chocolate Avenue, Hershey, PA 17033-1240, USA, www.igi-global.com. This series is composed of titles available for purchase individually; each title is edited to be contextually exclusive from any other title within the series. For pricing and ordering information please visit http://www.igi-global.com/book-series/advances-educational-technologies-instructional-design/73678. Postmaster: Send all address changes to above address.

Titles in this Series

For a list of additional titles in this series, please visit:
https://www.igi-global.com/book-series/advances-educational-technologies-instructional-design/73678

Handbook of Research on Diverse Teaching Strategies for the Technology-Rich Classroom
Lawrence A. Tomei (Robert Morris University, USA) and David D. Carbonara (Duquesne University, USA)
Information Science Reference • ©2020 • 450pp • H/C (ISBN: 9781799802389) • US $265.00

Emerging Technologies in Virtual Learning Environments
Kim Becnel (Appalachian State University, USA)
Information Science Reference • ©2019 • 348pp • H/C (ISBN: 9781522579878) • US $205.00

English Language Teaching in a Post-Method Paradigm
Paulette Joyce Feraria (The University of the West Indies, Jamaica)
Information Science Reference • ©2019 • 426pp • H/C (ISBN: 9781522592280) • US $175.00

Cases on Digital Learning and Teaching Transformations in Higher Education
Rebecca J. Blankenship (Florida Agricultural and Mechanical University, USA) and Charlotte Baker (Virginia Polytechnic Institute and State University, USA)
Information Science Reference • ©2019 • 266pp • H/C (ISBN: 9781522593317) • US $175.00

Handbook of Research on Innovative Digital Practices to Engage Learners
Prince Hycy Bull (Gardner-Webb University, USA) and Jared Keengwe (University of North Dakota, USA)
Information Science Reference • ©2019 • 468pp • H/C (ISBN: 9781522594383) • US $275.00

Blended Online Learning and Instructional Design for TPACK Emerging Research and Opportunities
Margaret L. Niess (Oregon State University, USA)
Information Science Reference • ©2019 • 213pp • H/C (ISBN: 9781522588795) • US $145.00

For an entire list of titles in this series, please visit:
https://www.igi-global.com/book-series/advances-educational-technologies-instructional-design/73678

IGI Global
DISSEMINATOR OF KNOWLEDGE

701 East Chocolate Avenue, Hershey, PA 17033, USA
Tel: 717-533-8845 x100 • Fax: 717-533-8661
E-Mail: cust@igi-global.com • www.igi-global.com

Table of Contents

Detailed Table of Contents

Out of a conversation between two long-time colleagues—each a science educator and practicing artist—emerged the question, "What does it mean to STEAMify a lesson, and why would a teacher actually choose to do such a thing, other than, say, for-grant-writing-purposes?" Their science selves really liked the idea of a STEAM system, acted upon by forces, both from the outside and from within, and with energy flowing and cycling, all the while transforming grey matter in ways that sustained the teaching/learning process. When it came to their art; however, their dialogue followed pathways grooved by long years of practice and hard work in their respective fields. One author is a seasoned vocalist, trained in the nuances of both individual and group vocal performance as well as the attendant dimensions of music, its composition and phraseology. The other is a painter, poet, and novelist, shaping words, color, and line to tell stories and communicate ideas. What was significant to each was that their artistic habits of mind had shaped their axiology, transforming their ways teaching.

Section 2
Art Practice

Focusing on the interdisciplinary connections between STEAM education and graphic novels, this chapter first establishes historical and educational contexts for the use of graphic novels in STEAM education. A literature review focusing on the use of graphic novels in a science course as well as how graphic novels have been used in secondary classrooms will be discussed. Literature that is reviewed includes books, articles, and edited volumes. The strategies for implementing graphic novels in STEAM education promote constructivist learning as students are asked to access their intellectual and cultural capital in order to ascertain meaning from given content. Additionally, multimodality and multiliteracies are highlighted throughout the chapter. Finally, this chapter ends with linking the use of graphic novels in STEAM education to the power of narrative inquiry in educational contexts.

Integrating high-quality children's tradebooks into elementary content areas has long been considered a best practice. When teachers choose to incorporate these texts into content area lessons, they are exposing students to art through the pictures and reaching an array of visual learners. There is a delicate balance between teaching the literacy strategies needed to read these texts and the actual content materials that students need to learn in the STEAM areas. This chapter explores how to incorporate texts appropriately into content area lessons so that students can focus on the content, as well as apply literacy strategies for comprehension.

This chapter connects the use of creative dramatics with engaging students previously disengaged with STEM classes. It elucidates a variety of reasons how creative dramatics allows STEM teachers and their students to succeed. Each of the assertions is backed by citations of research studies, classroom practice, and details of the theoretical underpinnings. Also included are detailed descriptions of three effective classroom methodologies using theater in STEM classes, along with specific examples of each that include student and teacher interactions.

This chapter explores the use of creative movement to extend meaning to inquiry-based science investigations. This process embraces the addition of A to STEM to realize the impact of STEAM. The chapter builds on the import of scientific and physical literacy, interdisciplinary learning, and the power of kinesthetic engagement. Students become active collaborative agents within a dynamic model using creative movement to bring meaning to the science of simple machines. The authors utilize working words into movement strategy to help students use their past experiences and motor memory to explore, interpret, and engage with as they seek understanding of simple machines. A Midwest urban elementary school provides the context for a unit plan culminating in a dance performance. The foundational ideas presented within this unit can be enacted within any classroom by creative movement (physical education or dance) specialists, science specialists, or classroom generalists. It follows with a presentation of science content on simple machines exploring the disciplinary core idea of force and motion.

This chapter describes the evolution of an arts-integrated approach to science curriculum inquiry which has been evolving since the 1990s—before the national science standards, the acronym STEM, much less STEAM, appeared across educational horizons. It reads as ethnography and has been performed in community, in association with the most caring of souls, with the goal of achieving a more inclusive/empowering, aesthetic science education, and a deep appreciation of the importance of the creative arts in the science learning process. It presents two research-based iterations of STEAM education in practice: 1) the creation of arts-integrated middle school ocean science curricula and 2) the development of a pedagogical tool called the "Know"tation as a way for teachers and students to make learning visible and integrate the languages of science throughout the process of inquiry. The cases described here apply many features of the STEAM model developed in Chapter 2 of this book.

Section 3
Theme-Based

The next generation science standards promote the teaching of engineering skills including the designing, testing, and building of models. Tower building can yield real world experience that not only provides the student with physics and mathematics through motion and stability but also through the explanation of the use of models and the engineering practice of design, redesign, and testing of these models. Tia Pliskow used the project of building a tower with her middle school students in order to provide a cooperative team long-term project. She focused first on the design, using background information on existing towers. She required each team to design their tower first using graph paper and scale. This process stressed the need for science, technology, engineering, art, and mathematics. The case included in this article expands her process by including a cost analysis attempting to promote real world engineering, links to more content, and final project photos. In addition, by building a shake platform, a test for the tower is added.

This is a case in which students build paper airplanes from templates provided by the instructor as well as those which they can design themselves. They extend their ideas on flight in by using the principles developed in the paper airplane race to power simulated rockets made out of balloons. They consider such variables as materials, mass, and design to see which combination of material design and mass are most effective in constructing an airplane or a rocket that flies the fastest and the furthest in a competition. Contestants are rated on team consideration of variables needing to be controlled in order to have a fair assessment of the designs. When the designs are agreed upon and constructed, a race is conducted. The ideas developed in the paper airplane competition are then used to design a rocket carrying a paper airplane capable of flying across the classroom in the fastest time with the most direct route. This is a simulation of the space shuttle flights. The parameters of the races are developed by the participants.

The purpose of the chapter is to provide an exemplar of an inquiry-based unit on pollination for designing and implementing constructivist instructional practices while simultaneously providing outstanding teacher preparation. The unit on pollination was developed by preservice teachers through a partnership between the Monmouth Conservation Foundation and the Monmouth University School of Education. Through collective action, these institutions were able to enhance student learning on a vital part of the science curriculum, provide a rich clinical experience for pre-service teachers, and to familiarize teachers with a more constructivist approach to pre-school science instruction.

The catapult was one of the most effective and deadly weapons in medieval siege warfare. They have been critical to warfare since ancient times and have been used by Greek, Roman, and Chinese warriors to vanquish their enemies. This case will present a problem-based scenario where students will assume the identity of an astronaut stranded on Mars. Like the character in the Disney film The Martian, the astronaut only has a small collection of "spare parts" at their disposal. In an effort to fend off predators, they must design and construct a catapult.

How do we keep hot drinks hot and cold drinks cold? Companies such as Tervis, YETI, and Thermos spend their time researching and designing products around that very question. In this lesson, students will discover, through mathematical modeling, which materials provide the best insulation and be tasked with designing their own insulator. This lesson has been designed at two different levels for students from grade three through high school with an optional extension activity for more advanced students. Students will use technology to explore the rate of change of the temperature of hot water over two minutes using different insulation materials.

After this exploration, students will use the data they have collected to determine the best materials for designing their own insulator. This insulator will then be judged based on the ability to keep a hot drink hot and on the aesthetic value.

This chapter discusses challenges faced by first year pre-engineering students. Also discussed are several topics taught within the curriculum including: the engineering design process, sketching, measurement, the elements and principles of design, and three-dimensional modeling. The chapter proposes a design project for engineering students that would tie all of these concepts together to provide an additional learning opportunity for students and more relevant practice of skills like isometric sketching, creating three-dimensional computer-aided design models, and measurement. Samples of student work from the project are included.

In history, geometry was founded more as a practical endeavor than a theoretical one. Early developments of the branch portray philosophers' attempts to make sense of their surroundings, including the measurement of distances on earth and in space. Such a link between earth and space sciences and geometry motivated us to develop and implement a multidisciplinary lesson focusing on the conceptual understanding of the law of cosines in the context of astronomy. In our content specific STEAM lesson, the authors aimed to facilitate an understanding of the law of cosines in ninth grade students, and then apply the law in a star map task to find approximate distances between stars. The second part of the lesson also included the use of an instructional technology to support students' work with the star map task. In the conclusion, the authors discuss possible ways to improve the quality of their STEAM education efforts for the given context.

This case summarizes two perspectives on inclusion of Arts in STEM/STEAM education and how they influence the modification of an existing STEM lesson. Teachers are encouraged to use many instruction models throughout their careers,

Preface

The research on STEAM education is still in its infancy. In education, research is generally based on either a testable model or an ability to identify and describe salient features of practice. This book presents case studies of practices and invites teachers to test, refine, and even refute models that are presented here. We welcome and invite critique. The primary purpose of this work, designed for the science teacher educator as well as the classroom teacher, is to present various ways to implement STEAM as pedagogy. This second edition features adaptations of previously shared cases as well as the introduction of new lessons and updated references.

The chapters in this book provide lessons that meet academic standards and offer educators case study strategies for developing, designing, and implementing integrated, collaborative, inclusive, personalized, and transdisciplinary STEAM teaching and learning lessons. A detailed Table of Contents describes the conceptual framework of each case study presented. Section 1 introduces STEAM as a theoretical framework of practice. Accordingly, the first chapter asks, "Why this book?" and "What is STEAM?"

Section 2 provides the readers with cases that include deeper descriptions of the art practices used within the lessons. The case chapters include information about graphic novels, trade books, theatre, movement, and fine art. Section 3 focuses on theme-based chapters. Lessons include topics from tower building, airplanes, bee pollination, catapults, insulators, engineering, *Star Maps,* marine science, to 'Imagineering.' Section 4 explores key strategies for differentiating lessons for special needs students. Finally, the conclusion sums up and highlights special features from the cases presented herein.

We would like to thank all of the case authors for their informative and inspiring work and also for the time they spent peer-reviewing their colleagues' chapters. The peer-review process strengthened every chapter in this book. May this book inspire you to eagerly consider how STEAM pedagogy might be integrated into your own teaching and learning practice!

Acknowledgment

The editors would like to acknowledge the hard work of editing by James Kelsey, College of Charleston graduate student, and Hannah Finley, College of Charleston undergraduate student. We also thank all of our authors and peer reviewers for their hard work and persistence in the chapter updating processes. Both Meta and Judith also thank Dr. Zambak, Monmouth University, for his contributory article.

Introduction

Creating catchy educational acronyms seems to be a national pastime. The acronyms STEM and STEAM are no exception. Before it was STEM (Science, Technology, Engineering, Math), a short-lived word jumble for the same disciplines was SMET. An interagency meeting on science education at the US National Science Foundation deemed SMET not memorable enough. The acronym STEM comes pre-loaded with multiple pre-existing significations. As a verb, the word "stem" can connote the possibility of branching out, supporting, or developing into something new. For example, inventions *stem* from an innovative idea. "Stem" as an action verb can also signify a cutting-off or stopping of something deleterious from happening. For instance, educators seek to *stem* the negative consequences of pedagogy that is not culturally responsive, inclusive, intellectually engaging, or empowering. Along these lines, we might also ask, "What factors are cutting off huge numbers of potential students from either seeking or succeeding in STEM careers?"

STEM careers remain significantly homogenous, and women and minorities hold only about 33% and 11%, respectively, of STEM jobs (National Science Board, 2018). The plan to diversify the STEM workforce so that it reflects multiple perspectives, knowledge, and experiences requires that we alter the way we think about and teach STEM. This effort "begins in elementary school and extends through the professorate and the rest of the workforce" (Educational Testing Service (ETS), 2015, p. 3).

The very idea or invention of STEAM as an educational acronym and theoretical framework *stems* from a perceived lack in STEM as inclusive and empowering pedagogy for *all* students (Koester, 2015). Support and interest in the possibilities of STEAM are growing. In 2011, Jim Langevin introduced a House Resolution 319 that "expresses the sense of the House of Representatives that adding "A", art and design, into federal programs that target STEM (science, technology, engineering and mathematics) fields encourages innovation and economic growth in the United States." The bill called for Secretaries of Commerce and Education, the Chairman of the National Endowment for the Arts (NEA), and the Director of the National Science Foundation (NSF) to develop a STEM to STEAM Council in order to facilitate a comprehensive approach to incorporate art and design into federal STEM programs.

NSF then funded three workshops called from STEM to STEAM (Maeda, 2012). The National Endowment of Arts hosted an event in order to promote STEAM (Maeda, 2012). The Innovation Collaborative, a national level think tank initiative, is now in its sixth year of research project to determine the most effective practices that "promote important workforce-related creative and innovative thinking skills at the intersections of the arts, STEM, and the humanities (STEAM) in K-12 learning settings" (Innovation Collaborative, 2018). We ourselves were called to develop and edit this book as a resource. To provide a strong foundation for the STEAM movement will require much evidence-based, action-based research, and case study.

Integrating the arts into STEM pedagogy asks for more time for the imagination, creative agency, and innovation to enter into the teaching-learning equation. Bringing in the arts addresses the concern that a profound kind of alienation is what is keeping students from pursuing higher levels of STEM education and the careers that could follow. According to Chan (2019):

Given the subjectivity of what constitutes 'good' education, neither STEM nor STEAM is necessarily better than the other. However, by ensuring art and design principles are not neglected in the lesson plan, student will enjoy a more holistic education that will prepare them for any future profession. After all, in the real world, everything is connected. We need the artists, designers and writers as much as we need the scientists, programmers and doctors. (p. 93)

These same early studies suggest STEAM-based curricula produce a higher percentage of students who remain interested in pursuing careers that require knowledge and skills from the fields of math and science (Masata, 2014). An overarching claim is that the arts afford students experiences that inspire them to explore and form relationships with academic content which may have before seemed irrelevant, inaccessible, and thus, uninteresting (Vanasupa et al., 2007). Martin and Calvert (2018) have provided strong evidence that intellectual engagement and perceived "social empowerment is higher when social change initiatives include exploration and expression through the arts" (p. 38). Koester (2015) persuasively argued that art connects, communicates, and breaks down barriers and believes that a STEAM approach can open windows to the world for English language learners, struggling readers, and others too commonly mislabeled as poor students because of their test scores. She emphasized that selecting an art form that matches the dynamics of the science phenomenon under study is key for an authentic synthesis of ideas to occur. Finding the time and space for developing the creative competencies associated with that art form (best achieved in collaboration with a proficient artist if the academic teacher does now already have these skills) is an important part of the planning process.

This book takes a curious stance toward the many ways in which the pedagogy now called STEAM can be enacted. It invites the reader to mine it for helpful teaching strategies that may spark the imagination and engagement of the student. We advocate for a constructivist, making approach that leads to the creation of artifacts (projects) that can be shared, critiqued, and celebrated. We also call for a kind of STEAM that is *transdisciplinary*, adopting Carillo's (2018) description of transdisciplinary learning as featuring a personally relevant concept, place-based issue or problem, while also integrating the perspectives of multiple disciplines. Thus, as a framework, we call for the STEAM pedagogy that incorporates (whenever possible) a PBL^3 (Problem, Place, and Project-based) approach (Williams & Houseal, 2018).

Which problems, which places, and which projects to investigate and about which to design, and implement lessons become the target of collaborative planning and discourse, with the collective goal of bringing the students into leading roles in the performance of the learning. At the core of the teaching practice, there would be an ethic of care, one that recognizes the power of personalized, agentic experiences, wherein the student can direct the process of his or her own learning. Students begin where they are and enter into a process of growth through formative, constructive feedback, feeling included as a diverse array of outcomes is not only expected but celebrated.

Hoping you enjoy the book:

Judith Bazler
Monmouth University, USA

Merrie Koester
University of South Carolina, USA

Meta Van Sickle
College of Charleston, USA

REFERENCES

Carillo, J. (2018). *Transdisciplinary education: Learning that is authentic and relevant.* Magellan International School. Retrieved from https://www.magellanschool.org/transdisciplinary-education-approach-learning-authentic-relevant-real-world/

Chan, K. (2019). *STEM vs STEAM, Essential Education*. Pressreader.com.

Educational Testing Service (ETS). (2015). *The STEM pipeline booklet.* Retrieved from https://www.insidehighered.com/booklets

Koester, M. (2015). *Science teachers who draw: The red is always there*. Blue Mounds, WI: Deep University Press.

Koester, M. (2016, March 10). *Imagination and the arts as antidotes for STEM education malaise and alienation*. Paper presented at the American Educational Research Association Conference. Retrieved from AERA Online Paper Repository.

Loewus, L. (2015). *When did science education become STEM?* Education Week's Blog.

Maeda, J. (2012). *STEM to STEAM: Art in K-12 is key to building a strong economy*. Edutopia.

Martin, B. H., & Calvert, A. (2018). Socially empowered learning in the classroom: Effects of arts integration and social enterprise in schools. *Journal of Teaching and Learning*, *11*(2), 27–42. doi:10.22329/jtl.v11i2.5057

National Science Board. (2018). *2018 Digest: Science and engineering indicators*. National Science Foundation.

Quigley, C., & Herro, D. (2014). *An educator's guide to STEAM: Engaging students using real-world problems*. New York: Teachers College Press.

Vanasupa, L., McCormick, K., Stefanco, C., Herter, R. & McDonald, M. (2011). Challenges in transdisciplinary, integrated projects: Reflections on the case of faculty members' failure to collaborate. *Innovations in Higher Education, 37*(3).

Williams, M., & Houseal, A. (2018). Composting. *Science Scope*, *41*(6), 36–44. doi:10.2505/4s18_041_06_36

Section 1

Chapter 1
Musing on Unanswered Questions

Meta Van Sickle
College of Charleston, USA

Merrie Koester
University of South Carolina, USA

EXECUTIVE SUMMARY

Out of a conversation between two long-time colleagues—each a science educator and practicing artist—emerged the question, "What does it mean to STEAMify a lesson, and why would a teacher actually choose to do such a thing, other than, say, for-grant-writing-purposes?" Their science selves really liked the idea of a STEAM system, acted upon by forces, both from the outside and from within, and with energy flowing and cycling, all the while transforming grey matter in ways that sustained the teaching/learning process. When it came to their art; however, their dialogue followed pathways grooved by long years of practice and hard work in their respective fields. One author is a seasoned vocalist, trained in the nuances of both individual and group vocal performance as well as the attendant dimensions of music, its composition and phraseology. The other is a painter, poet, and novelist, shaping words, color, and line to tell stories and communicate ideas. What was significant to each was that their artistic habits of mind had shaped their axiology, transforming their ways teaching.

DOI: 10.4018/978-1-5225-9631-8.ch001

INTRODUCTION

Why should just one more in an interminable line of educational acronyms or slogans actually matter? Is STEAM really *different*, or is it another fad? That the idea of STEAM has even emerged in the first place implies some underlying and global lack of creativity and imagination in both science and STEM education. And yet, we have personally observed many such teachers employing highly artistic teaching styles, flexibly improvising and adapting to changing situations and student needs all while driving home the message that in the end (as Eisner has said), education is the act of inventing yourself through hard work, practice, failing, and trying again. There is no kit with instructions for the making of the self. Such an effort begins with and is sustained by the individual. We are calling for a kind of STEAM education that develops the habits of mind associated with making art, which includes the practice of critique and care throughout the learning/making process. Such STEAM educators would recognize that what educators Hetland, Winner, Veenema, and Sheridan (2013) describe as "studio thinking" is highly congruent with the Science and Engineering Practices (NGSS Lead States, 2013, Figure 1).

We advocate for a kind of STEAM that builds in time for critique to help students learn to observe, interpret, explain, and evaluate (Hetland, Winner, Veenema, & Sheridan, 2013). Through the imagery of musicianship— be that improvisation, composition, or the interpretation of its performance—we will introduce a model of teaching science as aesthetic inquiry that we think can move students away from alienation and towards affiliation with science and STEM education. Our aesthetic model foregrounds knowing as communal and value-laden, rather than as a static, detached, objective process of naming and identifying. Aesthetic inquirers recognize that the creative process (always present at the heart of science) spirals outward through cacophony and chaos, and that neat, harmonious compositions may not

Figure 1. Comparison of art studio habits and NGSS science and engineering practices

Art Studio Habits of Mind	Science and engineering practices (NGSS Standards)
·OBSERVING	·Ask questions and define problems.
·ENVISIONING	·Develop and use models.
·INNOVATING THROUGH EXPLORATION AND DESIGN	·Design solutions.
·REFLECTIVE SELF-EVALUATION	·Obtain, evaluate, and communicate information.
·ENGAGING AND PERSISTING THROUGH MISTAKES	
·CRITIQUING	
Source: Hetland, Winner, Veenema, and Sheridan (2013)	Source: NGSS Lead States (2013)

always be the result, especially as new and yet unanswered questions emerge. Finally, we have conceptualized this form of STEAM education as a kind of communal *making*—one that unites the head, heart, and hands as ideas and learning artifacts are constructed. We fully recognize that our approach may evoke responses ranging from dissonance to resonance and that there is no one best way to *do* STEAM education.

THE RESONANCE/DISSONANCE DIALECTIC

When people are deeply connected positively with one another, they are often described as being *tuned into* one another, *on the same wavelength*, or in complete *harmony*. Their emotional and intellectual *frequencies* are in *resonance*, and they can often anticipate the other's next *phrase*. They listen to one another attentively and appreciatively, until and unless they are interrupted by a noise, which suddenly knocks them out of synch and forces them to attend to the distraction. They may even become irritable. How dare an intruder ruffle the smooth *texture* of their discourse and introduce *tension* into their relationship! Do they really have to *listen* to such noise? There are many educators who may criticize STEAM education as not-science or not-STEM. We recognize that we may not even approach equilibrium with these individuals. For them, even the idea of STEAM introduces too much cognitive *dissonance* and threatens their sense of should and ought. But we wonder, too, if the arts were a significant part of their own education, and if so, how were such experiences organized for them?

Liora Bresler (2005) is a musicologist who has devoted her career and scholarship to exploring what musicianship can teach educational researchers. She has reminded us of the "complexity of translating anything, but particularly a fluid form into a fixed one" and that the power of performances (be it an educational seminar or a musical performance) reside in the degree to which they establish connection and interaction (p. 169). How is the room arranged? Does it allow for movement? The expression of feelings? The development of empathy and trust? When conceptualized as *music* creation, what are the "forms, rhythms, textures, and timbres" of the performance (p. 170)?

Inspired by such writing, we asked, "What if science and STEM educators acquired the capacity to compose and perform lessons as would a musician? How might performances be enacted so that independent learning continued in the time and space outside the classroom, too? What science standards might be taught through this kind of teaching/learning performance?" Our musings were piling up, our notes strewn everywhere. We could feel ourselves toppling into entropy—until

Meta suddenly blurted out, "Why do we science teachers so often avoid the ethics and aesthetics (axiology) of teaching?"

A DIALOGUE COMMENCES

Merrie: Meta, I've often heard you talk about these things. What exactly do you mean?

Meta: That's just it. In science, we're zeroed in on the exactly. The philosopher in me is troubled by this approach. Philosophers view a topic from many different perspectives. They have conversations about how it is that we come to know anything and whether that knowing matters or not. When you teach, unless you marry the content with life—with what matters—deep learning cannot occur. But teaching in this way can be a very messy and complex process, because it has to do with feelings. Musicians work at shaping feelings into forms that touch the heart. To achieve the effects they want, they also know they must practice and persist through many revisions of their work as they experiment with different possibilities. Science educators have a lot to learn from musicians!

Merrie: Wow! That's a compelling premise. What if we were to show through a musical performance (that is, provide authentic evidence) that the claim you just made about employing musicianship to teach science can be supported? Better yet, what if we could figure out a way for our readers to interact with that performance and, at the same time, employ science and engineering practices? Do you know of a piece that could do those things?

Meta: I do! Have you ever heard of the composer Charles Ives?

Merrie: No, but I'm sure going to tell me!

Meta (laughing): Do you have your laptop? Pull up the PBS link for their Music Mavericks program called "The Unmixed Question": http://musicmavericks.publicradio.org/features/feature_unmixed_question.html (This file could take a long time to download, so readers, you may want to go get your coffee. Please do not skip over this part of this chapter!)

Merrie: Got it! It says, "Okay, let's play!" I love this already. [She clicks on PLAY and what looks like an audio mixing board pops up on screen.] WHAT???? It's a virtual mixing board! It says here in the written description that "the piece was created using binaural 3D positioning software to move the musical parts around in an imaginary hall." That sounds like music GPS!

Meta: That's essentially what it is. They use two microphones to create a 3D stereo sound that makes you feel like you're in the performance hall. Charles Ives accomplished a similar effect by positioning members of the orchestra in different physical spaces. He was a musical genius. He'd put the horn section offstage sometimes and the winds in places you'd never normally find them.

Merrie: So, he was engaging in science and engineering practices, intentionally changing one variable at a time to document the experimental effect(s).

Meta: Of course! This is what musicians have to do every day. A good choral director is constantly going after different effects, depending on the feelings and sounds he or she wants to evoke. For example, the arrangement of the singers would be different if the conductor wants the audience to feel surrounded by sound versus the room being filled with sound.

Merrie: So, here, in this simulation model, it looks like we have three different kinds of voices—a trumpet, the winds (clarinets, flutes, saxophones, etc.) and the strings (violin, viola, cello, bass, etc.). There's also this "miscellaneous" voice we can vary.

Meta: Right, but you get to control what you hear. Try it!

Merrie (listening and playing with the different control knobs): Here is where I would ask our readers to enter into this experience and ask, "How many different independent variables can be manipulated in this experience? How many experimental questions might be generated and then explored through science and engineering practices? The observational data might take the form of narrative writing and the use of descriptive adjectives or symbolic emoticons. They might make different kinds of expressive lines or body movements to communicate what's happening with them. Students might be asked to record their feelings when they hear each voice by itself or describe what happens when that voice is combined with others. Learners might work together to create different kinds of "noticing" charts to visually represent their observations. I can tell you that I personally wanted to silence those jabbering winds. They were so annoying! My emoticon would be scowling.

Meta (laughing): Just think of the conversations this experience might start among both teachers and students. For example, how many times do we teachers have to persist in the face of annoying distractions? Which ones do we pay attention to? What if those winds were students who were desperately trying to tell you that what you were doing wasn't working for them? Perhaps what you thought was perfectly suitable music was for them an ugly noise. These students would not feel cared for at all.

THE ROLE OF CARE

Merrie: OMG. You're right! This activity has powerful implications for the study I am doing with dyslexic science learners. Several have already confessed that they will go to great lengths to call attention away from the fact that they can't read. They say it's especially embarrassing when a science teacher asks them to read directions out loud or be a scribe in a group activity; so they'll act like the class clown, pretend to be sick, or even start a fight. When I asked what kind of science learning they did enjoy, they all said the kind where they get to solve the problem on their own.

They absolutely hate doing worksheets and claim that printed directions make them cringe! (Koester, in press).

Meta: Indeed, arts-integrated lessons tend not to marginalize the struggling reader. In this simulation, all students will likely enjoy playfully experimenting with sound energy and then personally arriving at the best combination of voices for their sensibilities. They'll also acquire an appreciation for how musical conductors use dynamics to achieve entirely different effects. Teachers can start dialogues about how sound engineers must work closely with the conductor and how both STEM and the arts knowledge, skills, and practices are required to make a musical performance work. A care for ideas will emerge (Noddings, 2005).

Merrie: This dialogue is getting more exciting by the minute! As it turns out, most of the dyslexic individuals in my study are also deeply involved in some kind of art, whether it's music, painting, theatre, or martial arts. They like the idea of making new worlds in which they play a starring role. One profoundly dyslexic art professor in my study shared that science had always made her feel stupid, but that her art made her feel powerful. I recently went to a community forum facilitated by an organization called Understood.org. I learned to my amazement that 1 in 5 students struggle with learning and attention issues, often making it difficult for them to read (http://www.adcouncil.org/Our-Campaigns/Education/Learning-Attention-Issues). This means that if we teach science through mostly talking, telling, and reading, we may be marginalizing fully 20% of our students! No wonder there has been a GREAT TURN OFF to science (Trefil, 2008).

Meta: That's another reason why arts-informed science education has such great potential. The arts offer other ways in, pathways to understanding and the freedom that comes with being included in the teaching/learning process. However, when you introduce the arts, you also enter into the world of feelings, aesthetics, and creating caring, empathic relationships—topics which can make a lot of science educators uncomfortable.

Merrie: What you have just said reminds me of a statement by Sartre (1976):

"Authenticity consists in having a true and lucid consciousness of the situation, in assuming responsibilities and risks that it involves."

We are most definitely asking science teachers to take risks and to teach toward outcomes that may not be measurable with a number or test score. For many (if not most) science teachers, lecturing is an acceptably authentic way of being a science teacher; whereas being asked to teach in an "artistic" way may provoke considerable concern, especially for the teacher who claims to have no artistic "talent." Furthermore, they may balk at the extra time an arts-integrated approach requires. In order for them to risk experimenting with an aesthetic paradigm, science teachers will want examples which show that it works. They'll need to see that their science standards can be taught; otherwise, we won't get much traction. Lederman & Lederman (2015) call on science education researchers to clearly define their theoretical frameworks. For us, this means we must answer two questions: 1) What problems might be solved by experimenting with this kind of approach, and 2) Why is our approach feasible?

Meta: Scientists like their numbers and their stats, but in a science teaching performance, things often happen that can't be measured. The simulation we just experienced was an example of a deeply aesthetic, open-ended, personalized form of inquiry—so unlike the prescriptive science education lab with its attendant work sheets and linear progression toward right answers. The simulation also taught the participant a lot about nature of science (NOS), through which discoveries emerge far less methodically (and more naturalistically) than is often presented in the science education canon (NGSS Lead States, 2013).

Merrie: This one STEAM lesson—through which we explore science and engineering standards through music-informed inquiry—could serve as a springboard into dialogue about energy transfer, properties of sound, form and function, cause and effect, and patterns—all cross-cutting principles in the NGSS standards (2013). If all these standards can be "covered," then the time spent on an aesthetic approach will have been well invested indeed! By "noticing what there is to be noticed"—as we engage in participatory arts exploration of content—Green (2001) has reminded us that "we create new connections in experience, allow new patterns to be perceived and open new vistas of understanding" (p. 6). The teacher does not necessarily need any artistic talent, but collaboration with talented students or a creative arts specialist can produce some phenomenal results.

Meta: I completely agree! An aesthetic approach is not only feasible but desirable in the science classroom. In Bresler's (2005) model of music-informed inquiry, a musicianship approach opens spaces for "improvisation, empathy, and embodiment" (p. 169). The Ives spatial music simulation is not just about playing with the controls on the sound mixer. Otherwise, it's just another fun technology toy, app, or game. You have to have dialogue about what is happening! For example, what does manipulating the technology signify for the one at the controls? For the student, those jabbering winds could signify anything from a nagging mother to a little brother who won't be quiet. You just said yourself that they reminded you of the student who ceaselessly interrupted you, possibly to call attention away from the fact that he or she could not read.

Merrie: These matters of signification are important points of dialogue and reflection. In the midst of a learning experience, the "image" that takes shape in my "mind's eye" can be completely different than what's in yours. Finson and Pederson (2011) provide a comprehensive argument for why and how, for example, visual data are so important in science education. They are a way for teachers to see student thinking and the types of connections they have made—or not.

Meta: Yes! In our simulation, spatial music becomes both a visual and aural metaphor for the classroom, which is a highly complex and dynamic setting where many voices are competing for attention. In the "Unanswered Question" performance, there are serene moments throughout, played by the strings. The trumpet sounds at interspersed moments. Did you hear the clarion call of the question? Just as you likely picked up on many sound patterns in the performance of this spatial music, caring, perceptive teachers also pick up on patterns of discourse and behavior of their students.

Merrie: We keep returning to the role of care in the teaching/learning process. Care is a significant component of an aesthetic approach to education (Siegesmund, 2010). In your own writing on the caring science teacher (Van Sickle & Spector, 1996), you state that caring teachers notice how their students are responding to what's happening in the classroom.

Meta: That's right. You were talking earlier about your research on the dyslexic learner. Caring teachers realize that if they teach only through talking and telling, they may lose, or marginalize some 20% of their students. The caring teacher finds ways to move all students forward, finding ways to challenge them, and offers constructive critique to validate them. They identify each child's strengths and what matters to him or her and then use that information as on-ramps into the learning. Noddings (2005) talked a lot about this.

Merrie: And so, each learner can shape a different type of experience for him or herself…

Meta: Yes, but only if we don't over-value instant success and right answers — something we do all too often in science classes.

Merrie: So often, we hardly wait at all before we just tell them the right answers, too. Why should they bother to do the work of making their own ideas? No one enjoys being told what to think all the time (Duckworth, 2006).

Meta: Yes! Emergent, creative thinking is characteristic of the nature of science and art. If you look back at the chart you made that compares artistic habits of mind with science and engineering practices, you'll see that artists and scientists all make a lot of observations before they develop their questions. When you read Nobel laureate literature, they talk about "playing" with ideas, just as artists play with materials, notes, dance moves, voice modulation, etc. They all do thought experiments and mess around with notions. They all get worked up, angry, defensive, happy, or despondent when they fail, but we rarely hear about the feelings, the fun, or the fear of failure when we read science papers.

Merrie: I think this is a really important point you're making. It's as if we're dehumanizing the nature of science in our classrooms by making it seem less messy and more predictable than it is in authentic practice. By contrast, art teachers preach persistence in the face of failure, which they conceptualize as an integral and essential part of the learning process.

Meta: In our choral group, we learn to enjoy the process, make a ton of mistakes, make improvements, and persist through hours of rigorous practice. And we do that so that the performance moment can be all that it can be. That's pure joy, and it's at the heart of what it means to learn in an aesthetic space.

Merrie: You're right! My athletic coaches' favorite expression was always, "No pain. No gain." And, I've been blessed to have deeply caring coaches. I always wanted to work harder to improve my performance because they made me feel like I mattered. Coleman's (2003) idea of human capital is very relevant here because it's acquired when one learns a new skill or gains new knowledge. However, Adjibolosoo's (1995) construct of aesthetic capital is even more to our point. Adjibolosoo explained that "aesthetic or artistic capital brightens or enriches people's' lives" (p. 181). According to Uhrmacher (2010), aesthetic capital may cause a student to feel or act differently and undergo a radically different way of being in the world (in Koester, 2015).

Meta: It's a teacher or coach's ethical responsibility to set high expectations and keep you moving forward and upward, all the while making it clear that it's your responsibility to get there. You can't produce an aesthetic transformation—the pleasure and even the feeling of power derived from a beautiful experience—without also having an ethic of care. When you practice an ethic of care, it's comparable to the Hippocratic oath of doing no harm. An ethic of care always asks the question of what happens in a relationship. When there is an ethic of care, a system can thrive. When students, the primary actors in an educational system, don't feel cared for, they stop interacting, and they stop learning (Newmann, 1981). They become alienated from the adults and the ideas. They may even retaliate against the equipment and each other. They'll either act out or shut down…

Merrie: …like so many students who struggle to understand the language of science.

Meta: Yes! When they feel included, the winds will settle down and stop interrupting so much. There is a movement toward affiliation with each other and the content, and there will be resonance in the classroom. Tenets of care will have been met.

Merrie: Can you explain more what you mean by tenets of care?

Meta: I think the best explanations can be found in Nel Noddings's work (2003). She talks about the difference between caring about and caring for. I can care about many things, and take no action. If I'm going to care for something or someone, however, I actually have to do something. I can care for others, in this case, our students. I can care for other living things. I can care for objects, too. This latter tenet is very important in your classroom because you need your students to take care of the equipment you're using…

Merrie: …what about care for the self?

Meta: Caring for self is truly important, because the mind and body have to be healthy in order to grow and develop. If you don't care for your mind, you are, by default, limiting yourself and your possibilities of becoming. You certainly can't pass tests and college entrance exams if your mind isn't prepared.

Merrie: This is amazing and beautiful to me, Meta. In my workshops with teachers, I often bring in an inflatable brain. You saw me pull one out at a conference recently.

Meta: I do remember! You had us imagine a huge, brain-shaped hot air balloon, which could only lift off if the teaching practices (in the basket below the balloon) did not bring the system down…

Merrie: …Yes, and the question was, "What's in your science teaching basket?" (See Figure 2).

Meta: An ethic of care and aesthetic transformation cannot be separated from one another in practice…

Merrie: …so one's the balloon, and the other's the air in the balloon. Without both, there's no semiotic lift off, that is to say, no deep meaning-making and no upward and outward movement in the system.

Meta: Now, you're talking!

Merrie: It seems to me that you're also talking about flow. I've felt this so many times when I'm painting or writing.

Meta: Yes! Csíkszentmihályi Mihályi, an educational psychologist, writes about flow (1997). In flow, emotions are aligned with the challenge at hand. When flow happens in a classroom, the best thing you can do is get out the students' way and let it happen. It results in deep, joyful learning. It's the place where ethics and aesthetics merge into a beautiful teaching/learning performance.

Merrie: It seems that we're talking about a process that moves in spirals. Making art happens in the same way. You spiral in and out of flow, but you keep on moving, attending to what's happening in the moment, until a problem is solved. **Meta:** And this takes time, work, and persistence. It takes grit. Angela Duckworth gave a TED talk on grit that I share with all my students. Here's the URL: https://www.ted.com/talks/angela_lee_duckworth_grit_the_power_of_pas-sion_and_perseverance#t-34479

Merrie: Wait. Let's go to it now. (We hope our readers will do the same.)

Meta: The most fascinating thing Angela Duckworth and her colleagues uncovered in their research was that grit has very little to do with IQ or talent. Instead, it begins with passion and moves forward through perseverance. The "gritty" individual does not quit when bored or tired, or when they fail (2007).

Merrie: Yes! And hard work can require a tremendous expenditure of energy. It's as if the teacher (acting as a conductor) is a heat source that initiates and sustains a convective process suffused with aesthetic energy (Koester, 2015).

Figure 2. What's in your teaching basket?

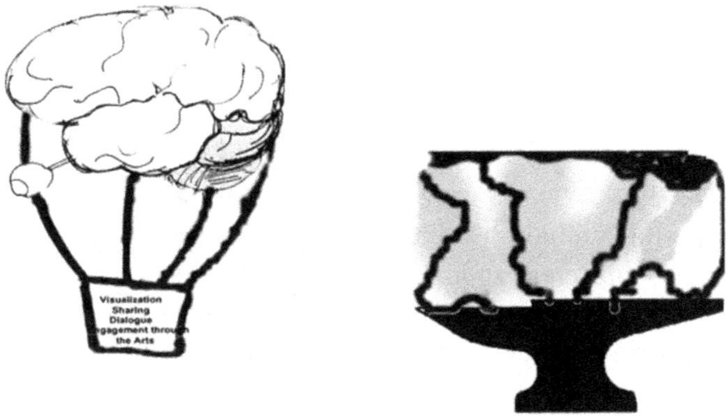

Meta: Agreed. I think the concept of energy transfer and flow is a topic about which the arts have a lot to teach us all. In the arts, energy is transferred from the performers to an audience, who are transformed, transported, and moved. Knowing that they will be the source of this energy, performers are told to rest before they go onstage…

Merrie: This is why caring teaching can be so exhausting.

Meta: Yes, and I would add that even more energy is required when the teacher has the courage to say to the student(s) working on a project or learning a new skill: "You're not there yet" (Dweck, 2006). Persistence and hard work are key mindsets for learning, and there can be harm done if praise is given too often for minimal improvement. On the other hand, the worst thing a teacher/coach/ conductor can do is allow a student to just give up. Caring is complicated.

Merrie: Several times in our conversation, you have moved us forward by saying "Yes, and" instead of "Yes, but." I recently heard Erica Halverson give a talk at AERA. A practicing actor herself, she explained that these "Yes, and" affirmatory communications are the mainstay of excellent improvisational theatre. In her work, she advocates for the use of improvisation techniques as a "new framework for the design of learning environments," all the while recognizing that doing so may make many educators somewhat uncomfortable (2016). In my own participatory action research, I often encounter teachers who say they are worried that they draw like nine-year-olds. But then, they plunge right in. Significantly, they find that making themselves vulnerable is perceived by their students as a profound act of care.

Meta: Once again, you've described an ethic of care—the honest relationship to the other.

VISUALIZING THE PROCESS OF STEAMIFICATION

Merrie: This book is about STEAM teaching. This chapter is about how doing STEAM might look and feel when that happens. We've brought our readers into a music simulation that allows them to experiment with different features of a spatial music performance, and each has likely experienced the simulation in his or her own way as a factor of which variables they chose to manipulate and the sounds they hypothesized they might be able to achieve. As sound engineers, they used technology to shape different audio outcomes. And, if they wanted to re-produce a particular sound effect, they needed to carefully document their experimental procedures. Depending on their technological and mathematical skills, they might have been able to deduce multiple relationships between variables.

Meta: Indeed. I think we've made an excellent case for why taking the time to employ an arts-integrated approach to STEM teaching/learning is worth the investment. Music is a powerful medium for teaching principles of energy transfer and experimental processes, but I think it has even broader applications. I personally see music composition as a metaphor for the art of teaching. [Picking up a pen and starting to draw]. (See Figure 3).

Figure 3. Thinking through drawing

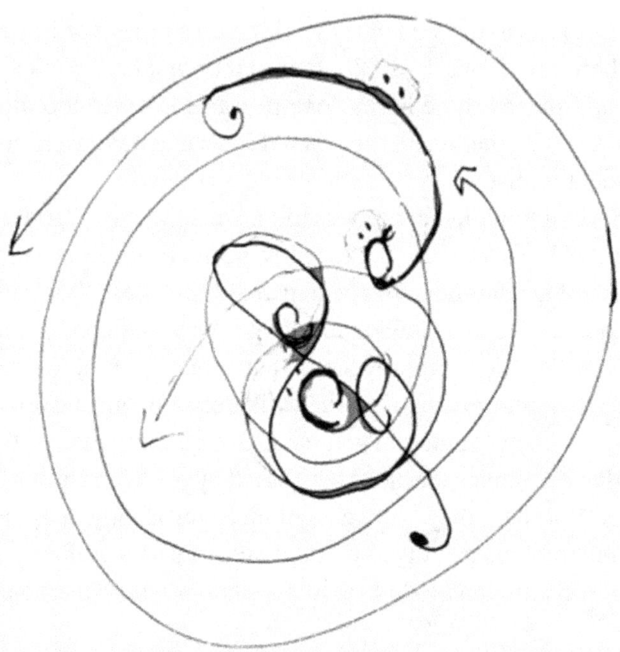

If you look at this drawing, everywhere you see two dots (a colon), I drew a new spiral. Each colon signifies the beginning of an individual student's experience—wherever he or she may be at the moment. The process is organic, temporal, and spatial.

Merrie: …which means that both time and space have to be factored in by the teacher for growth and development to occur. It may seem like not much is happening (that is, there may be a lot of noise and even confusion to the casual observer), while, in fact, many possible pathways for learning are opening up…

Meta: …and, in a musical performance, there are many different voices, which can be instruments, humans, and even the building itself. There are times for individual voices to be heard and other times when they must blend in harmony, or be altogether silent.

Merrie: Liora Bresler (2005) talked about this when she described music dynamics, or the perception of loud and soft. She explained that context is essential. A loud sound seems louder if it arrives on the heels of silence or a soft tone. Likewise, those interruptions by the winds in the simulation contrasted so vigorously with the strings and the horn that I, at least, had to lower their volume even to hear the other voices.

Meta: That's why music is such a great metaphor for the teaching/learning process. Its dimensions of form, melody, dynamics, and texture, as well as the sheer amount of work that must go into its performance, can be applied in the classroom. But none of this can take place in the absence of an ethic of care…

Merrie: …which is at the heart of all aesthetic inquiry (Siegesmund, 2010). And so, we've spiraled back to our beginning premise: Aesthetic inquirers recognize that the creative process (always present at the heart of science) spirals outward through cacophony and chaos, and that neat, harmonious compositions may not always be the result, especially as new and yet unanswered questions emerge. This process is not a linear, inorganic, or prescriptive one. Any number of different outcomes are possible…

Meta: …which should then be critiqued in a way that is empowering and moves the process forward and deepens the learning.

Merrie: There is nothing at all static about the STEAMification process we're describing.

STEAM AS CRITICAL MAKING

Meta: Yes. We want our model to exude possibilities and motion and only be limited by the imagination. With regard to care, it's even more important for a teacher to enact this than to say it (Noddings, 2005). From the very beginning, the teacher needs to communicate (model) that his/her students are expected to engage with and deeply explore a concept, like the energy transfer that takes place during this simulation. This models caring for ideas. Noddings talks a lot about this, too. As we've said earlier, while this is happening, the caring teacher also challenges students, offering constructive critique to keep them moving forward.

Merrie: For my part, I'd like to see our model include features of aesthetic inquiry as described by Pestalozzi (1898), Dewey (1938), Greene (2001), and Eisner (2002)—a participatory and inclusive, sense-based, interactive, connecting, and above all, purposeful engagement with the arts, as we did here with this spatial music simulation. I don't see STEAM achieving deep learning unless the dynamics of the art form are in synch with that of the STEM content being explored. When students perceive the arts as an add-on or a frill or prescriptive craft activity, they don't enter into an appreciation of how it makes it their minds better and stronger. They don't get how rigorous the arts can be, especially when critique is valued and practiced.

Meta: That's huge! For example, making an artbot from a kit can then evolve into designing new versions of artbots that can draw in different ways, which students then share with one another.

Merrie: Exactly. I think today's Maker Spaces can epitomize STEAM learning when explicit connections to both arts and STEM knowledge, skills, and practices are made. There needs to be time for students to notice what's happening and then to improvise with the materials. Within reason and safety measures, they can ask, "What if I try to manipulate a different variable than the one directed by the kit instructions? What will happen then?"

Meta: And then you critique again. The process keeps spiraling, and the performance can better and better as skills are developed through practice (Noddings, 2005).

Merrie: We are clearly both channeling Seymour Papert's (1991) and Kafai's (2006) kind of making, in which dialogue and critique go hand-in-hand as part of the overall practice, establishing the relationship between all the components of the teaching/learning system. His theory of constructionism helps us to see that making a public artifact (a presentation or poem, a musical, theatre, or dance performance, a painting, robot, a new piece of software, etc.) informs and transforms learning.

Meta: Yes. Skills are built and knowledge is constructed. Through practice, understanding emerges as you begin to lean into the learning. For example, if I have to learn to sing a specific run in a song, I have to practice it again and again until it's in me. The muscles controlling my vocal cords and diaphragm have literally learned what to do so that, when it comes time for the performance, I don't have to think about it.

Merrie: I so get what you're telling me. We can't expect our students to be prepared for STEM jobs if we don't give them time to learn and practice STEM skills, which include the ability to creatively and collaboratively identify and solve a problem as well as use the equipment and perform the skills necessary for work. I can remember having to throw away countless electrophoresis gels before I made even a single one that could be properly used in an experiment.

Meta: That's what we mean by practice. Once you've got it, it's yours!

Merrie: So often, we just march through a lab activity, follow the steps, and don't allow time for playing with the materials and seeing what else might happen. The music simulation allowed for the important aesthetic component of taking time for noticing. It's a use of technology that has real educational value. When we use technology, I think we need to ask, "How is it moving the lesson forward? Is it part of the problem-solving process and facilitating deeper learning, or is it just another distraction?"

Meta: Exactly! For the Ives simulation to be a meaningful form of arts integration (STEAM), you can't just play with the controls. Otherwise, it's just another fun technology toy or game. You have to have dialogue about what is happening. For example, what does manipulating the technology signify for the one at the controls? For the student, those jabbering winds could signify anything from a nagging mother to a little brother who won't be quiet. You yourself said they reminded you of the student who ceaselessly interrupted you, possibly to call attention away from the fact that he or she could not read.

Merrie: You're right. That's because my own work is about helping the struggling reader find a way into meaning-making in the science classroom. For me, the matter of signification is really important. I think a lot of what we are talking about resides at the intersection of communication and signification. What kinds of messages are being sent and how are they interpreted? How has the experience been organized so that it promotes inclusion rather than exclusion? Francois Tochon, a world language scholar and semiotician, has stated that the form communication symbols take and the way they are presented is the key to whether the educational system functions as intended and deep learning occurs (2013). All too often, students feel left out of community because they can't understand STEM content languages. These languages are not only foreign, they don't seem relevant to their lives. Thus, there's no incentive to do the work to learn them. The arts can readily serve as a common language and as a bridge to meaning-making and deep learning. They can act as media for both confirmation and affirmation.

Meta: This is both empowering and freeing. AND, you can't take this power away once they have it. What we are talking about gives rise to a pedagogy of possibilities, one that is in constant motion and nurtured, with feeling, by building caring relationships of many kinds. This takes courage and energy.

CONCLUSION

We recognize that what we have proposed here may seem like a rather *avant-garde* approach to the teaching of science. In this epilogue, written for the second edition of this text, we seek to briefly show how this aesthetic way of teaching is not as "strange" as it appears *and* that it is strongly aligned with the national call for "three-dimensional" (3D) science teaching and learning (NGSS Lead States, 2013). At a presentation for the Association of Science Teacher Educators (ASTE), we guided participants through the simulation described in this chapter. We explicitly showed how what seemed like "playing" with sound and stage position controls, all the while talking about and documenting what was happening, helped make connections with disciplinary core science content, science and engineering practices, crosscutting concepts, as well as with the nature of science–the three dimensions. We believe an aesthetic form of science teaching also meets the call for students to "make sense of phenomena using the three dimensions, construct their own explanations and arguments, and evaluate these explanations and arguments based on evidence" (NSTA position statement, 2018). Most importantly, the work, conceptualized as a "pedagogy of possibilities," seeks to align with the long-held national mandate for achieving science literacy for all students (Lee, 2003; National Research Council, 1996; NGSS Lead States, 2013).

As we have claimed, emergent, creative thinking is characteristic of the nature of science and art. Both artists and scientists all make a *lot* of observations before they develop their questions and then ultimately design, develop, test, and finally share the artifacts and/or performances they make. Science and engineering practices abound in this method. As stated, teachers can start dialogues about how sound engineers must work closely with the conductor and how both STEM and the arts knowledge, skills, and practices are required to make a musical performance work. A care for ideas will emerge (Noddings, 1992).

We make the claim that 3D science learning can move forward (and be personally meaningful) only when experiences are designed and implemented that afford time for aesthetic noticing and the making of conceptual relationships. The ways in which this objective might be achieved is a phenomenon in and of itself.

REFERENCES

Adjibolosoo, S. (1995). *The human factor in developing Africa*. Westport, CT: Praeger Publishers.

Ainsworth, S. E. (2006). DeFT: A conceptual framework for learning with multiple representations. *Learning and Instruction*, *16*(3), 183–198. doi:10.1016/j.learninstruc.2006.03.001

Bordieu, P. (2003). The forms of capital. In A. H. Halsey, H. Lauder, P. Brown, & A. S. Wells (Eds.), *Education: Culture, economy, society* (pp. 46–58). Oxford, UK: University Press.

Bresler, L. (2005). What musicianship can teach educational research. *Music Education Research*, *7*(2), 169–183. doi:10.1080/14613800500169399

Coleman, J. S. (2003). Social capital in the creation of human capital. In A. H. Halsey, H. Lauder, P. Brown, & A. S. Wells (Eds.), *Education: Culture, economy, society* (pp. 80–95). Oxford, UK: University Press.

Csikszentmihalyi, M. (1997). *Finding flow: The psychology of engagement with everyday life*. New York, NY: Basic Books.

Dewey, J. (1938). *Experience and education*. New York, NY: Kappa Delta Pi.

Duckworth, A. L., Peterson, C., Matthews, M. D., & Kelly, D. R. (2007). Grit: Perseverance and passion for long-term goals. *Journal of Personality and Social Psychology*, *92*(6), 1087–1101. doi:10.1037/0022-3514.92.6.1087 PMID:17547490

Duckworth, E. (2006). *The having of wonderful ideas: And other essays on teaching and learning* (3rd ed.). New York, NY: Teachers College Press.

Dweck, C. (2006). *Mindset: The new psychology of success*. New York, NY: Ballantine Books.

Eisner, E. (2002). *The arts and the creation of mind*. New Haven, CT: Yale University Press.

Finson, K., & Pederson, J. (2011). What are visual data and what utility do they have for science education? *Journal of Visual Literacy*, *30*(1), 66–85. doi:10.1080/23796529.2011.11674685

Greene, M. (2001). *Variations on a blue guitar: The Lincoln Center Institute lectures on aesthetic education*. New York, NY: Teachers College Press, Columbia University.

Halverson, E. (2016). *The art of improvisation in teaching and learning*. Paper presented at the American Educational Research Association Conference, Washington, DC.

Hetland, L., Winner, E., Veenema, S., & Sheridan, K. (2013). *Studio thinking 2: The real benefits of visual arts education*. New York, NY: Teachers College Press.

Kafai, Y. B. (2006). Constructionism. In K. Sawyer (Ed.), *Cambridge handbook of the learning sciences* (pp. 35–46). Cambridge, MA: Cambridge University Press.

Koester, M. (2015). *Science teachers who draw: The red is always there*. Blue Mounds, WI: Deep University Press.

Koester, M. (in press). Deep learning in science: An argument for an aesthetic paradigm of science education. In F. V. Tochon & D. Busciglio (Eds.), *Deep Education Across the Discipline sand Beyond: A 21st Century Transdisciplinary Breakthrough*. Blue Mounds, WI: Deep University Press.

Lederman, N. G., & Lederman, J. S. (2015). What is a theoretical framework? A practical answer. *Journal of Science Teacher Education*, *26*(7), 593–597. doi:10.100710972-015-9443-2

Lee, O. (2003). Equity for culturally and linguistically diverse students in science education: Recommendations for a research agenda. *Teachers College Record*, *105*(3), 465–489. doi:10.1111/1467-9620.00247

National Research Council. (1996). *National science education standards*. National Academies Press.

Newmann, F. (1981). Reducing student alienation in high schools: Implications of theory. *Harvard Educational Review*, *51*(4), 546–564. doi:10.17763/haer.51.4.xj67887u87l5t66t

NGSS Lead States. (2013). *Next generation science standards: For states, by states*. Washington, DC: The National Academies Press.

Noddings, N. (2003). *Caring: A feminine approach to ethics and moral education*. Berkeley, CA: University of California Press.

Noddings, N. (2005). *The challenge to care in schools: An alternative approach to education* (2nd ed.). New York, NY: Teachers College Press.

NSTA Position Statements. (2018). Retrieved from https://bit.ly/2TiXolx

Papert, S., & Harel, I. (1991). Situating constructionism. In I. Harel & S. Papert (Eds.), *Constructionism* (pp. 1–11). Norwood, NJ: Ablex.

Pestalozzi, J. H. (1898). How Gertrude teaches her children (2nd ed.). Syracuse, NY: Bardeen.

Sartre, J.-P. (1976). *Anti-Semite and Jew* (G. J. Becker, Trans.). New York, NY: Schocken. (Original work published 1948)

Siegesmund, R. (2010). Aesthetics as a curriculum of care and responsible choice. In T. Constantino & B. White (Eds.), *Essays on aesthetic education for the 21st century* (pp. 81–92). Rotterdam: Sense Publishers.

Tochon, F. V. (2013). *Educational semiotics: Signs and symbols in education.* Blue Mounds, WI: Deep University Press.

Trefil, J. (2008). *Why Science?* New York: Teachers College Press.

Uhrmacher, P. B., Conrad, B., & Lindquist, C. (2010). The power to transform: Implementation as aesthetic awakening. In T. Constantino & B. White (Eds.), *Essays on aesthetic education for the 21st century* (pp. 183–203). Rotterdam: Sense Publishers.

Van Sickle, M., & Spector, B. (1996). Caring relationships in science classrooms: A symbolic interaction study. *Journal of Research in Science Teaching*, *33*(4), 433–453. doi:10.1002/(SICI)1098-2736(199604)33:4<433::AID-TEA5>3.0.CO;2-T

Wright, P. (2016). *The unmixed question: the spatial music of Charles Ives.* American Mavericks. Retrieved from http://musicmavericks.publicradio.org/features/ feature_unmixed_question.html

KEY TERMS AND DEFINITIONS

Aesthetic Inquiry: We employ Maxine Greene's conceptualization of this concept, which features a curriculum that begins with and is sustained by noticing what there is to be noticed. The lesson progresses through a participatory engagement with the arts, as new connections are made and conceptual relationships evolve.

Aesthetics of Care: Using the arts to develop an appreciation of how learning through them can create multisensory ways into learning. In this work, we try to show how the aesthetics of care is directly linked to an ethic of care.

Axiology: The philosophical study of value. It typically involves the discussion of both ethics and aesthetics.

Constructionism: A learning theory developed by Seymour Papert, in which he contends that you have to make it to learn it. Learning is achieved through the making of a public artifact, which is then critiqued. This shareable construction may take the form of a robot, musical composition, paper-mache volcano, poem, conversation, or new hypothesis. We believe that this theory holds a key to understanding and developing the potential of what is now being called a Maker Movement.

Critique: A process described by Yasmin Kafai as the ways in which a learning artifact is analyzed and reviewed in a social context. In the process, connections between old and new knowledge can be made.

Dialectic: The exploration of seemingly contradictory ideas and the relationships between them. It employs logical argumentation techniques to search for a unity of opposites.

Dissonance: In music, the sine waves of pitch interfere with each other to the point that the sound may cause the human ear to react with discomfort. As a metaphor for teaching and learning, being off-pitch can have a deadening effect on the lesson.

Dynamics: Musical conductors use dynamics to achieve entirely different effects. Dynamics can include anything from loudness and softness (volume) to fast and slow (tempo).

Resonance: In music, resonance is the compounding of sound waves so that what you get is more than the sum of the parts. As a metaphor for teaching and learning, resonance is the achievement of synergy—when all the parts of the system unite to produce a level of learning not possible if the elements were separated. It often manifests as joy and exuberance. It's a beautiful thing.

Semiotics: The study of the meaning-making process, or semiosis. In Peirce's theory of signs, semiosis depends on the triadic interaction between an object in the world, a sign or symbol that re-presents that object (or phenomenon), and the way(s) a person in a given culture or situation interpret(s) that contextual sign.

Spatial Music: A modern music composition technique that purposefully uses the location of the instruments to achieve different effect. Charles Ives's "Unanswered Question" of 1906 was the first ever 20th century composition to feature spatial separation as a major element.

Voice: At a minimum, this is the utterance of the possible sounds a human can make. At a maximum, voice is the ability to express personal ideas.

Section 2

Art Practice

Chapter 2
Graphic Novels and STEAM:
Strategies and Texts for Utilization in STEAM Education - Graphic Novels and STEAM

Alex Romagnoli
Monmouth University, USA

EXECUTIVE SUMMARY

Focusing on the interdisciplinary connections between STEAM education and graphic novels, this chapter first establishes historical and educational contexts for the use of graphic novels in STEAM education. A literature review focusing on the use of graphic novels in a science course as well as how graphic novels have been used in secondary classrooms will be discussed. Literature that is reviewed includes books, articles, and edited volumes. The strategies for implementing graphic novels in STEAM education promote constructivist learning as students are asked to access their intellectual and cultural capital in order to ascertain meaning from given content. Additionally, multimodality and multiliteracies are highlighted throughout the chapter. Finally, this chapter ends with linking the use of graphic novels in STEAM education to the power of narrative inquiry in educational contexts.

INTRODUCTION

In *All-Star Superman* (Morrison and Quitely, 2011), the superhero Superman finds himself in a laboratory full of amazing and bizarre science experiments. After having saved pioneer astronauts who were exploring the outer atmosphere of the sun, Superman tests his strength levels in this futuristic menagerie of scientific

DOI: 10.4018/978-1-5225-9631-8.ch002

extravagance. The head scientist tells Superman, "You're now pushing against the equivalent of **200 quintillion tons**, Superman" (p. 20). However, Superman's trip to the sun to save the astronauts had a side-effect:

Your trip to the sun exposed you to critical levels of stellar radiation, more raw energy than your cells are able to process efficiently. Apoptosis has begun. Cell death. There can be only one outcome even for You. (p. 20)

Superman's exposure to the sun is more than he can take, and within one year, The Man of Steel will meet his end. This is the premise of Morrison and Quitely's acclaimed graphic novel *All-Star Superman*, which has been used as a text to teach composition (Romagnoli, 2015).

The value of this text doesn't end with the story, though. Looking back at the scenario that was just presented, unmistakable teaching moments have arisen. 200 quintillion tons? Solar atmosphere? Apoptosis? Cells processing solar energy? A science lesson has just taken place.

LITERATURE REVIEW

The STEAM and Comic Connection

Comic books have traditionally been associated with science and technology, and Jones (2004) even traced the connection of science and popular literature to the "pulps" of the early twentieth century:

In the archaeology of popular culture, the August 1928 issue of Amazing Stories appears again and again as a pivotal memory of a generation of moviemakers, science fiction writers, cartoonists, astronomers, futurists, and rocket engineers. Against a lucid yellow background floats a man in a skintight red costume, leather pilot's helmet, and sleek black boots, his body prone but angling upward in an attitude of nascent flight. (p. 30)

Most important to note from Jones' historical study of comic books is his implicit connection of various and diverse disciplines: cartoonists, astronomers, moviemakers, and rocket engineers. This is the essence of STEAM (science, technology, engineering, arts, and mathematics) education: the idea that the logistics of scientific realities are informed by the aesthetics of the scientists' and engineers' imaginations who are responsible for the creation and implementation of new machines. Most importantly, this has been the case for quite a long time.

The graphic novel, "a comic style narrative, fictional or nonfictional, that is either a collection of individual comic book issues or an extended, original work" (Romagnoli, 2013), has transcended its pulp roots to inform culture, popular media, and education. Tatalovic (2009) has presented a proverbial call to action for both educators and scientists about incorporating graphic novels and comics into science education. This incorporation is meant to help bridge the gap between those who work in the sciences (engineers, scientists, mechanics, educators, etc.) and those who do not (students, general public). Recognition of the benefits that comics possess for education is becoming more apparent in science education:

from my exploratory research of science comics, it appears that scientists and educators are becoming more aware of the appeal that comics hold for young people and are starting to use them more as a vehicle to communicate scientific ideas. (p. 4)

The benefits of comics lie in their very structure which incorporates multiple modes of interpretation with which readers interact.

The unique combination of visual stimuli with narration and dialogue allows for students to access multiple modes of interpretation and comprehension that might otherwise be missing in a monomodal text. An interesting way of approaching graphic novels is to think of them as traditional novels that have the setting and the action drawn instead of written. Some educators may view that as blasphemous, but graphic novels' visual nature benefits students who are becoming increasingly versatile in their abilities to interpret meaning from varying sources, both traditional and academically unestablished.

Even though comics have traditionally been relegated to the proverbial slums of literary canon because of their sophomoric content and/or emphasis on superheroes, their visual characteristics can benefit students. Specifically, graphic novels access students' "multiliteracies" (New London Group, 1996) which have become a mainstay in literary studies over the past twenty years.

Multimodality, Graphic Novels, and the STEAM Connection

Even though comic books have been a mainstay of popular media since the early twentieth century, it wasn't until 1978 when comic book artist Will Eisner wrote *A Contract with God* which became what is considered the first graphic novel. This text informed what became Will Eisner's professional thesis on comic art and storytelling, mainly *Comics and Sequential Art* (1985). Eisner's focus of what he termed "sequential art" was on the relationship between text and images which he said were, "irrevocably interwoven" (p. 122). The connection that Eisner speaks of is symbiotic as, "the words become welded to the image and no longer serve to describe but rather to provide sound, dialogue and connective passages" (p. 122).

McCloud's seminal work *Understanding Comics* (1992) deconstructed comic storytelling and further highlighted the important connection between text and art. However, McCloud's commentary on this relationship informs Tatalovic's (2009) assertion that comics can help bridge knowledge gaps. McCloud sums up the popular argument against comics this way: "Words and pictures **together** are considered, at best a diversion for the masses, at worst a product of crass commercialism" (p. 140). The irony of this statement is that as technology evolved, the importance of visuals only increased because, "words and **moving** pictures have half the world in thrall to their charms…" (p. 141).

The New London Group (1996), which was comprised the world's most elite compositionists, deemed it necessary to coin a new term to address the evolution and eventual dominance of media, the Internet, and computers as the premier forms of communication and entertainment. "Multiliteracies" was established, and it embodied a new set of views on how audiences interpret media. Traditionally, literacy, "has referred to reading and writing of printed text" (Westby, 2010, p. 65). The privileging of printed text over more dynamic and visual forms of media had been powerful in schools, and that prevailing notion of accepting printed text over forms of media is still present. The New London Group saw literacy as being multifaceted especially in terms of pedagogy:

A pedagogy of multiliteracies, by contrast, focuses on modes of representation much broader than language alone. These differ according to culture and context, and have specific cognitive, cultural, and social effects. (p. 64)

For STEAM education, this has profound repercussions. Jolly (2014) states, "The purpose of STEAM should not be so much to teach art but to apply art in real situations. Applied knowledge leads to deeper learning." This mirrors The New London Group's view on multiliteracies: readers possess multiple literacies that work in tandem with one another to achieve a holistic understanding of a given text.

The majority of research and literature on incorporating graphic novels into educational contexts has been in language arts (Carter, 2007; Eckard, 2015; Romagnoli & Pagnucci, 2013). Social studies also has become a popular subject in which to utilize graphic novels. However, mathematics and the sciences are just beginning to access the potential of graphic novels.

Brozo, Moorman, and Meyer (2014) have also outlined ways in which graphic novels can be used in STEAM education. The benefit of using graphic novels in science courses is that "with their blend of dialogue and images, [they] allow students to gain and interpret science concepts and information through a unique and accessible format, opening a window to science for even the most reluctant learners" (p. 77). The "blend of dialogue and images" is akin to "multiliteracies," and it highlights the importance of the multiple forms of interpretation and creation that graphic novels embrace. With mathematics, the authors offer a unique view of how mathematics can benefit from graphic novels: "We would like to propose a vision for mathematical teaching and learning that moves beyond routine knowledge to something closer to socially constructed mathematical inquiry" (p. 100). Interestingly, Brozo, Mooreman, and Meyer are not the only educators who feel this way about mathematics and graphic novels.

Reilly (2015) presents a way in which a middle school mathematics course can utilize graphic novels to increase students' understanding. Reilly had students in an eighth-grade math course create "three comic books with various themes. All of the comic book adventure themes involved heroes working together to prevent some disaster from occurring" (p. 67). Students were specifically tasked with writing a story "which focused on the overall theme of fully developing one's talents in the face of societal pressure to conform" (p. 66). Reilly connects this socially conscious angle of her assignment to the superhero team The X-Men who are faced with fear by humans who do not fully understand them. This assignment provides three significant forms of literacy:

1. Social Literacy – Students are required to consider social issues while creating the mathematics themed graphic novel.
2. Mathematical Literacy – Students are required to fully explain a mathematical concept in the context of their story.
3. Creative Literacy – Students are required to create a comic book with both text and visuals.

This assignment contains all of the components of STEAM education and blend the various disciplines into a cohesive and powerful unit for learning mathematics. Additionally, as The New London Group (1996) asserts, the students are able to access their multiliteracies to ensure understanding of complicated mathematical concepts.

In science education, the graphic novel has become a means to instill interest in the subject and make complicated scientific principles intellectually accessible for all students. Tatalovic (2009), whose call to action to bridge this proverbial gap was cited earlier, has more concrete commentary on the benefits of utilizing graphic novels in scientific education. The benefits of using graphic novels in science relate to issues of content accessibility and generating interest. However, Tatalovic also warns against incorporating graphic novels or comics that may be perceived as being too loaded in their agendas: "Similarly, some have even suggested replacing textbooks with comics. But if the text in comics is overly didactic, it may easily lose connection with the story that is being told resulting in a less interesting read" (p. 13). There needs to be a balance in the way graphic novels are used in STEAM education.

Reilly (2015), whose mathematics class was discussed earlier, maintains this balance by having students create their own graphic novels. Doing so caused Reilly's participants to have intellectual and cultural capitals invested in both the story *and* the content. This is the key to integrating graphic novels into STEAM education: balancing the entertainment of the story with the scientific/mathematic principles that are expected to be understood given a particular curriculum.

Texts and Activities for STEAM Education and Graphic Novels

The overarching benefits that useful activities and graphic novels in STEAM have on students are the multiple forms of interpretation they provide for complicated content. The combination of the visuals and text, which graphic novels are revered for, is the prime benefit for students. Traditionally, much of the content students studied in their courses would be monomodal: one form of interpretation. For example, a traditional prose text would be monomodal because it only employs text to convey meaning. Alternatively, an instruction manual could be multimodal because it employs both text and images to convey meaning. Providing multiple ways to interpret meaning from a text allows readers to access their multiliteracies. Having established the importance of multiliteracies and multimodality in STEAM education, the following are examples of powerful texts teachers can utilize for various disciplines. This is by no means an exhaustive list, but it is a significant starting point for STEAM educators.

Biology and Evolution

Evolution can be a startlingly complicated concept for students to grasp, and evolution's place in curriculums around the country is sometimes challenged for socio-political and socio-cultural reasons. These issues notwithstanding, Charles Darwin's *The Origin of Species*, which was originally published in 1859, serves as the premiere text for studying both the historical context for this scientific theory and the science behind the theory.

The 502-paged original text may prove unwieldy to some students, and that is understandable given the language of the original text:

It need not be supposed that all varieties or incipient species attain the rank of species. They may become extinct, or they may endure as varieties for very long

Figure 1.

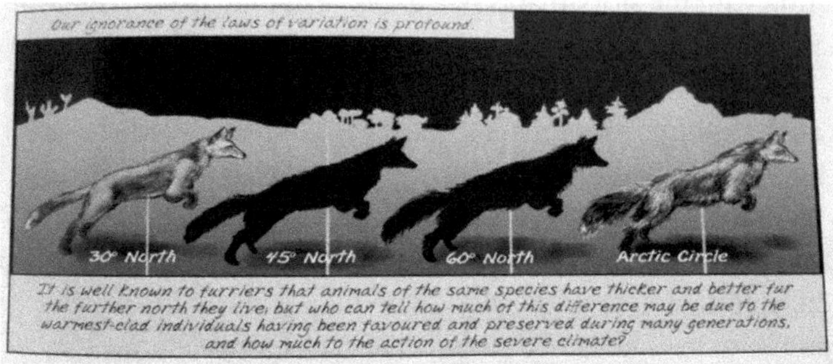

periods, as has been shown to be the case by Mr. Wollaston with the plants by Gaston de Saporta. (Darwin, 1909, p. 68-69)

Citing this language is not to criticize Darwin's work or downplay the text's significance. Rather, citing this language presents what students would be tasked with if they were required to read Darwin's text. This leads to how graphic novels may help students with access to Darwin's important and seminal work.

Charles Darwin's On the Origin of Species: A Graphic Interpretation (2009) is written and illustrated by Michael Keller and Nicolle Rager Fuller. It presents Charles Darwin's original story and scientific observations with beautiful illustrations that provide multiple forms of interpretation for students to analyze.

One particularly powerful example the book's utility is when Darwin's Laws of Variation are explored. The Laws of Variation, which look to clarify why certain species vary in appearances and builds, is particularly interesting. This law helps students understand why all animals of a given species have similar characteristics (i.e. birds and their wings) but look decidedly different and possess varying degrees of attributes (i.e. which birds can fly and which cannot). Keller and Fuller present the following panels to help explain this:

The visual of the canine's fur changing as it goes farther north, coupled with the original narration from Darwin, helps solidify the concept of variation. In short, the canine needs less fur if it lives in a warmer climate. Complementing this is the next set of panels which elaborates on the effects of climate on particular species' evolution.

This section's multimodality is a highlight of a text that is rife with these multimodal explanations of Charles Darwin's theory of evolution. This section blends Darwin's life story as well as his evolutionary theory. Charles Darwin's story, which provides even more context for students as they read about his evolutionary theories, encapsulates

Figure 2.

the very spirit of STEAM education. Incorporating a given scientist's story while he/she was developing ground breaking discoveries contextualizes the struggles, both scientific and social, that many brilliant minds faced throughout history.

To complement the study of this text, a teacher could have students create their own sections on Charles Darwin's evolutionary theories. The graphic adaptation is very extensive, but it is by no means complete. Reading at 189 pages, *Charles Darwin's On the Origin of Species: A Graphic Interpretation* leaves parts of Darwin's theory out; this is where the students come in. A teacher could explore the original text on his/her own and provide short excerpts for which students to create their own graphic novels. Not only would this encourage students to interpret a text using multiliteracies, it extends the discussion about Darwin and evolution through student created content. Additionally, the creation of an original graphic novel section informs the very nature of STEAM education, particularly as it relates to the "A" (Art).

Having students create original artistic sections about Darwin's work follows Bloom's Taxonomy (Bloom, B.S. et al., 1956), a staple of pedagogical and assessment instruction for the past fifty years. Specifically, students apply their knowledge of the content and synthesize the learned content by creating visual representations of learned content. By having students create original sections, they are simultaneously exploring advanced scientific concepts whilst creating stories through their art.

Mathematics and the Moon Landing

One of the most significant events in American (and world) history was when Apollo 11 successfully landed on the moon, and astronauts walked on the surface. This historical event had implications that extended beyond the science and math that was necessary for astronauts to safely make it to the moon and back again. With the backdrop of the Cold War and the Space Race, The United States and The Soviet

Figure 3.

Union competed for political and military dominance on Earth with the battlefield being beyond the planet.

The scientific lessons associated with the moon landing are apparent, and the study of the moon landing serves as the basis for studies of gravity, the moon itself, and how the moon's proximity to Earth affects the planet. However, the math behind this human achievement is unmistakably important for students to explore in some high school classrooms and beyond.

T-Minus: The Race to the Moon (Ottaviani, Cannon, and Cannon, 2009) tells the stories of the astronauts, scientists, politicians, and mathematicians who were responsible for getting the United States to the Moon with Apollo 11 in 1969. Throughout the graphic novel, there are sequences that outline both the scientific and mathematical concepts that both scientists and mathematicians had to navigate throughout the preparation process.

One of the major factors that needed to be addressed for astronauts and rocket builders was the amount of g-force(s) an astronaut could withstand. This becomes both a scientific and mathematical problem for students to explore. Scientifically, understanding how gravity affects the body informs the study of space travel for humans in general. Mathematically, understanding how to calculate g-forces addresses the reality of space travel and how understanding mathematics plays into that reality. One poignant scene from the graphic novel is when a cosmonaut is put into a centrifuge for training purposes:

Below the image of the cosmonaut who is clearly distressed by the training is a definition and example of "g-force" from which students can begin to explore further. However, the image of the cosmonaut in fear is incredibly powerful; it helps students connect the mathematics of gravity to the insanity of space flight immediately.

Figure 4.

One page later, this same cosmonaut is put in a pressure chamber to test his ability to concentrate during a space-travel crisis. This test involves mathematics:

The extreme conditions the cosmonaut is put through are intended to mimic what a space traveler would have to navigate if s/he were actually in space and an emergency arose. Additionally, having to solve math problems in rapid succession during a given space crisis makes sense. This would particularly be plausible should the ship's trajectory need to be adjusted or the cosmonaut had to make quick, life-saving decisions using data from the ship's telemetry. However, one major factor that students reading this text in a math class have to explore is the absence of calculators. Not having access to a calculator should become a discussion point in class which has socio-cultural implications. In a math class, where conversation is stereotypically associated with quantitative reasoning, students are able to apply experiences to support assertions concerning the use of a calculator.

Additionally, *T-Minus'* multidisciplinary content encourages the cross-curricular exploration of themes. This particular graphic novel has literary, scientific, political, and mathematical themes embedded throughout, and this highlights the symbiotic relationship that all disciplines share. Chemistry, one of the major disciplines that is cited in *T-Minus* is also present in other graphic novels.

Chemistry and the Graphic Novel

The prominent visual for chemistry is the Periodic Table of Elements. The table, which displays the atomic details of all known elements that are the building blocks for all matter, is a mainstay in chemistry and science classrooms around the world. Each element on that table is a story unto itself. The story of how those elements were discovered and the discovery's implications are as interesting as the element itself. In short, chemistry involves both the elements' technicalities and the stories behind those elements.

Gonick and Criddle (2005) tell the story of chemistry's history and evolution in *The Cartoon Guide to Chemistry*. This graphic novel is a unique way to explore chemistry with humorous illustrations and comments. Within the first two pages of the graphic novel, Gonick and Criddle summarize the field of chemistry:

Fire – and those other processes – revealed hidden features of matter. If you heat a piece of wood, all you get is a hot piece of wood, at first...but suddenly, at some point, the wood bursts into flame. Where did that come from? Chemistry is the science that answers that question, and chemical reactions are the strange transformations that reveal matter's hidden properties. (p. 2)

Accompanying this explanation is a humorous cartoon of two cave people confusedly observing wood burning. *The Cartoon Guide to Chemistry* illustrates complicated concepts whilst explaining the concept in a humorous and accessible way.

The three stages of matter are explored as well. Understanding the differences between solids, liquids, and gasses is a fundamental concept that involves complicated science for students. Identifying solids, liquids, and gasses is easy enough; taking an ice cube and placing it in a hot pan shows the process occur. However, the science of these states is something else entirely. Gonick and Criddle describe the states using the following image:

(p. 105)

Placing the characters in oversized jars and explaining how the states of matter work is an effective visual for students. It also serves as a great discussion piece about why the characters are reacting the way they are. Even before exploring

Figure 5.

temperature, pressure, and other factors that influence states of matter, students are free to propose their own hypotheses as to how states of matter work.

This is the prime benefit of this particular text: it encourages discussion through the images. More so than the other graphic novels that have been cited throughout this chapter, *The Cartoon Guide to Chemistry* serves as a discussion starter. Throughout the text, there are cartoons that depict chemical processes and facilitate class discussions.

Using this text as an introduction throughout early chemistry lessons would benefit students. With chemistry, the instructor's expertise and experience in the field serves as the primary resource, but *The Cartoon Guide to Chemistry* is excellent for beginning conversations and accessing students' intellectual capital when it comes to chemistry.

Ethics and Science Education

The beauty and majesty of space exploration has been an exotic concept for people of all ages going as far back as the early twentieth century. Magazines and pulps of the time were inundated with images of flying men, alien contraptions, ray guns, and impractical flying spaceships. Stories of fearless astronauts traversing the stars fueled the minds of young people who would one day be responsible for sending others to the moon and exploring the stars. It was all a beautiful fantasy in richly colored pages.

Even after the beginning of the space race in the mid-1960s but before man ever stepped foot on the moon, children and adults alike watched *Star Trek* (1966) on their televisions: a progressive and promising view of science fiction as a genre. However, a mere nine years before *Star Trek's* influence, the world witnessed the reality of space travel. The reality of space travel, as it turns out, is far less romantic.

The graphic novel *Laika* by Nick Abadzis (2007) tells the story of the first living animal in space. Launched in November of 1957, Sputnik II had a dog named Laika aboard who became an international sensation. She also became a symbol of Soviet strength at the genesis of the space race. The launch was a major victory in the space race. Man had successfully sent a living being into space and could monitor a mammal's reactions to the harshness of space and the violence of a rocket launch.

Laika chronicles this momentous achievement in science, but it presents the story from a unique perspective: Laika and her caretaker. Yelena, Laika's caretaker, struggles with preparing the stray dog for space flight and completing the duties expected of her. Emotional attachment to *Laika* consumes Yelena, and the narrative throughout tests her commitment to the mission as well as her allegiance.

The graphic novel addresses scientific matters regarding preparation for space travel. Laika's physical limits are tested in a centrifuge designed to mimic g-forces, and she experiences weightlessness via an airplane performing parabolic dives. For students, there is science throughout the graphic novel that is worthy of being examined. However, the true power of the graphic novel comes from its emphasis on ethics in scientific research.

Balancing the benefits and deficits of sending an animal into space is a cornerstone of the graphic novel, and the emotions of the scientists are apparent throughout. Of

Figure 6.

particular note is Yelena who begins to interpret the dog's barking as communicating her frustrations and fears.

For students working in STEAM environments, *Laika* can serve as a final and significant discussion point in a science course. In *Laika*, all the logistics of launching a rocket, calculating orbits, developing appropriate test subjects, and developing protocols for space travel ultimately result in one profound question: does the life of a single, stray dog matter when attempting to advance humanity's exploration of space? The answer is not easy, and scientific advancement rarely offers easy questions. Instead, humanity is forced to question not only the task at hand, but its ethical implications as well.

This question, whether it is right to sacrifice life for scientific advancement, is STEAM education. It is the totality of human experience questioned at a singular point in time, a singular human achievement that can simultaneously be celebrated and mourned. Using *Laika* as a reflective practice can encourage students to consider their expanding scientific knowledge and how to apply it. It is a deep question, and one that needs significant contemplation, but that is what educators need to consider when implementing STEAM education. Educators are responsible for disseminating scientific knowledge as well as instilling a conscience in young scientists who will one day lead humanity to its next great discovery. *Laika* serves as a reminder to future scientists that one generation had a similar struggle and explains what that struggle meant for humanity. In an afterword, Abadzis cites a quote from one of the mission scientists, Oleg Georgivitch Gazenko: "Work with animals is a source of suffering to all of us. We treat them like babies who cannot speak. The more time passes, the more I'm sorry about it. We did not learn enough from the mission to justify the death of the dog" (p. 201).

CONCLUSION

When looking back on the graphic novels that have been cited in this chapter, it becomes apparent that all of the graphic novels have one characteristic in common that transcends all of the disciplines; they are stories. While stories and storytelling are often associated with English or history classes, they are the core of human experience. Those experiences account for the scientific and mathematical discoveries that students explore in STEAM education.

For STEAM education, making the content accessible to students of all experiences and backgrounds is of paramount importance. This mentality reflects the reality of what it means to be a scientist. Kakalios describes what it means to be a physicist in the book *The Physics of Superheroes* (2006):

Despite the impression gleaned from popular movies, being a physicist does not require an encyclopedic knowledge of equations and fundamental constants, coupled with the ability to perform complex arithmetic in one's head with robotic speed and precision. Physics is not about having memorized all the answers, but rather about asking the right questions. For when the right question is posed of a phenomenon, either the answer becomes clear, or at least the manner as to how one should go about obtaining the answer is revealed. (p. 16)

In short, science is about experience and sharing those experiences with each other through stories that enlighten, educate, and connect. Graphic novels serve as a positive and multimodal option for educators and students alike.

Pagnucci (2004) has stated that, "Stories connect us to other people, and we have a responsibility to respond to those stories" (p. 76). When students are accessing their multiliteracies and engaging with multimodal texts to explore complicated and significant scientific and mathematical principles, they are taking part in the discovery of these principles. Their own discoveries are occurring through the text and the images, and graphic novels facilitate this discovery.

This is an important distinction to mark when it comes to STEAM education and graphic novels. Utilizing graphic novels in class is about the balance of literacies and recognizing diverse ways to understand concepts; that is all very important. However, a pattern has started to emerge where the perceived coldness and rigidity of science and mathematics is being dispelled. Science is not just lab coats, hypotheses, beakers, and experiments. Mathematics is not just equations, proofs, memorized operations, and prime numbers. Science and mathematics are influenced and shaped by people, and the people's experiences affect the sciences and mathematics. More so than the multimodality of the comic books, the stories within this literature help to humanize what are sometimes perceived as robotic subjects. There is beauty in constellations, exploding stars, sea life, and cellular reproduction. There is beauty in perfect proofs, statistics, and pi. There is true beauty in all of the people who sacrificed their lives, their fortunes, and sometimes their well-being in the noble efforts to simply inform the world's people about how the universe works and why they should care. There is beauty in knowing these stories and reflecting on not just the discoveries, but the lives of these scientists who made it their missions in life to better the world through discovery and experimentation.

REFERENCES

Abadzis, N. (2007). *Laika*. New York: First Second.

Bloom, B. S., Engelhart, M. D., Furst, E. J., Hill, W. H., & Krathwohl, D. R. (1956). *Taxonomy of educational objectives: The classification of educational goals*. New York: David McKay Company.

Brozo, W. G., Moorman, G., & Meyer, C. K. (2014). *Wham! Teaching with graphic novels across the curriculum*. New York: Teachers College Press.

Carter, J. B. (Ed.). (2007). *Building literacy connections with graphic novels: Page by page, panel by panel*. Urbana, IL: National Council of Teachers of English.

Cazden, C., Cope, B., Fairclough, N., & Gee, J. (1996). A pedagogy of multiliteracies: Designing social futures. *Harvard Educational Review*, *66*(1), 60–92. doi:10.17763/haer.66.1.17370n67v22j160u

Darwin, C. (1909). *The Origin of Species*. New York: P.F. Collier & Son Company.

Eckard, S. (Ed.). (2015). *Yin and Yang in the English Classroom: Teaching with Popular Culture Texts*. Lanham, MD: Rowman & Littlefield.

Eisner, W. (1978). *A contract with God*. New York: Baronet Books.

Eisner, W. (1985). *Comics & Sequential Art: Principles & Practice of the World's Most Popular Art Form*. Poorhouse Press.

Gonick, L., & Criddle, C. (2005). *The Cartoon Guide to Chemistry*. New York: Collins Reference.

Jolly, A. (2014). *STEM vs. STEAM: Do the Arts Belong?* Retrieved from http://www.edweek.org/tm/articles/2014/11/18/ctq-jolly-stem-vs-steam.html

Jones, G. (2004). *Men of tomorrow: Geeks, gangsters and the birth of the comic book*. New York: Basic Books.

Kakalios, J. (2005). *The Physics of Superheroes*. New York: Gotham Books.

Keller, M., & Fuller, N. R. (2009). *Charles Darwin's On the Origin of Species: A Graphic Interpretation*. Rodale.

McCloud, S. (1993). *Understanding comics*. Tundra Publications.

Morrison, G., & Quitely, F. (2011). *All Star Superman*. New York: DC Comics.

Ottaviani, J., Cannon, Z., & Cannon, K. (2009). *T-Minus: The Race to the Moon*. New York: Aladdin.

Pagnucci, G. S. (2004). *Living the narrative life: Stories as a tool for meaning making*. Portsmouth: Heinemann.

Reilly, E. (2015). Superheroes in Math Class: Using Comics to Teach Diversity Awareness. *Works & Days*, *32*(1 & 2), 61–74.

Romagnoli, A. (2015). Up, Up, and Away: Superman in the Composition Classroom. In S. Eckard (Ed.), *Yin and Yang in the English Classroom: Teaching with Popular Culture Texts* (pp. 123–135). Lanham, MD: Rowman & Littlefield.

Romagnoli, A., & Pagnucci, G. (2013). *Enter the Superheroes: American Values, Culture, and the Canon of Superhero Literature*. Lanham, MD: Scarecrow Press.

Tatalovic, M. (2009). Science comics as tools for science education and communication: A brief, exploratory study. *JCOM: Journal of Science Communication*, *8*(4), 1–17. doi:10.22323/2.08040202

Westby, C. (2010). Multiliteracies: The changing world of communication. *Topics in Language Disorders*, *30*(1), 64–71. doi:10.1097/TLD.0b013e3181d0a0ab

KEY TERMS AND DEFINITIONS

Accessibility: The theory that all students are able to read and interact with a given text so as to be able to understand a given and/or participate in the learning process.

Comic(s): A shorter comic style narrative that serves as a short story or part of an ongoing serial or series.

Graphic Novels: A comic style narrative, fictional or nonfictional, that is either a collection of individual comic book issues or an extended, original work.

Monomodal Text: Texts that contain a single mode of interpretation (novel, poem, etc.).

Multiliteracies: The theory proposed by The New London Group that readers and learners possess multiple literacies that work in tandem in order to ascertain understanding or synthesize meaning from a multimodal text.

Multimodal Text (Multimodality): Texts that contain multiples modes of interpretation that work in tandem to present a story or concept.

Chapter 3
High-Quality Trade Books and Content Areas:
Planning Accordingly for Rich Instruction

Carolyn A. Groff
Independent Researcher, Switzerland

EXECUTIVE SUMMARY

Integrating high-quality children's tradebooks into elementary content areas has long been considered a best practice. When teachers choose to incorporate these texts into content area lessons, they are exposing students to art through the pictures and reaching an array of visual learners. There is a delicate balance between teaching the literacy strategies needed to read these texts and the actual content materials that students need to learn in the STEAM areas. This chapter explores how to incorporate texts appropriately into content area lessons so that students can focus on the content, as well as apply literacy strategies for comprehension.

LITERATURE REVIEW

Literacy and content area learning are in a constant state of change. According to Adams and Pegg (2012), "one dimension relates to shifting understandings regarding student learning, and the other dimension involves the relationship between content and literacy" (p. 151). Citing the work of previous researchers, Adams and Pegg argue that discipline-specific discourse can be taught through literacy practices; that is, the way the students talk about science, technology, engineering, and math (STEM) can be learned through literacy practices in the classroom. Moreover, because

DOI: 10.4018/978-1-5225-9631-8.ch003

discipline-specific discourse is used by authors to create informational texts, using such texts can lead to increased understanding in content areas.

Integrating high-quality children's trade books into elementary content areas has long been considered a best-practice (Olness, 2007). More specifically, high-quality informational texts can assist students in learning STEM concepts. Using these texts with rich visuals and vocabulary forges a link between the arts and the STEM focus, creating the use of STEAM. Yopp and Yopp (2012) state that students' engagement with informational texts more richly promotes their knowledge of content areas by teaching them vocabulary, text structure (such as compare and contrast), and text features (such as charts and tables). Teachers must be able to merge their selection of high-quality informational texts and scaffolded instructional models in order for students' experiences to be successful (Fisher & Frey, 2015). Fisher and Frey (2015) state that "as students invest themselves in these informational texts, look for opportunities to encourage ways to inspire further investigation, discussion, and writing" (p. 529). However, for pre-service teacher candidates and novice teachers, planning for the blend of literacy and other content areas may not come naturally (Hoffman, Collins, & Schickedanz, 2015). Citing prior research, Hoffman, Collins, and Schickedanz (2015) state the following with regard to science instruction:

Research on effectively integrating science and literacy instruction to develop conceptual understandings uses informational text as a complement to hands- on scientific investigations in the science curriculum, not as a replacement for such experiences (Palincsar & Magnusson, 2001; Pappas & Varelas, 2004; Varelas & Pappas, 2006). Thus, informational texts are most meaningfully integrated in literacy instruction when they align with the ongoing science curriculum (i.e., the texts are related to the scientific concepts investigated in hands- on experiences). (pp. 368-9).

The lesson plan template provided in this chapter is meant to assist teacher candidates and practicing teachers in preparing for the integration of high-quality informational texts into STEM content area lessons, focusing on the integration of visual information and concept development.

According to Mallett (2010) who draws on the work of Piaget and Vygotsky, there are three principles for effectively teaching and learning with informational texts. The first principle is that the act of teaching concepts always begins with active exploration (Mallett, 2010). Children are able to become naturally curious about concepts presented to them, and teachers can draw on that excitement by allowing children to actively explore the concept in combination with evidence from informational texts (Mallett, 2010). Mallett (2010) writes that "experiments in science, field trips in geography, outings and museum visits to enliven history and English all help to sustain and deepen interest, and all nourish learning from texts" (p. 232). Further, Mallett continues to explain what she contends is the second principle, "this exploration is not done in isolation, but rather through dialogue with peers and teachers." Therefore, during instructional time, teachers should allow students to talk to one another. Teachers should also prompt students to share their ideas about concepts (Mallett, 2010). This communication helps children forge critical links between what they are actively exploring in a hands-on manner and what they are reading in texts (Mallett, 2010). Mallett's third principle is that the teacher's role in the use of informational text and concept development is central. Effective teachers understand how to choose appropriate texts, plan effective lessons that promote active engagement, and model how to learn from text and first-hand experiences so that children can emulate these behaviors (Mallett, 2010). It is critical to choose only high-quality informational texts.

Texts come in many different forms, including digital texts which can be displayed on a screen. However, even in traditional printed texts, the layouts task the reader to think about the content in new ways (Hassett & Schieble, 2010). According to Hassett and Schieble, "The computerization of type design and the photomechanical printing technologies available today make it possible for alphabetic text to mingle with graphics, extending the ways in which thought might be represented" (2010, p. 325). Hassett and Schieble argue that the linguistic cueing systems, graphophonic, semantic and syntactic, traditionally used to understand how reading works are no longer enough to navigate these new types of texts (including social media, technology, icons, etc.). They suggest that "we might begin by updating our reading cueing systems to encompass a greater scope. For example, visual cues clearly extend beyond the sign systems of print and now include images, graphics, and the look of the word on the page" (Hassett & Schieble, 2010, p. 330). Despite the challenges of doing so, the opportunities that this provides in turning STEM lessons into STEAM lessons are many. Choosing books containing rich vocabulary, complex structure, and thought-provoking and exciting visuals, will demonstrate to students that art and literacy can be integrated into any content area.

CONTENT INFORMATION

This chapter's lesson plans for integrating informational texts in STEAM lessons is designed to work with the Common Core College and Career Readiness Standards as well as The Next Generation Science Standards. The specific grade level standards can be found on the following website: corestandards.org/ELA-Literacy/CCRA/R/. In most situations, teachers may be using informational trade books to expose students to concepts; therefore, it is important for teachers to look for the specific grade level under *Reading: Informational Text* found on the same webpage. Depending on the content area being addressed, teachers also will have to choose grade-appropriate standards in the STEAM areas. Because high-quality informational texts can be used in several different ways, the standards addressed vary depending on the area of integration. For example, if an instructional book about writing is used during a science lesson to teach procedural writing in a visually artistic way, then the corresponding science and English/language arts standards would be invoked at the appropriate grade level.

CASE DESCRIPTION

The following lesson plan format can be used to integrate trade books into STEAM lessons, forging a link between the content areas, literacy, and the arts. As Yopp and Yopp (2012) suggest, the way teachers share an informational text with the students is essential to the lesson's success. They recommend a "before reading, during reading, after reading" format, which is followed here.

Lesson Plan Information

Activation of Background Knowledge and Hands-On Activity

Elementary school children often carry misconceptions about STEAM concepts because they have not developed enough knowledge to understand basic principles (Hoffman, Collins, & Schickedanz, 2015). As a teacher, it is important to discover these misconceptions so that they might be corrected through the use of literacy strategies, and especially through the use of trade books. Having students discuss and hypothesize through an activity can serve as an informal assessment of the misconceptions students carry. With a given STEAM concept in mind, develop an activity and corresponding questions. For example, a popular science activity for young children is determining whether objects will sink or float. Let the students experiment with objects, each time predicting whether an object will sink or float and explanations as to why it occurs. If students inadequately or incompletely articulate why certain objects might sink or float, take note of their misconceptions. Further, prompt them with questioning such as, "What made you give the reason why that object could float (or sink)?" Observing students is extremely important here. Carefully document (through video if possible) their responses and dialogue as they proceed with their hands-on activity.

Strategy Choice/Book Research

With misconceptions documented, students' background knowledge activated, and students having acquired new understandings in the hands-on activity, educators can begin to address the central focus of their respective lesson. Consider which literacy strategies would help the students understand the concept through reading trade books and writing. Choose several informational or hybrid trade books on the concept. Make sure to choose books at varying reading levels to accommodate all readers. One of the books should be the mentor text that is shared with the entire class. Align an appropriate reading strategy with your text; for example:

- Making connections (e.g., text to visuals)
- Visualizing
- Inferring
- Asking Questions
- Determining Importance
- Synthesizing

Aligning reading strategies with books gives the students a purpose for digesting the book's information, especially for the concepts that relate to the central STEAM focus of the lesson. For example, by modeling how to ask questions while reading text, and allowing students to do the same, teachers demonstrate how asking questions can lead to a wealth of information discovery that might have been overlooked had not questions been asked.

It is helpful to create a chart that aligns the chosen books with the strategies, as well as the content area standards being covered and the content area(s) to which the books relate.

Using the Books to Foster Understandings

Before Reading: Before beginning the read-aloud of the mentor text with the whole class, review the discussions that took place during the hands-on activity. Remind students of their predictions and explanations for why something occurred. For example, in the case of having objects sink or float, a teacher might say, "Group 1, you said that the rubber duck could float because there was nothing weighing it down." After reviewing the discussion and activity, introduce the book by doing a picture walk. In most cases, the book will be an informational book, so be certain to point out the text features (photographs, charts, graphs, bold words, captions). Critical to the picture walk is the introduction of vocabulary that students may not have used when describing the experiments or activities. Providing the students with access

Table 1. Related books

Book title/author	Reading Strategy	Content Area Standards	Content Area(s)
Example *Sky Tree* Thomas Locker with Candace Christiansen	Making connections between text and visuals	K-ESS2-1 (Weather conditions)	Earth Science Language Arts/Reading

to that vocabulary will help them better understand the text and its relationship to the concepts at hand. Finally, state and model the chosen strategy which will help them think about the text as you read. For example, if "*Determining Importance*" was chosen, explain that not every detail in the text may be critical to understanding the main idea; therefore, it is crucial to choose the most important details to remember. It is helpful to guide the students in a brief practice of this strategy, using a paragraph from a previously read informational text on a STEAM concept that they have already mastered. Teachers should allow the students to assist them in determining the most important ideas in the paragraph and highlight them.

During Reading: There are key components to this phase. First, check continuously for understanding by questioning and prompting students for comments. Instead of having the students always answer the teacher, they can tell the answer to a friend next to them as the teacher listens. If the text is appropriate for making predictions, allow students to do so, and then confirm whether or not their predictions were correct. Second, continue the focus on the chosen strategy by modeling it frequently and having the students assist in using the strategy. Third, be certain to point out the textual features such as pictures, photographs, charts, graphs, bold words, and captions that assist in understanding, especially for ELL students. If possible, have enough copies of the text per two students. If these are not available, teachers should be sure to place students who need extra support towards the front of the classroom or project the book on a document viewer. It is at the teacher's discretion whether to read the entire text or not, depending on the rest of the lesson and the difficulty of the text. In the event that a text aligns with most students' reading levels, a teacher might read half of the book, and then send the students in partners to fill out a graphic organizer while reading the rest of the book together. They would then come back to share their findings. Students who need extra support can be provided with easier texts on the same topic so that they can also participate in the discussion. The mentor text and the other books can then be placed in various content area centers so that students can refer to them as they go about their work.

After Reading: When the text has been read, it is time to assess whether or not the reading increased understanding and cleared up misconceptions. Return to the discussion held before reading about statements made during the activity and ask students to confirm or deny those statements using textual evidence. Repeat the activity from the beginning of the lesson or a similar activity. Ask students to describe the events using vocabulary and information from the book. If a copy of the book is available to students, they can write down the page number where they found that information in observational logs or lab report books.

Post-Lesson Professional Development

It is critical for teacher candidates and practicing teachers to analyze their integration of informational trade books into STEAM content areas because this integration may not always come naturally. Provided below are some reflective questions that can be used for professional development purposes.

- What do you feel were the strengths and weaknesses of your book choices?
- How could you embed trade books differently in a future content area lesson?
- What high-quality characteristics did your mentor text have that made you choose it for your whole group reading?
- How will reading this text help your students master the Content Area standards listed on your chart?
- How did the text help the students better understand the STEAM concept that was the central focus of the lesson and clear up any misconceptions?

CURRENT CHALLENGES FACING TEACHERS

Choosing high-quality informational texts for elementary grades can be overwhelming and time consuming. Often, curricula materials may come with their own texts which do not involve as much planning. However, these texts can be dull, artistically lacking, and often do not accommodate the range of concepts which need to be conveyed in a unit (Hoffman, Collins, & Schickedanz, 2015). While teachers may possess excellent pedagogical skills, their lack of practice in cross-curricular integration may prevent them from attempting such endeavors, especially given the high-stakes environments in which many teachers work (Hoffman, Collins, & Schickedanz, 2015). According to Hoffman, Collins, and Schickedanz, "even highly skilled, experienced teachers found discussion of concepts during read-alouds challenging in practice" (p. 365). Teachers may become too focused on the literacy skills and strategies involved in reading informational text, causing them to overlook the possibilities for engaging in rich discussions about the concepts (Hoffman, Collins, & Schickedanz, 2015). With many school districts focusing on preparing students for standardized tests in literacy and math, other content areas can be overlooked, and teachers are forced to weave this knowledge into literacy instruction time. Therefore, some teachers may lack practice providing full lessons in other content areas such as science.

SOLUTIONS AND RECOMMENDATIONS

During clinical practice and professional development, teacher candidates and teachers need practice with the following:

- Choosing high-quality informational texts that present, both in words and visually, a rich conceptual development of the topics they wish to cover;
- Creating hands-on experiences that will link to the reading of the text;
- Forming lesson plans in the STEAM content areas that strike a balance between concept development and the literacy skills/strategies needed for text comprehension (Hoffman, Collins, & Schickedanz, 2015).

In sum, teacher candidates and teachers need to be provided with the time to practice teaching in this manner. To maintain the status quo of only integrating content areas into literacy instruction is to rob both students and teachers of opportunities to deeply engage with concepts and make important connections to the world around them.

ACKNOWLEDGMENT

The author wishes to gratefully acknowledge the help of Brian Edgar, Ph.D. for his assistance and feedback during the writing of this chapter.

REFERENCES

Adams, A. E., & Pegg, J. (2012). Teachers enactment of content literacy strategies in secondary science and mathematics classes. *Journal of Adolescent & Adult Literacy*, *2*(56), 151–161. doi:10.1002/JAAL.00116

Fisher, D., & Frey, N. (2015). Selecting texts and tasks for content area reading and learning. *The Reading Teacher*, *68*(7), 524–529. doi:10.1002/trtr.1344

Hassett, D. D., & Scheible, M. B. (2010). Finding space and time for the visual in K-12 literacy instruction. In M. Cappello & B. Moss (Eds.), *Contemporary Readings in Literacy Education* (pp. 325–331). Thousand Oaks, CA: SAGE Publications.

Hoffman, J. L., Collins, M. F., & Schickedanz, J. A. (2015). Instructional challenges in developing young childrens science concepts. *The Reading Teacher*, *68*(5), 363–372. doi:10.1002/trtr.1325

Mallett, M. (2010). Choosing and using fiction and non-fiction 3-11: A comprehensive guide for teachers and student teachers. New York, NY: Routledge.

Olness, R. (2007). *Using literature to enhance content area instruction: A guide for K-5 teachers*. Newark, DE: International Reading Association.

Yopp, R. H., & Yopp, H. K. (2012). Young childrens limited and narrow exposure to informational text. *The Reading Teacher*, *65*(7), 480–490. doi:10.1002/TRTR.01072

ADDITIONAL READING

Avgerinou, M., & Ericson, J. (1997). Are view of the concept of visual literacy. *British Journal of Educational Technology*, *28*(4), 280–291. doi:10.1111/1467-8535.00035

Burmark, L. (2002). *Visual literacy: Learn to see, see to learn*. Alexandria, VA: Association for Supervision and Curriculum Development.

Coiro, J., Knobel, M., Lankshear, C., & Leu, D. J. (Eds.). (2008). *Handbook of research on new literacies*. London: Routledge.

Considine, D. M., & Haley, G. E. (1999). *Visual messages: Integrating imagery into instruction. A media literacy resource for teachers*. Santa Barbara, CA: Libraries Unlimited.

Cope, B., & Kalantzis, M. (2009). Multiliteracies: New literacies, new learning. *Pedagogies*, *4*(3), 164–195. doi:10.1080/15544800903076044

Dondis, D. A. (1974). A primer of visual literacy. Cambridge, MA: The MIT Press.
Elkins, J. (Ed.). (2008). Visual literacy. New York, NY: Routledge.

Frey, N., & Fisher, D. (2008). *Teaching visual literacy: Using comic books, graphic novels, anime, cartoons, and more to develop comprehension and thinking skills*. Thousand Oaks, CA: Corwin Press.

Keene, E. O., & Zimmermann, S. (1997). *Mosaic of thought: Teaching comprehension in a reader's workshop*. Portsmouth, NH: Heinemann.

Kiefer, B. Z. (1995). *The potential of picture books: From visual literacy to aesthetic understanding*. Englewood Cliffs, NJ: Merrill.

Lankshear, C., & Knobel, M. (2003). *New literacies: Changing knowledge and classroom learning*. Buckingham, UK: Open University Press.

Leu, D. J., Kinzer, C. K., Coiro, J. L., & Cammack, D. W. (2004). Toward a theory of new literacies emerging from the Internet and other information and communication technologies. In R. B. Ruddell & N. J. Unrau (Eds.), *Theoretical models and processes of reading* (5th ed.; pp. 1570–1613). Newark, DE: International Reading Association.

Messaris, P. (1994). *Visual literacy: Image, mind, and reality*. Boulder, CO: Westview Press.

Schönborn, K. J., & Anderson, T. R. (2006). The importance of visual literacy in the education of biochemists. *Biochemistry and Molecular Biology Education*, *34*(2), 94–102. doi:10.1002/bmb.2006.49403402094 PMID:21638645

Sinatra, R. (1986). *Visual literacy connections to thinking, reading and writing*. Springfield, IL: Charles C. Thomas.

Stokes, S. (2002). Visual literacy in teaching and learning: A literature perspective. *Electronic Journal for the Integration of Technology in Education*, *1*(1), 10–19.

KEY TERMS AND DEFINITIONS

Content Area Integration: The combining of two or more elementary content areas during a lesson plan or unit.

Elementary Content Areas: All subjects taught in elementary grades (language arts, math, science, social studies, world language, music, art, health, and physical education).

High-Quality Children's Trade Books: Books with rich vocabulary and complex structure, as well as thought-provoking and exciting visuals.

Linguistic Cueing Systems: Sound-symbol, structural, grammatical and meaning relationships that aid a reader, writer, speaker or listener in processing language.

Mentor Text: The main text used to teach a whole-group lesson to students of varying ability levels.

Misconceptions: Ideas that students hold about concepts which are inaccurate or false, especially in the scientific sense.

Reading and Writing Strategies: The application of purposeful and metacognitive ways of thinking to assist one with language arts tasks.

Visual Learning: Learning through viewing pictorial representations.

APPENDIX

Sample Lesson Plans for Grades K-2 and 3-5

INTRODUCTION

The goal of these lessons is to help students comprehend scientific and technical concepts by applying strategies that use images to assist in comprehension. In this sample lesson, we begin by looking and noticing. A strategy that can be used with any book is beginning with the title and assessing the images on the cover. Book covers are designed with care, and in the case of picture books, artists spend hours drawing, painting, or using computers to consider and design the images on the covers. We can then read the book and make evaluations about how the story or information is represented visually. During and after reading, we can consider how the images contribute to the understanding of the text.

Lesson Plan 1

1. Into the Text: Activation of Background Knowledge and Hands-On Activity

To assess prior knowledge and misconceptions about the seasons, have the students draw a picture of a tree in the four different seasons. Ask them to describe the changes in their trees (e.g., Why did you draw the autumn tree differently from the spring tree?). Drawings can be displayed or shared in pairs or groups. Students should be asked about the decisions they make with different colors and scenes.

After they describe their changes, ask them why those changes occur. Document their answers in a way that can be displayed for the whole class.

To prepare students to read the text and be introduced to the strategy of making connections between text and images, we might ask students to look at the cover of *Sky Tree* and ask them questions such as:

- What do you notice?
- What objects do you see?
- What colors do you see?

We might begin by noticing the tree. The blue sky between the branches and the clouds that form the boundary. We might notice the grass, river, and mountains in

Table 2. Simple lesson place; k-2 science: changes in the season; next generation science standard

K-ESS2-1.	Use and share observations of local weather conditions to describe patterns over time. [Clarification Statement: Examples of qualitative observations could include descriptions of the weather (such as sunny, cloudy, rainy, and warm); examples of quantitative observations could include numbers of sunny, windy, and rainy days in a month. Examples of patterns could include that it is usually cooler in the morning than in the afternoon and the number of sunny days versus cloudy days in different months.] [*Assessment Boundary: Assessment of quantitative observations limited to whole numbers and relative measures such as warmer/cooler.*]

Book: Sky Tree by Thomas Locker with Candace Christiansen. HarperCollins Publishers, 1995

the background. We might see a leaf that is formed by the shape of the clouds and formed in the blue sky. We can notice, areas of light and dark. The tree is the focus and is atop a small hill. Other details might stand out as well. For instance, there are no leaves on the tree. From this observation we can make inferences about the text's setting including the weather and season. Next, we may ask students to think about the title. We might ask students what *Sky Tree* means.

2. Through the Text: During Reading

At the beginning of the read-aloud, tell students that there will be stops throughout the reading so that they can make connections between the text and the pictures, commenting on the images and details the same way that was modeled with the class while looking at the cover. Explain that making connections between the ideas and the pictures in the book is an important strategy that will help them understand the seasons thoroughly.

Model this by using the first page: "The text is telling me that the leaves are fluttering in the summer breeze, and when I look at this picture, I can see that the tree is full of green leaves, which is true of trees in the summer. Moreover, the grass and the bushes are green, too, which tells me the season is one of the warmer seasons." Turn to the next page and have the students make connections between the text and the picture. Begin the reading of the mentor text, *Sky Tree*, by reminding the students about their tree descriptions and the reasons that the trees might change through the seasons. Show the students a few pages in the book and ask them what they are seeing. Through modeling, encourage them to use vocabulary relevant to the seasons; these can be written down to display. For example, "sprout" or "sprouting" would be an appropriate word to describe plant life in early spring. As the reading progresses, *Sky Tree* will also raise questions about the images such as: how does this make you feel? or "what season is it?" and so forth. We are doing two things here. One is building skills and strategies in reading images that students can apply to any texts. With the *Sky Tree* lesson, we are also helping students better understand the science concepts of changing seasons and the natural cycles of nature.

3. Beyond the Text: After Reading

After the read-aloud, teachers may use their K-W-L charts and ask students: What did you learn? What did you notice? This can be done orally or in writing, think-pair-share, or any other preferred method. Here is a point to assess the students' understanding and then deepen it by developing conceptual knowledge about the seasons and the cycles of the year. Ask students if the book begins in the fall or autumn? Where does it end? What is the process that occurred?

Once the reading is finished, instructors should return to the display of the statements the class made at the beginning of the lesson. Read the statements and ask the students to use examples from the text and pictures to confirm their validity. For example, if at the beginning of class a student said the leaves fall off the trees when it gets cold, ask the students to show an example from the book.

As a closing activity, students (in pairs or in centers) can match visual depictions of the seasons with vocabulary words or simple sentences by cutting out provided (pre-made) pictures or drawing them. They will then have to orally explain their reasoning for matching a picture with a word or sentence.

Lesson Plan 2

1. Activation of Background Knowledge and Hands-On Activity

To assess prior knowledge of coral reefs, have the students in pairs or groups sort pictures (pre-made by the teachers) of animals and plants by whether or not they could be found in coral reefs. Ask the student pairs or groups to explain their sort: why did they think that some could survive in coral reefs and some could not? Document their explanations in a manner that can be displayed for the whole class to see.

2. Before Reading

Begin the reading of the mentor text, *Coral Reefs*, by reminding the students about their explanations as to why certain life forms survive (or don't) in coral reefs. Turn to one of the pages in the book and ask them what they are seeing. The two-page spread thatshows the child swimming with a turtle is an excellent visual. Through modeling, encourage them to use vocabulary relevant to the topic; these can be

Table 3. Simple lesson place; grades 5-8 coral reefs; next generation science standard

3-LS4-3.	Construct an argument with evidence that in a particular habitat some organisms can survive well, some survive less well, and some cannot survive at all. [Clarification Statement: Examples of evidence could include needs and characteristics of the organisms and habitats involved. The organisms and their habitat make up a system in which the parts depend on each other.] (retrieved from http://nextgenscience.org/pe/3-ls4-3-biological-evolution-unity-and-diversity)

Book: Coral Reefs by Jason Chin. Roaring Brook Press, 2011.

written down to display. For example, "ocean floor" or "food chains" would be appropriate words to describe a coral reef habitat.

Tell students that as the book progresses, they will not be seeing pictures so that they can visualize the information instead. They will see pictures once the read-aloud is finished. Explain that visualizing the book's ideas is an important strategy that will help them understand the coral reefs thoroughly. Model this by using the first page which was previously covered with paper, leaving only the text exposed. "When I read this, I try to imagine in my mind an animal that looks like a plant moving in the water. I am going to try to draw a picture of that on my paper." Draw a picture and ask students to comment on whether they were imagining the same picture in their minds.

3. During Reading

Continue reading the book and engaging the students in visualizing. Provide each student with a piece of paper so that they can sketch what they are visualizing. You can also have the students "turn and talk" to classmates in order to share their visualizations and sketches. Check continuously for their ability to make visualizations by asking what evidence in the text made them visualize an idea in a certain way.

4. After Reading

Once the reading is complete, show students the book's pictures one page spread at a time, and have students explain how their visualizations and sketches compare to the actual pictures. Ask how visualizing helped them understand ideas about coral reefs, and make a list of new understandings they gained. Return to the display of the statements the class made about why different plants and animals could or could not survive in coral reefs at the beginning of the lesson. Read the statements and ask the students to use examples from the text and pictures to confirm their validity. Take note of the animals and plants about which students are confused, and tell the students that they will be researching these to confirm whether they are part of a coral reef habitat in a future lesson.

To extend the learning to cover all elements of the lesson standard, students (in pairs or in centers) should fully draw out their sketches of animals and plants only found in the coral reefs and cut them out. They will then make a web of how these organisms are related (for example, predator/prey) within the coral reef habitat on a large sheet of paper. They will then have to explain in writing their reasoning for constructing the web they way they did.

Chapter 4
Theater as the STEAM Engine for Engaging Those Previously Disengaged

Paul C. Jablon
Independent Researcher, USA

EXECUTIVE SUMMARY

This chapter connects the use of creative dramatics with engaging students previously disengaged with STEM classes. It elucidates a variety of reasons how creative dramatics allows STEM teachers and their students to succeed. Each of the assertions is backed by citations of research studies, classroom practice, and details of the theoretical underpinnings. Also included are detailed descriptions of three effective classroom methodologies using theater in STEM classes, along with specific examples of each that include student and teacher interactions.

INTRODUCTION

Skilled inquiry-based elementary teachers who engage students in daily materials-based STEM investigations maintain students' interest throughout elementary school. However, for much of the older school age population, science, technology, engineering, and mathematics (STEM) are seen as "cold" subjects that lack emotion and humanity (Heaverlo, 2011; Burke & Mattis, 2007), and by the time students reach early adolescence they have essentially disengaged from these STEM subjects (Grocholski, 2018; Kesar, 2018; USDOE, 2017: USDOE, 2006; Osborne, et al., 2003).

Why do so many adolescents who previously enjoyed STEM in elementary school disengage from it as they get older? In order to answer this question, it is

DOI: 10.4018/978-1-5225-9631-8.ch004

important to investigate what is relevant to adolescents. Perhaps a better way to ask the question is, "what matters to teenagers?" Before reading on, please take a moment to answer this question.

Over the years I have done this exercise with thousands of teachers and parents, and in virtually every case the consensus has been the same. I have also done it with hundreds of teenagers, and they have also reached the same conclusions. Everyone agrees that they care most about being *accepted by a group of pee*rs who matter to them. They also care about being able to *demonstrate their competence* in things they and society find important. This differs from the former need in that in the latter they are *respected* for who they are and what they can accomplish.

STEM education becomes a natural, organic process when teachers can meet these needs while engaging students. It becomes a task that students want to do, not one that they must be forced to do. Students may practice STEM even if they weren't required. The humanistic psychologists' explanation of human needs, which is supported by current understandings of brain neurobiology, may put this in a better conceptual perspective.

In the 1950s and 60s, Abraham Maslow (1998) and his colleagues studied individuals who many recognized as having incredible mental health. From this study the researchers deduced a hierarchical set of human needs. Until one need is reasonably met, the others are not of consequence to the person. After testing this on thousands of individuals globally, researchers have found that this sequence of needs is *genetically programmed* into all humans. This aligns with neurobiological knowledge that the frontal cortex and other reasoning parts of the brain surround the amygdala, the most ancient part of our evolutionary brain.

Figure 1 (page 2) illustrates the hierarchy of needs (Maslow, 1998). The figure shows that survival and emotional needs serve as the hierarchy's foundation and therefore always override the rational aspects towards the hierarchy's top. Once people have met their physiological needs (food, clothing and shelter), then they have to focus on their safety needs. Homeless children keep slipping back into this category and can't much focus readily on anything happening in school. Safety needs comprise the hierarchy's second tier. Safety means the ability to regularly predict life's events. Safety reaches beyond physical safety; it includes psychological wellness. This explains why kids raised in abusive or dysfunctional households struggle to focus on school. Teenagers whose safety needs are fulfilled focus mostly on the two hierarchy tiers that reflect aspects of adolescence that teens care about most. The next two levels of belonging (acceptance in a group) and esteem (respect for competence) are where *genetically* teens must be. Society often demands that teens reach the top level, self-actualization, functioning with all preceding needs sufficiently fulfilled, but adolescents typically don't have these levels filled. Nevertheless, schools generally ask them to do things because they will need it in

Figure 1. Maslow's hierarchy of needs

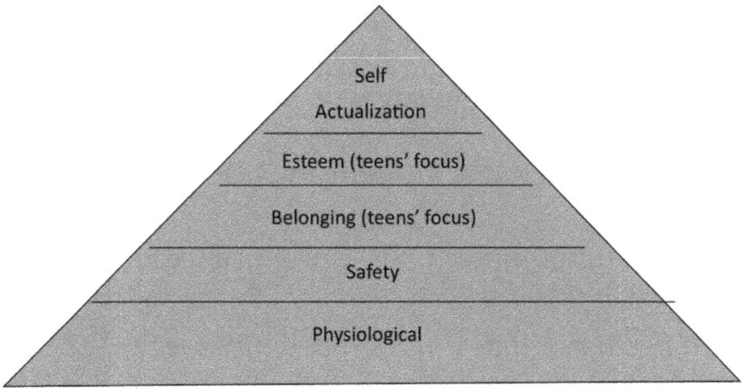

the future, or not to worry about what their friends think. Yet, these are sentiments that teenagers generally struggle to adopt.

If teachers can operate their classrooms so as to meet these needs while engaging students in academic explorations, then students will be more willing to be there. They will put more effort into learning than most teachers have had the opportunity to experience. Examples of how their needs can be met while studying STEM concepts will be demonstrated in detail in the next section.

Thirteen years into my public-school teaching career, I turned to a colleague, and, pointing at my classroom, said, "Look at *that*. I never imagined it could be this self-sustaining. They are so intellectually and socially engaged. It is beyond anything I could imagine."

The students had done role-play earlier in the class about a piece of literature's underlying theme related to biological concepts they were investigating. I currently assigned them to create short theater pieces in groups demonstrating their understandings of the relationship between science and social science in relation to *evolution and revolution*. Each group's self-motivation as they intellectually argued about the intricacies of these complex concepts and then worked equally hard to dramatically convey their intended meaning was incredibly impressive. The previous term a number of these students were cutting their didactic transmission style classes. It boggled my mind (Jablon, 2014).

Therefore, as students progress into early adolescence, their daily attention centers around the drama of personal relationships; their struggles with the unfairness of the world; and their confrontation with rules, regulations, and ethical issues. The pre-teen and subsequent "teen in angst" rarely finds relevance in anything outside of their daily drama, and the previously engaging mysteries investigated in inquiry and materials-based science, engineering, and math fall to the wayside (Kuenzi, et al., 2006; Bennett and Hogarth, 2009). Until these inter-relationships are brought into the STEM classroom, as Maslow noted, most adolescents cannot deeply engage. Using drama as a tool in the STEM classroom can assist teachers to accomplish this task.

Effective Models for Using Theater in STEM Classrooms

Three approaches to using theater, and creative dramatics in particular, are outlined in this section. There are other ways that theater can be utilized in STEM classrooms. One of these alternative approaches is where students role-play physical STEM systems such as the example mentioned in a latter section about students physically representing the chemicals in a redox equation. This can easily be done, as can representing parts of mathematical algorithms. However, it is not nearly as powerful a tool in deeply engaging students in STEM understandings, especially those previously disengaged, as are the three models explained below.

Each approach is explained in step-by-step detail so that educators can understand how to utilize it in their classrooms. Each explanation is followed by an example that has worked effectively in classrooms.

Role-Playing of Actual STEM Professional Situations

These are scenarios in which students role-play various constituencies and the STEM professionals that are involved in government, legal, medical, engineering design, or other professional situations where decisions need to be made. In these role-plays, students are given criteria for their characters, and they then improvise these situations. Some students play the STEM professionals utilizing their subject area knowledge, while others play lay people who interact with the professionals. Each role utilizes decision-making skills. These role-plays are usually brief, lasting less than five minutes. However, the whole-class, follow-up analysis is more extensive.

Although some of the role-play situations involve teenage characters, they usually do not. Rather they are adult situations involving STEM professionals in governmental hearings or local community meetings. Another common scenario includes individuals who are in conversation with personal advisors like health professionals.

This approach is most commonly used by STEM educators and is a good starting point for those who have never utilized creative drama in their classrooms. Because it has teenagers playing adults, rather than teenagers playing teenagers making decisions about issues they deal with daily, they are not as engaging to the students. Nevertheless, the method is still a powerful tool.

The scenarios can be used as a motivational tool to introduce new topics by engaging students and sparking curiosity about STEM knowledge. The same role-play can be utilized by the teacher in the middle of a unit where the discussion of the role-play can be a formative assessment tool or used at the end as a summative assessment tool. At the beginning of the term students are told that everyone will be asked to improvise and act out a role explained to them by the teacher. The teacher tells each character individually how to react and what to believe. The characters don't know what the others have been told. The rest of the class and the two or three role-players are given a bare-bones outline of the situation. The role-play begins with the students improvising the dialogue. The rest of the students are asked to remain silent during the role-play and to refrain from laughing unless something is very funny since it makes it hard for the role-players to focus.

The rest of the students are told that although they will each get credit for role-playing when it is their turn, most of the credit will come from the whole-class discussion following the role-play, so it behooves them to be attentive. The teacher watches how the role-play progresses and notices if it does not hit upon enough of the issues and STEM content. The teacher then asks the role-players to stop for a minute and asks if the viewers think a particular aspect is real to try and move the dialogue in a more useful direction. Surprisingly to teachers who are novice to using creative dramatics, this almost always works, and the role-play resumes in a more useful way.

However, the scene often unfolds on its own for a few minutes, fulfilling its goal. The teacher then asks students what is real about this situation. This allows the rest of the class to note what they perceived as genuine interaction or as accurate math, science, or technological content. Then the viewers have a chance to speak to what they see as staged, unusual, or inaccurate. They can suggest solutions not presented, alternative decisions to those made by characters, or perhaps comment on societal situations that they think are unfair and should be altered. All of these need to be supported by evidence from the content of the class or things they have experienced in their lives. Sometimes the teacher will ask the student who says there is another way for a character to deal with the situation to take over the role of that character, and the improv continues.

Usually the teacher thinks of these situations ahead of time while writing lesson plans. Many times, there are STEM societal scenarios related to the content being studied written in books for students just to read and comment on. These are great sources for teachers to use as the basis for a role-play. There are a few on the National Science Teachers Association (NSTA) website and many more elsewhere. However, once teachers become proficient doing this, they can make up their own role-play situations to fit given content. I have seen teachers only have students read the scenarios and have discussions and I have seen the same scenario first role-played and then discussed. The level of participation, passion, deep involved complex thought, and application of STEM content is always exponentially higher when the students have experienced the dramatic improvisation version in class.

Example 1: A Role-Play of an Actual STEM Professional Situation Focusing on Sickle Cell Anemia Testing on a College Campus

The teacher asked for two volunteers to role-play. One volunteer was told that the character they'd be playing was Frankie Fulcher, an African-American college student. About a month ago Frankie and some friends had been walking across campus when they saw a sign advertising free Sickle Cell Anemia testing in the student center. One friend convinced the others that this was really important. They all had their blood tested. Frankie had pretty much forgotten about it when she received a letter from a laboratory saying that her test results had come back, and she could set up an appointment in the student center to get the results. Even though she didn't know much about sickle cell, she made an appointment. The teacher also told the student that Frankie is engaged to be married in the near future. She is told that the genetic counselor might say she is a carrier of sickle cell. She should act skeptical since she is a great athlete and is completely healthy. She should keep referring to that. If she suggests that she tell her fiancé she should be hesitant and then say she's probably not going to do that, because she might think she is defective and then won't want to marry her.

The other student is told that she is a genetic counselor, Pat Nash, who interprets genetic test results and assists people in making decisions. Since it is the beginning of the term, the teacher has to coach the student on what to say. The teacher explains that sickle cell is a disease that only manifests if the zygote is homozygous for the trait. She explains that if two of the trait's genes are absent, someone may be a carrier, but may not necessarily have the disease, which he explains is a really debilitating disease. Therefore, both mother and father need to be carriers, and then there is a one in four chance of the baby having the disease. That is why when Frankie is told that she is a carrier for sickle cell, she must be asked if she is married. If she plans on having a child, the partner should be told and tested.

The role-play of Frankie coming to talk with the genetic counselor begins. Frankie plays awkwardly astonished when she is told that she is a carrier, and protests that she is not only healthy, but fit. Pat, the counselor, does a pretty good job of explaining the genetics that she has memorized. When she asks about being married and Frankie explains about her fiancé, she suggests she tell him, but she rejects the idea. She talks about the fiancé loving her and how he will respect her for being so mature, but Frankie is afraid of rejection, and says it is only a one in four chance and his fiancé is rich and probably isn't a carrier, equating wealth with health.

At this point, the teacher stops the improv, and the class applauds. The teacher then asks what is real. The students mention they have heard about sickle cell, but like Frankie they don't know much. They talked about how her fear of losing her fiancé could be real. They talk about themselves, their friends, and their failed relationships. They say that the genetic counselor tries to explain things, but the scientific terms make it difficult to understand.

The teacher then asks the class, would they want to marry Frankie given she doesn't want to tell her fiancé? This raises a heated discussion, but also has many students in the class asking questions like, "How can someone have the gene, but not have the disease?" and "How can this counselor know that there is a 25% chance of the child getting the disease?"

The teacher tells students to be attentive to the upcoming investigations in order to discover the answers to these questions. As the class engages in genetics investigations over the next few weeks, both the teacher and the students keep referring to Frankie Fulcher as they study Punnet Squares, recessive and dominant traits, and how a disease like Sickle Cell Anemia can actually be a positive thing in a race or ethnic group genetic pool if Malaria is common among the group.

In this case the role-play was used as a motivation for students by staging a captivating scenario that makes them eager to understand the science, math, and technology behind it and then to make informed decisions. Sickle Cell Anemia is not limited to individuals of African descent. As a positive adaptation to fight malaria, therefore it is most common among people of African descent but is found in any population with more than a few generations of living with malaria exposure.

The multidimensional ethical issues here are abundant. Students need a deep understanding of them, and this scenario can facilitate extensive discussions about genetic testing as an emerging and informative field. There is a great deal of misinformation being perpetuated related to socially constructed groups and genetic correlation (Roberts, 2012). This is a grand opportunity to not only wrestle with the Next Generation Science Standards (NGSS) disciplinary core ideas of *Inheritance of Traits, Variation of Traits, Adaptation, and Biodiversity and Humans,* but also a grand opportunity to practice the core skills of *Asking Questions and Defining problems*, and even more so *Engaging in Arguments from Evidence*. Likewise, as the students engage in the ethical decision-making that this scenario demands, they will also touch upon the Core Science, Tecnologoy, and Society (STS) idea of *The Influence of Engineering, Technology and Science on Society and the Natural World.*

Analogous Interrelationship Role-Plays

These are role-plays created by teachers that utilize typical teenage situations but are analogous to governmental or corporate ethical and economic dilemmas. Often, they can help students understand the complex interplay of ethics, economics, and subject area knowledge when they cannot comprehend it when looking at a large-scale governmental or corporate problem.

For the past forty years I have observed adolescents in a wide range of classes. For some of these classes I was the teacher but since 1990 I have been blessed with the opportunity to be an outside observer of a cross-section of classes in various disciplines. I've observed in both urban and suburban locations, and in small, non-traditional schools and large, fairly traditional schools. In each of these classrooms there were at least three or four students who were completely disengaged from any learning activities no matter how much inquiry-learning and interdisciplinary format was used. If we multiply these students by the number of classes in the school, this number becomes substantial. In many urban schools this number was much higher.

During the twenty-one years—1970-1990 and 2001-2002—that I taught adolescents, my colleagues and our students created an approach to truly engage these disengaged students in learning and in becoming full participants in our community. Inquiry-based learning was a major component, but it was not sufficient by itself.

Action-based classes that connect academic subjects to societal issues appeal to the innate desires of young people *to make a difference* in their world. Giving attention to social conscience issues enables students to act upon their decisions about the issues through letter writing, direct community involvement, and more. However, many times when teachers introduce societal issues such as solid waste management, genetic engineering, and such, they are surprised to find that many students are not motivated by such matters. For those students who initially demonstrated an interest, their attention either is sporadic or wanes completely.

Yet there are some programs utilizing societal issues that have the ability to sustain a deep student interest throughout the semester. Some of these programs have utilized societal issues with great success with students who have been labeled "at-risk" (Jablon and Born, 1988). If we look at the nature of human beings and the nature of adolescents in particular, some of the underlying reasons for this success in engaging students with societal issues become clear.

As outlined earlier, most adolescents, but especially at-risk students, do not have their security, love, or self-esteem needs met—as defined by Maslow's hierarchy of needs (1998). Yet most of what we do in schools is to ask students to work within the realm of self-actualization, saying that these things they are asked to study will be useful to them some day or to the betterment of society. Rarely are they engaged in any activities in the academic classroom that will fill their immediate needs for these lower-level foundational elements. These needs, when not addressed, prevent students from engaging in the self-actualized activity we present. Many students whose more emotional needs aren't addressed in their academic classes seek them in their anti-school, negative peer groups (Pulkkinen, 1982).

Underlying these societal issues are psychological and sociological relationships. Over the years, teachers found that most of these big ideas and concepts that they wish students to acquire, whether in social studies, science, technology, or in the relationships of characters in novels, are universal relationships between individuals and groups, individuals and other individuals, or between groups and groups. They are in many ways based on the same psychological and sociological foundations as the relationships these students have with their peers and parents, the two primary areas of concern to adolescents. They are *analogous.*

Once these needs are met, students will not only engage in study of the analogous societal relationships but delve into the deepest esoteric questions that are posed. In addition, they acquire conflict resolution skills and coping mechanisms that they can utilize in their everyday relationships. Personalizing the subject matter by having students investigate these *analogous interrelationships* between their own personal interrelationships and those that exist at a societal level is one of the most powerful tools that I have ever experienced in engaging students previously disengaged from schools (Jablon 2010). If the psychological needs of the students are met as they are engaged in activities with the other students in the STEM class, then a positive alternative peer group is created that students can go to as an alternative to their negative neighborhood peer groups where they normally turn in their search for fulfillment.

Teachers facilitate the role-plays the same way as with the previously described STEM professional role-plays, but the result is far more effective in engaging these emotionally needy students. I could not think of a more powerful pedagogical tool to utilize in a STEM classroom.

Example 2: An Analogous Interrelationship Role-Play Focusing on the Greenhouse Effect and the Brazilian Rainforest's Deforestation

After a few weeks of studying the effects of cutting the Brazilian rainforest in order to sell hardwoods and create grazing land for cattle to be sold as beef. The biology students had done a myriad of investigations with stream tables in order to understand the relationship between planted and unplanted soil and water erosion. They had done greenhouse effect experiments with jars and thermometers. They'd also done several bromothymol blue photosynthesis experiments. They really had a handle on the science and had been especially engaged because of the societal issues raised about the plight of those living in the rainforests. They were very aware of how Brazil was up against a wall in need of exports to avoid inflation and the World Bank calling in loans. However, they were constantly perplexed that, given what ecologists knew as the long-term negative effects on the local and global ecosystems, adults in Brazil just didn't stop the devastation. Teachers' explanations didn't seem to help with these complex real-world interactions.

In order to have students see the interrelationships in both the country of Brazil and in the world groups with whom they had a relationship, the following role-play was created and done in class the following day.

Three students assume the roles of the following characters:

- Marie Sussman, a fifteen-year-old high school student who must work long hours to add to her family's income because a parent has recently died
- Ms. Pape, Marie's English teacher who is upset by the drop in quality of the student's work in her class
- Marie's mother, Diane Sussman, who is caught between wanting the best for the student and the rest of the family's welfare

The students playing the high school student and her mother are given the information about their roles. Marie is the oldest child, having a four-year old brother and an eight-month-old brother. Up until the summer, her mother worked part time, helped Marie do great on homework, test, and quizzes, and her mother spent many hours at home taking care of the kids. Her father worked full-time and earned a decent salary. Unfortunately, her father died during the summer and, being a "private" family, they felt no obligation to notify the school. Her mother went back to work more hours, mostly on the weekends. The family is on food stamps, but having grown up in a welfare family, the mother will not put her family through that "same humiliation." In September, Marie was doing fabulous in Ms. Pape's class. She loved the novels and the class discussions. But her mother came to her and said the family wasn't making enough money to pay rent, buy food, and keep her siblings in Pampers. Thus, she needed to get a job to assist with these expenses. They have a family friend who gets Marie a job doing mostly filing work at a doctor's office from 4:00 to 8:00 every weeknight. Marie's grades have been dropping drastically. By the time she gets home, she's exhausted. She has six subjects' homework to do, and she is way behind in everything, including English class. On weekends she babysits her siblings while her mom is at work. She and Ms. Pape have spoken before about her grades dropping, poor quality of work, and the lateness of her assignments, and Marie has assured her that she plans on doing better. Marie has even asked the doctor if she can cut back some hours, but the doctor says that he is already doing her a favor, and he can't have more part-time employees, so no cut back in hours. Ms. Sussman is told that if she gets a call from any of Marie's teachers, to be huffy thinking that these teachers just don't get how difficult a position she is in. Marie is told that in her family, the family's business is kept within the family, so she would be hesitant to talk about it when asked by teachers. When pressed, she can explain a little bit, and then finally the whole story.

The student playing Ms. Pape has not heard any of this. She is told that Marie was a great student in her class in September, but her work has dropped off so much that she is on the brink of failing. Occasionally, Marie submits an amazing piece of writing, so she suspects there is some reason that Marie's not keeping up. She has spoken to Marie before, who keeps assuring her she will get in the work and improve, but that she is just busy. She is also told that she has asked Marie to come after class, and she will really confront her with how close she is to failing. She is also told she can call Marie's mother if she can't resolve this with Marie.

The class is simply told that Marie is a ninth-grade student in Ms. Pape's English class, and Ms. Pape has asked Marie to see her after class. They are also told that Marie's mother is at home and can be reached by phone.

Marie doesn't tell the teacher much to start with, although Ms. Pape is probing. Finally, Marie admits that she has a job and it is hard for her to do homework when she gets home at almost 9:00 every night. When pushed further by Ms. Pape, Marie explains about her father's death and her need to help support the family. Ms. Pape is helpful, asking can she cut back hours (no), asking what about the weekends (babysitting siblings), and offering to stay after school and help Marie (can't because need to travel to work). Finally, Marie needs to go to work and Ms. Pape calls Marie's mother, Diane Sussman, who is at home with her other children. At first Diane is pleasant and saying she is also concerned about Marie's grades, but as Ms. Pape continues speaking about Marie's future and how she is almost failing, Diane pushes back. She says she is responsible for her whole family and she tries to give Marie as much time as possible to do her schoolwork, but they don't eat and pay the rent if Marie does not hold down this job.

The teacher stops the role-play at this point and asks the class what is real. Most hands in the class go up with comments about how the teacher is caring but only cares about grades. Other comments are about that parents don't realize how much pressure there is on high school kids, or that it is true that there are lots of kids like Marie who are hesitant to talk about their family issues with teachers. When asked what is not real, a student says that the situation where Marie really has to work, not just to earn spending money, isn't real. Three students all fire back that either they, or friends, or brothers or sisters are in that situation right now. Others say that teachers really don't care that much, and many students give examples of teachers who bend over backwards caring, or teachers who wouldn't even go so far as calling Marie's mother. Other kids say that Marie needs to stand up more for herself with her mother, and the class gets into a long and serious discussion about relationships between parents and teens, the power differential, and other issues.

The teacher asks the class to think for a few minutes before answering the next question, "Who is to blame?" At first there is a lot of back and forth discussion and disagreement that Ms. Pape is to blame, and she should give Marie less homework and a break on her tests. Then the same thing happens about the mother or Marie being to blame. Finally, after about five minutes a student puts forth the notion that, "none of them are to blame." When asked by others in the class how can that be, someone needs to be wrong, the student replies, "the system in this community, city, or state that they live in is to blame. Each of them is doing the best they can do at that moment." The teacher asks if any students can think of a time when they were caught in something like that, and one example is given, and then another that is not so clearly the fault of the situation.

Up until this point, this is about social issues that teenagers are grappling with in their everyday lives with teachers and parents. However, the teacher asks the whole class to use wait time before they answer, "Think back to the cutting of the rainforest in Brazil. Who here is like the ecologist? Who is like the government of Brazil with the World Bank on their back? Who is like the folks living in the rainforest and cutting it down to get by?"

There is a very long silence and then various opinions are given with evidence from the role-play and from what they studied about the greenhouse effect and cutting the rainforest to back them up. There were these "aha!" moments as students said things like, "ecologists only care about the future, just like Marie's teacher Ms. Pape who kept talking about her future in college."

This role-play allowed this class of students studying deforestation to understand the complex relationship between the poor native inhabitants cutting the rainforests (the student), the government of the third world country who is encumbered with a large international debt (the parent), and the environmentalists trying to help both the native and the future world ecological situation (the teacher). It was a clear *analogous interrelationship.* Students not only had a unique insight into world economics but into their own personal relationships and therefore identify positively with their academic classes.

For days following this long role-play and discussion, students were deeply engaged in the science related to this topic. Why? Their Maslowian needs were met with serious discussions about their daily dilemmas with parents, teachers, and responsibilities, with many coping mechanisms suggested along the way. Although it helps all students, for those students who are most lacking emotionally, the more of these types of role-plays done, the more they will not cut class and the more they will attend to the content as well. This example described an actual classroom experience that I encountered in my teaching career and how analogous interrelationship role-plays that were created helped solve some student lack of understanding of a complex topic. However, if a teacher was just to introduce this in a generic deforestation/ecosystem unit, it might be more effective and create less of a them-versus-us mentality if the teacher was to use local or US national examples such as the deforestation of the Pacific Northwest. Using the Brazil, or an Indonesia example, there is the potential for teaching that deforestation and other topics impacting the environment is something that happens "elsewhere" may limit student ownership and agency as well as reinforce a narrative of "us: good, smart, stewardly" and "them: unaware, abusive, ignorant, destructive." Just as in example one, here is another grand opportunity to not only wrestle with the NGSS (2013) disciplinary core ideas of *Global Climate Change, Ecosystems: Dynamics and Resilience,* and *Interdependent Relationships in Ecosystems*, but also to practice the Core Skills of *Asking Questions and Defining problems*, and even more so *Engaging in Arguments from Evidence.*

Writing Scripts and Producing Plays Utilizing the Subject Areas Studied

This third model for using theater goes well beyond creative dramatics and role-plays, although they both are utilized in the model. It usually involves other subject area teachers working with one of the STEM teachers, is much more complex and difficult to maintain, and is best explained by citing an example. Therefore, a brief description will be given here, and the model will be elucidated in greater detail in *Example #3* where a full description of a program is given.

After a certain amount of subject area material has been uncovered through inquiry teaching approaches, the students as a class decide on the important STEM and societal or ethical issues they want to have others consider and then as a class write, rehearse, and produce a dramatic piece that creatively embeds these issues in the plot. This piece is then performed for a public audience. This process is usually done in collaboration with other subject area teachers.

The various subject area teachers working together with the same group of students usually select some powerful universal theme to help integrate the various cross-disciplinary content, which simultaneously assists students make sense of complex world issues and decisions. Such themes as "power over versus power with" or "the right of the individual versus the right of society" allow students to investigate some of the greatest of human dilemmas and ideas, while pulling together seemingly unrelated subject areas.

Example 3: Using Student Written Theater Productions to Engage Previously Disengaged Students

Nellie McCaslin, the grande dame of American creative dramatics for over sixty years, highlighted a particularly successful program in her *Creative Drama in the Classroom and Beyond* (2006, pp. 264-5). What follows is a detailed description of that program in a public high school in New York City for students who were either regularly truant or had previously dropped out of school. It shows how an interdisciplinary program with drama components had these same students attending daily, succeeding in science and other subjects, and finally graduating at amazingly high numbers. A ten-year mixed method study demonstrated how having these students use role-plays addressing STEM-Societal topics, utilizing interdisciplinary team teaching, and, most especially, writing original full-length plays based on the subject area content and important themes taught throughout the term completely transformed the academic and social-emotional achievement of a vast majority of the students enrolled. This BONGO program started by Terry Born, an English teacher and me, a science teacher, has been replicated in many other schools in many other states, but this particular term described below was from one of our groups of students at the original Middle College High School.

The program was three periods of a seven-class day where students studied rigorous subject matter in three state-mandated subjects. The students who completed whole-class dependent projects with public presentations—many with social advocacy themes—were enrolled for one semester. However, since the mix of subjects changed each semester, many students re-enrolled during a subsequent year for a semester when the classes they needed were available as they realized their need for the *positive alternative peer group* they had previously experienced.

The students in this semester were enrolled in a Biology I, physical education, and English III class as part of BONGO. During the term they read Orwell's *1984*, Huxley's *Brave New World*, and More's *Utopia*. The students also studied a bit about American government structures and engaged in physical education where small bicycling experiences twice a week led to an understanding of bicycle safety. Simultaneously, students were engaging in *doing* biology by designing and completing daily inquiry lab investigations related to genetics, cellular mechanisms, and ecological relationships. Genetic engineering, and its related ethical issues, became a lens through which the students studied various biological concepts and connected these to the novels they were reading. Students also employed many of these biological concepts to make sense of the physiological demands on their bodies during bicycling.

For the first two-thirds of the term, small cooperative teams completed many small interdisciplinary projects and daily *analogous interrelationship role-plays* that used and demonstrated mastery of science, English, and social studies knowledge and skills. These role-plays always involved adolescent characters interacting with other teens, their parents, teachers, or other significant adults in their lives. The personal relationships in the role-plays were analogous to the larger relationships being studied in an academic perspective in science, history or literature. Students had an opportunity to simultaneously confront and discuss the adolescent issues that made their lives so unpredictable and stunted them emotionally while studying important academic concepts.

During the final six weeks, students began to synthesize some of the larger concepts they had identified as being *vital driving questions* that they found significant and were wrestling with during the term. As adolescents they identified what was the most important social conscience issue they encountered as the term progressed so that they could then engage their peers, parents, and others who would see the play they were about to write as a whole class. Like the little projects all term long, the final project they created was group dependent for success. As a class they outlined, wrote, revised, and produced a two-act play that synthesized complex ethical, scientific, and legal issues into a compelling drama that spanned four decades in present and hypothetical future history. The first act of the play written by the students in 1983 took place in a hypothetical 1983 United States where a constitutional convention had been called to change the constitution so that a law could be passed that allowed the government to override the rights of individuals found in the Bill of Rights. A court scene takes place where a couple refuses to obey the law that would affect their unborn baby. The second act occurs in 2020 where the students devised a culture based on the extension of the idea of the government making decisions for individuals. The baby of the original couple, now a young man who has been hidden away in the mountains, emerges into this new culture with the ethics of pre-1983 Americans. This play was toured to surrounding schools and also presented in a local college theater (Jablon and Born, 1988).

During the thirteen years that Terry and I facilitated the BONGO Program, we were astounded by what our previously disengaged students accomplished. It was beyond my wildest dreams about what teaching and learning could be. We got to watch one class reinvent existentialism over a two-day period. The students created photographs and sculptures that were displayed in college galleries. They won the educational award from the National Hunger Project for a play they wrote and toured about world hunger and the politics involved. Four times they won first prize for young adults' playwriting contest from the Dramatist Guild in New York City. Groups of two or three BONGO students have taught over fifty classes, with adults only as observers, in other public schools before they presented their original plays the next day.

That term was six years into a three-period-a-day program that Terry and I co-taught and directed for thirteen years and that in some form or another has been continued by teachers around the country. A ten-year quantitative and qualitative research study of the program uncovered some fairly amazing results. Despite the almost universal agreement among the students in the program that the coursework was the most challenging they ever encountered in school, a ten-year study of the program demonstrated a remarkable success with a set of previously disengaged students. This structure accounted for some fairly compelling results with a population of urban African-American, Latino, and white students who almost all had a history of at-risk behavior. The results showed:

- Ninety-two percent did not cut class during the time they were in the BONGO Program. Seventy-seven percent of those students had a history of cutting classes before they were in the program, which dropped to zero cuts while in BONGO.
- Attendance also increased, both for students who were mostly truant, and for students who attended school fairly regularly, but had significant absences. Sixty-five percent of the students experienced increased attendance, thirty-one percent showing a massive increase. For any of these students to have above ninety-five percent attendance was unbelievable for their parents, their other teachers, and themselves. The number of students who showed either perfect or ninety-five percent attendance doubled (perfect attendance went from three to five percent; ninety-five percent attendance went from twenty-one to forty-five percent).While students were in BONGO, fifty percent had ninety-five percent attendance or better.
- Eighty-nine percent of the BONGO students either increased or remained constant in their academic performance with fifty-one percent showing a major increase. Thirty-six percent failed science before BONGO but passed during. Fifty-four percent had an increase in science grade.
- The most startling finding was that thirty-two percent of the students went from passing no courses, or one or two courses, to passing *all* of their subjects, including those classes outside of the three BONGO periods.

The Watson-Glaser critical thinking test administered to BONGO students demonstrated that sixty percent of those students significantly raised their *critical thinking abilities*, which was four times greater than their peers in control classes. The Process Skills and Science, Society, and Technology (PSST) test of process skills in science and abilities in understanding Science, Technology, and Society (STS) relationships revealed that forty-one percent of the BONGO group increased their science reasoning skills and STS awareness over the course of one term as compared to nine percent within the control group (Jablon, 1989).

Finally, of these students who almost all had a history of dropping out of another high school, 97% graduated. Although this is a much more complicated version of incorporating theater into STEM education, the compelling statistics about how it turned around the lives of hundreds of students should motivate educators to implement similar programs in their schools and university STEM education professors to include it in preparation and professional development programs.

Theoretical Underpinnings and a Review of the Literature of Drama in the STEM Classroom

For some students, the utilitarian application of designing useful things for our everyday needs while using scientific and mathematical knowledge brings STEM subjects closer to relevance. However, for most early adolescents, this practice of engineering and technology is devoid of the optimization and ethical drama that accompanies real-world applications. It is ironic that many of the societal issues raised by technological innovation are analogous to the issues that affect these very adolescents in their daily emotional lives, but these parallel emotional, social, and ethical issues are rarely at the core of the technology and engineering programs.

When some modestly effective STEM programs have infused these social and ethical issues into the curriculum, often they are introduced through readings about international environmental situations with countries or large corporations being the units of conflict. Even when the issues are locally based, all too often the societal issues are simplistically seen as one dimensional, black and white, right and wrong without the complex ethical and monetary dimensions being introduced. Some students can deal at this theoretical and formal thinking "someday I will need to know this" level, but most are still focused on their daily interpersonal interactions as they try to navigate their way to adulthood.

Over the past two years since the first edition of this book was published, I have seen a few other STEAM books written for teachers and a bunch of articles in teacher magazines. Very few focus on using theater. As I updated the research bibliography on using theater in STEAM, I realized that there were no new long-term studies done, and the handful of brief studies published do not present any new approaches beyond those already elucidated in the original edition. It is heartening to see that the drama educators have taken on the STEAM approach, but it is equally disheartening that few of the university and secondary STEM educators have truly embraced it even in the least complex ways. It is such an effective tool to engage adolescents in STEM that a concerted, and well-funded, effort by the STEM community needs to occur sooner than later.

For decades, teachers have found that using creative dramatics in their classrooms has allowed them to tap into these emotional needs of their students in order to engage them in the previously unlearned subject area conceptual understandings (McCaslin, 1984, 1996; Heathcoate & Bolton, 1995; Bracha, 2007). A vast majority of the teachers who utilized this effective approach did so with social studies or language arts (Salisbury, 1996; Duveen, & Solomon, 1994; Mardirosian and Lewis, 2009). Science, math, engineering, and computer science teachers have been loath to use these creative dramatics techniques (Kelner, 1993; Dorion, 2009), perhaps because of the way we as adults have experienced theater in our lives. Most of us are familiar with going to the theater or reading famous plays in school. Unlike theater, creative dramatics in the classroom focuses on those who are participating, rather than on an end product or a final performance. Creative drama is defined by the Children's Theatre Association of America as "…an improvisational, non-exhibitional, process-centered form of drama in which participants are guided by a leader to imagine, enact, and reflect upon human experiences" (McCaslin, 1981). It is almost always an improvised performance where lines are not written or memorized. Many members of the group get to play various parts. Participants are guided by a teacher, and no decorations, costumes, or special equipment is needed.

Using drama in the classroom builds community, engages students in content, enhances focus and concentration, and often creates meaningful experiences that help students understand abstract concepts (Berk & Treiber, 2009). Pieczura wrote that,

Creative dramatics, a highly effective method for integrating arts education into core curriculum, produces a positive and lasting impact on student learning, in terms of creative and critical thinking, language development, listening, comprehension, retention, cooperation, and empathy and awareness of others. Creative dramatics not only has the power to bring curriculum to life, but also to stimulate active involvement in the development of conceptual understandings. (2013).

Well-planned, engaging experiences with creative drama guides and supports student's problem solving and decision-making skills, while at the same time, works at increasing awareness of how to solve issues at hand (Duckert & DeStasio, 2016). Instead of the STEM classroom just being a place where students are being taught and told what to think and feel, drama turns this into a deeper experience in thinking, further motivating students to question, respond, and explain what they themselves are feeling and thinking (Segarra, et al., 2018; Ødegaard, 2003). Creative drama can be a useful tool for all stages of the learning cycle, providing the STEM teacher an authentic assessment measure and an excellent tool for engaging, explaining, exploring, elaborating and evaluating (BSCS, 2005; Llewellyn, 2004).

STEM subjects are still generally taught in didactic, teacher-dominated classrooms by teaching methods that do not consider students' Piagetian cognitive developmental stages, learning styles, or multiple intelligences (Goodnough, 2001; Weld, 2000). Creative drama activities can help students bring abstract concepts into a concrete context by exploring and constructing an understanding of abstract ideas during improvisational interaction (Bailey, 1993; McCaslin, 1996). Students are motivated to build concepts, negotiate ideas, and solve problems when these are relevant and meaningful for them (Dockett, 1995). Effective learning occurs when students construct their understanding by active learning and by building on their prior knowledge (Yager, 1993). In 2007, Bracha wrote that

Students who are usually shy, quiet, self- conscious or seem troubled, "wake up" or "come to life" when they are free to use their imagination. Teachers are often surprised at the knowledge and the vocabulary that students have internalized but had not shared through traditional forms of learning and assessment (p. 57).

Participation in drama activities provides the teacher with another way to assess the student (Davidson & Simms, 2017). Through dramatic play, students reveal how they organize ideas, solve problems, work in a group, deal with conflict, and use their imagination (Fels & Meyer, 1997). Observing how students dramatize an event offers valuable insight into how they perceive, interpret, understand, and analyze the material at the core of the lesson (Caldwell, 2016). Buchanan wrote that,

A student can, if only for a few moments, become another, explore a new role, try out and experiment with various personal choices and solutions to very real problems-problems from their own life, or problems faced by characters in literature or historical figures. This can happen in a safe atmosphere, where actions and consequences can be examined, discussed, and in a very real sense experienced without the dangers and pitfalls that such experimentation would obviously lead to in the "real" world. (2016).

Although the use of creative dramatics to tap into the emotional connection of subject matter to students' daily lives is the most powerful use of theater in schools, it is also useful to simply have students physically role-play various complex physical or biological scientific/technical systems to give insight into these interactions and concepts related to these systems. In addition to using creative dramatics, it can also be an effective practice to actually have students in classes write and produce short or full-length plays that are final public projects that allow a STEM class to express their science, technology and society understandings to those of importance in their own personal lives.

It is also important to note that the integrative thinking that occurs in the right frontal cortex that is predominant in the arts is the same type of thinking necessary for novel insights in all STEM fields. Far too often, STEM education focuses on using linear, logical thinking devoid of any practice of using a Gestalt approach to creating novel insights utilizing concepts and data from seemingly unrelated science and engineering fields. There are a whole set of students who are drawn to the arts for this creativity. Many can actually do well in science, math or engineering, but choose not to pursue a STEM related career because they never get to see this side of these professions. However, when drama and STEM are intertwined, several of those previously disinterested in STEM professions realize a possible compatibility.

Various Reasons to Use Theater, Especially Creative Dramatics, in STEM Classrooms

The following is a list of reasons that creative dramatics is an effective tool to be used in STEM classrooms. Although each of these reasons stands on its own, it is the synergy among these schemas that makes creative dramatics such a powerful tool.

It Gives Students Insight Into the Scientific, Mathematical, Social, Economic, and Ethical Issues of Users of Technology

Students may not be intrinsically interested in codons, transcription, or mitochondrial RNA, but exposing students to role-play about future parents, religious leaders, and doctors asked to make decisions about DNA testing and genetic editing, almost instantly intrigues students with the ethical dilemmas in a way that reading about such issues does not engender. Once they are hooked on debating the "right and wrong" of issues, they begin to want to know the science and technologies used in order to make sense of the difficult ethical dilemmas. Using drama to teach about societal STEM interactions has been noted to improve pupils' empathy and identification in socio-political situations of science (Cobern, 1993). In South Africa lessons have been evaluated where role-plays helped raise the conceptual insight and level of argumentation about the ethics of organ-trade for xenotransplantation (Braund, Lubben, Scholtz, Sadeck, & Hodges, 2007). Hence, creative drama can assist in the development of argumentation in science by anchoring scientific debate and concepts in pupils' real-world experiences (Papacosta, 2006; Duveen & Solomon, 1994).

It Gives Students Insight Into the Social, Psychological, and Economic Dimensions of the Design Process

My first teaching job was over forty years ago in a "street academy" in EL Barrio, Spanish Harlem, created by the Young Lords, a group who were trying to bring the young adult Puerto Ricans into places of power, knowledge, and authority. By the time they created this school, they were college graduates, but formerly many were members of the Puerto Rican street gangs made famous by the movie *West Side Story*. In addition to the school, another young Puerto Rican leader who had a degree in architecture ran a studio where a number of Columbia University graduate architecture students chose to do their internships. This gave these mostly white, middle-class architects-in-training an opportunity to work on designs for real projects to be built in the neighborhood. They would interview the people who would use the future community center or housing project and then create designs to be evaluated by the staff in the studio. Invariably these students would create designs that were aesthetically beautiful to them and their Columbia faculty, while having met some of the criteria set by both the engineers and the local residential users. However, most of their designs were not utilized. The mothers who asked for enclosed courtyards in public housing so their children could be safe outside playing while they cooked in kitchens that faced these courtyards could not be heard. They couldn't hear that the fathers asked that Spanish Colonial architectural components be included in the designs for this matched their cultural aesthetics. The city engineers said thirty-story buildings were more affordable, but then mothers couldn't keep track of their kids like they could in an eight-story building.

Therefore, it is vital to have STEM students role-play community board meetings, with various students taking on the roles of parents, local physicians, engineers, politicians, local storeowners, and perhaps epidemiologists. The board is having a hearing on some design for a local structure that allows students to hear the passion that each representative brings to the table. Mathematics, science, and engineering when thrown into this cauldron of everyday public debate brings the subject matter involved into the emotional realm of everyday life and the importance it plays in effectively planning our *designed environment*. This approach comes from a long history of educational drama and related approaches to education (e.g. Boal, 2002; Freire, 2000; Heathcoate, 1984; O'Neil, 1995; Grady, 2000; Miller & Saxton, 2004). These strategies support a variety of learning styles that keep students actively engaged in the learning process while also allowing for differentiated instruction for diverse learners, such as those who are English Language Learners or who have disabilities.

These creative drama techniques are useful, helpful, and effective in many curriculum areas, including STEM subjects (Bailey, 1993; Kase-Polisini & Spector, 1992; McCaslin, 1986, 1996; Cottrell, 1987). Science researchers and educators in particular have promoted and researched this approach (Butler, 1989; Duveen & Solomon, 1994; Kentish, 1995; Steinert, 1993; Stencel & Barkoff, 1993). Butler (1989) and Webb (1980) see a major benefit of using creative drama in STEM classrooms because it enables students to learn openly from and with each other. Their studies demonstrate that drama is a stimulating alternative to class discussions, lectures, or question-and-answer sessions. STEM becomes more than learning a collection of isolated facts when drama is involved, where application and synthesis of conceptual understandings occur naturally.

These creative drama approaches do not only exist in secondary school settings. Kelly, Carlsen, and Cunningham (1993) strongly recommend that the sociological aspects of a genuine STEM culture should be included in school curricula. Early childhood science education seems an ideal place to begin this incorporation through sociodramatic play. It is evident that children utilize play to clarify their scientific thinking (Bailey, 1993; McCaslin, 1996; Smilansky, 1968, Smilansky, 1990). They mentally play with the concepts they are dealing with, incorporating them into an imaginary setting. They use play in order to make sense of puzzling, stimulating, and familiar contextual ideas. The book *Dramatic Science - Inspired ideas for teaching science using drama ages 5–11* by Debra McGregor and Wendy Precious (2015) provides teachers with a range of tried and tested techniques to use drama as a support and aid early childhood STEM education. All the strategies in this book have been shown to work effectively in a wide range of primary schools. There are three studies that closely examine successful use of drama in elementary STEM classrooms (Kamen, 1991; Metcalfe, Abbott, Bray, Exley, & Wisnia, 1984; Varelas, M., et al. 2010). This last one demonstrates enhanced STEM learning in urban first through third grade classes using drama enactments.

It Gives Teachers Insight Into the Decision-Making Criteria of Their Students and the Origins of These Criteria From Their Student's Lives

In the engineering portion of STEM, it is an overt part of the curriculum to actually teach the decision-making process (Alenljung, Person, & Wohlin, 2003; Gomes, 2016). However, if science and mathematics are taught using effective inquiry teaching methodologies (Jablon, 2014), then a major focus of the pedagogy is also to work on students' decision-making models in their process of doing math and science. In order to be successful at having students metacognitively be aware of their thinking within a STEM subject, teachers need to uncover students' decision-making patterns. Teachers and students can then compare and contrast the students' everyday decision-making models with ones used in the math and science communities. When students in STEM classrooms engage in STEM-societal role-plays, students, as they improvise their roles, overtly demonstrate the way they make decisions by the interactive dialogue they create for their characters. Not only does it become clear to the teacher and the other students watching the role-play, but also the student questioned about the role they created gets insight into how they utilize STEM understandings to make decisions in their lives.

In classrooms where teachers effectively facilitate learning through the use of creative drama, students actively co-construct their learning process. They take on the "mantle of the expert" through the scaffolded safety of drama-based activities (Cawthon, Dawson, & Ihorn, 2011), trying on new language (first modeled by the instructor) both in verbal and written form (Heathcote & Bolton, 1995). This dynamic process demands higher-order thinking skills and increases emotional intelligence. In some cases, this shift in roles and knowledge making through drama-based work can illuminate where socially constructed meanings exist and create space for self-efficacy to emerge (Thompson, 2012). Kentish (1995) notes that when students use creative drama strategies, they develop greater ownership of their learning, which is reflected by their development of problem-solving and decision-making skills and abilities.

In addition, creative dramatics not only gives insights into students' decision-making and problem-solving skills, but also can teach, or at least reinforce, the same STEM skills in parallel. Similarities exist between STEM and creative drama processes. In both processes students are active, solve problems, ask questions, learn through inquiry, and construct knowledge built on prior experience and information (Bailey, 1993; McCaslin, 1996; Kase- Polisini & Spector, 1992). When integrated into teaching STEM, creative drama can assist students in developing STEM skills as well as affective skills, such as cooperative work, empathy, communication, listening, and reasoning (Cristofi and Davis, 1991, Hodson & Reid, 1988; Kase-Polisini & Spector, 1992).

It Gives Teachers Insight Into the Conceptual Understanding of Their Students of Subject Area Conceptual Knowledge as They Apply It in Dramatic Role-Plays

Just as with their decision-making skills, creative drama helps learners to make their STEM conceptual understandings *explicit* (Onwukwe, Akuezuilo, & Ijioma, 2011; Venkateswaran, 2008, HMI, 1989; ArtsCouncil, 1992). Certainly, role-plays are generally more spontaneous and more flexible than end-of-chapter questions or discussion led by the teacher (Steinert, 1993). Cornett and Smithrim wrote that,

Drama is an alternative way to assess by observing. When teaching new lessons, we always depend on prior knowledge. We start with what students know, which

further guides us with the next step to take in our teaching. It is difficult for some of our students to make sense of specific things, which is simply where drama fits in. Drama can be used to preview or review a lesson; further allowing teachers to assess what students already know or have learned. (2001).

Whether the role-play has students playing roles in societal situations related to decision-making related to STEM subject matter, or the students are actually role-playing the parts of a scientific or technological system using their bodies, the teacher and students still see and hear a physical analogy manifestation that can then be discussed concretely, and possibly then moved into a more theoretical realm. In *The Drama of Science* (2009), Bernhard Sturm offers a selection of drama-based activities to act or re-enact science in the chemistry and physics classroom. For example, he suggests a method to visualize the exchange of oxygen between different metal atoms within the redox series (after Lavoisier). Each student wears a cap in one of three colors, representing atoms of oxygen and two different metals (there should be equal numbers of each color). In groups of about 8-10, the students enact simple redox reactions with 1:1 stoichiometry, such as $CuO + Fe \rightarrow Cu + FeO$, and afterwards present them to the entire class. Students often find creative ways to represent both the activation energy and the release of energy in these reactions. Although these role-plays of actual physical reactions are not as powerful a way to engage students in science as are social decision-making situations, they are still somewhat useful to uncover students' understandings and misunderstandings of STEM subject matter (Ceylan, et al. 2015; Close & Scherr, 2015; Stroupe & Kramer, 2014).

Role-play not only allows students to demonstrate their understanding, but also to explore their views and develop deeper understanding of phenomena by combining new facts and concepts with previously acquired knowledge and applying it to solving real-life problems (Hendrix, et al., 2012; Butler, 1989). Role-play permits students to verbalize abstract science ideas in terms that they can understand (Peleg & Baram-Tsabari, 2011). Role-play has the potential to help students develop and create their own mental models. Gardner (2011) suggested, "in trying to understand science, students draw on available mental models" (p. 157). Often in science these mental models create images of a microscopic world, which is inferred from observations of theories while trying to explain phenomena (Aubusson, et al., 1997; Aubusson & Fogwill, 2006). Creative drama is an effective strategy to teach abstract concepts. According to Stencel and Barkoff (1993), drama is very useful in teaching abstract concepts, especially those at the molecular level that are often difficult for students to understand. Bachelis, et al. (1994) demonstrate that students can best understand an algorithm if they can see the process tangibly at work, and creative drama activities are suggested as a means of accomplishing this goal.

One of the great gurus of drama education, Dorothy Heathcote, went even further than just seeing drama as a milieu in which students can demonstrate and modify their understandings of STEM concepts. She asserts that the challenges involved in creating drama stimulate left-brain activity and hence scientific and mathematical functioning (Wagner, 1979). Evidence from neurological research into brain function supports the view that creative (right hemispheric) activity contributes to general cognitive ability (Gazzaninga, 1989). Inoa, et al., (2014) demonstrates the relationship between theater arts and mathematics achievement.

It Gives Teachers Insights Into Issues Affecting the Emotional Lives of Their Students

As educators we all know that the most defining criteria for successful learning to take place in a classroom is the relationship between the teacher and the student, followed only by the relationships between the student and other students in the class. For over three decades I have utilized role-plays related to the STEM content and skills in my classes, and as a result of students improvising their roles, or of other students speaking about the improvised situations, have learned much about events taking place in students' lives that have an impact on their engagement as learners. This is one of the reasons that my colleagues and I have had such success with students who were previously disengaged from school and STEM classes (Jablon, 2014).

In addition to uncovering emotional issues in students' lives, creative drama can also directly enhance students' psychological well-being. Under different characters, students can express their true feelings or sense of personality without fear of being judged or criticized. They can work on personal issues or solve personal problems while in character, which can simply help their overall well-being. Essentially, what this does is allow students to get things off their mind, further releasing emotion and tension and allowing students to be who they are (Cornett & Smithrim., 1999).

It Meets Many of the Emotional Needs of Their Students (Maslow), so These Students Can Focus on the Concepts and Thinking Skills Involved in Their Stem Subjects

If students are deeply engaged in school lessons where they are actively negotiating ideas and creating concepts, then deep understanding is the result. Therefore, it is simple. We just need to get students to come to school every day, then come to STEM classes every day, and then, when they are there, be completely focused on thinking and meaning making. We know that student engagement is tied to improvements in student behavior (including both an increase in pro-social behavior and a decrease in antisocial behaviors), improved learning outcomes, and greater respect for the rights of others (Covell & Howe, 2001; Decoene & De Cock, 1996; Howe & Covell, 2007). High levels of student engagement can also reliably predict lower high school dropout rates (Archambault, Janosza, Fallu, & Pagania, 2008). Student engagement has positive effects for teachers, as well; Covell, et al. (2009) found that low engagement is a contributing factor in teacher burnout and that increasing engagement can reduce teacher burnout and improve teacher self-efficacy.

In middle-grade and secondary STEM classes, most of the topics studied don't immediately appear relevant to students' lives and inherent psychological needs. The study of acceleration, subduction, or osmosis doesn't appear immediately relevant. Calculating centrifugal force seems so unimportant to almost every sixteen-year-old. Likewise, uncountable numbers of students have asked when they will ever need to use a quadratic equation. Yet each of these topics, if embedded in the right explorations, contextualized in challenging local projects, with the appropriate large conceptual objectives and effective themes can be made engaging and deeply relevant to many of the learners' lives. However, for the most disengaged students whose lives are unpredictable, unless what occurs in the classroom can meet their immediate psychological needs, they cannot focus on any of the STEM content.

It Meets the Needs of Those Students Whose Way of Learning STEM Is Through Creative Engagement in the Arts

Twelve years into my university career as a science and interdisciplinary education professor, I had an opportunity to spend a year again teaching in an urban high school. This particular school was fairly representative of the traditional pedagogy practiced in American schools in the early twenty-first century. However, in my classroom I engaged students in daily inquiry investigations, working in small research teams, focused on large conceptual understandings, problem solving, and students owning their learning. Although this was a whole new classroom paradigm for these students, many adapted well and had a higher level of success than in their other classes (Jablon, 2013). However, there were a number of students who were not thriving, and others who were not succeeding in this STEM class. That is, until I started using creative dramatics' role-plays on a regular basis. There are two students I remember in particular. One boy was there every day, but rarely participated in his group and didn't bother to study any of the material we were investigating. The other was a girl who attended sporadically, and when present feigned indifference to all that many others were engaged in creating.

One day the class was in the middle of a mini-unit on issues dealing with solid waste in our communities when I suggested we were going to do a role-play of a community board meeting about a proposal to build a solid waste incinerator. Within two days both of these students became animated, taking lead roles, and became engaged in learning the science concepts and technology applications. They not only intuitively understood the societal dilemmas, but also led other students to understand the complexities of the ethical, economic, psychological dimensions of the community roles they played. They were naturals at learning and helping others learn within this milieu. They physically attended class every day, but more importantly mentally attended to the material even when we did regular inquiry investigations – as long as we did some role-playing now and then as part of the pedagogical mix.

This was one of many examples from my STEM classes where, "drama does engage multi- dimensional learning styles including verbal-linguistic, interpersonal, intrapersonal, kinesthetic, spatial, and logical" (Ashton-Hay, 2005). As teachers, we understand that our students learn differently and at different paces because of their biological and psychological differences (Reiff, 1992). Learning styles not only comprise the cognitive domain, but also the affective and physiological domains (Oxford, Hollaway & Horton-Murillo, 1992; Kinsella, 1996). Strong, Silver and Perini (2001) divide verbal-linguistic intelligence into four learning preferences including mastery (knowing), interpersonal (connecting with people; social skills), understanding (discovery and reasoning), and self-expressive (creativity). Using drama in a classroom naturally matches the learning styles of a variety of students, thus their motivation, performance, and achievement is more likely to increase making the STEM content readily accessible to them.

CONCLUSION

Do I believe that teachers and professors reading this will be so informed and empowered that they will begin using this in their classrooms and professional development workshops starting next term? Actually, for most, they probably will not. However, for some, what has been presented here makes so much sense, and has enough studies to demonstrate its effectiveness, that they feel compelled to try to reach the rest of their STEM students that they are not currently deeply engaging. Therefore, what is a good path for them to take?

If you are currently a STEM teacher, find someone in your school or district who already does creative dramatics in their classroom. Ask if you can attend their class on a day when they will be doing some role-plays. Watch everything they do closely. Then ask if you can come again, but if the next time you could facilitate a role-play in their class, even if it is an English or social studies class. You will want to meet with them beforehand for a couple of minutes, so you know the purpose of the improv and where they expect it to go. That way, if it doesn't go exactly as planned when you are facilitating in their class, they are there to help you out if absolutely necessary. With this success under your belt, ask if they could meet with you after school or on a weekend and help you think up one or two scenarios that will work in your STEM class. Now it is time to jump into the pool. Try it in your STEM class. Pick kids who you think will love to do this and ask them the class before. You will most likely be successful, and the students will be engaged. You will also get better at it as you go along. If you are a STEM professor at a university, find the arts in education faculty and do the same protocol with them. Ironically, it is the theater educators who in 2017 had the focus of their national conference as *The Rise of STEAM: Real World Theater Education for the 21st Century* (Educational Theater Association, 2017). As STEM educators we need to join with them and utilize their expertise in our teaching.

I have always taught these approaches to creative drama in my science methods classes, and convinced colleagues in various university education departments to have a required class where these play a prominent role for all of our education majors. I have then seen many first-year teachers use these very successfully in their classes, and now years later have many who teach others in their schools how to use creative dramatics. I also run workshops in school districts throughout the country and at national conferences with STEM teachers at every grade level from Pre-K to graduate school, so there are a cadre of STEM professionals utilizing these techniques in their classrooms.

Before you have done these regularly in your classroom, you might think there will be a lot of resistance from a substantial number of students. There are some who initially bristle at the thought of doing it, but invariably after the class has been doing them for a month or so, most everyone is regularly volunteering when given the opportunity. You know it is working because students are eagerly asking if the class is going to be doing role-plays today. It also really changes the culture of the STEM class in a very positive way, bringing humanity right into the core of everything studied. Students are more creative, think more deeply, and are usually truly engaged in most everything the class attempts to solve.

Now it is your turn to raise the curtain on learning!

REFERENCES

Alenljung, B., Persson, A., & Wohlin, A. A. (2003). The fundamental nature of requirements engineering activities as a decision-making process. *Information and Software Technology*, *45*(14), 945–954. doi:10.1016/S0950-5849(03)00096-X

Archambault, I., Janosza, M., Fallu, J., & Pagania, L. S. (2008). Student engagement and its relationship with early high school dropout. *Journal of Adolescence*, *32*(3), 651–670. doi:10.1016/j.adolescence.2008.06.007 PMID:18708246

Arts Council. (2003). *Arts council guidance on drama education: Drama in schools*. London: The Arts Council of Great Britain.

Ashton-Hay, S. (2005). Drama: Engaging all learning styles. In *Proceedings 9th International INGED (Turkish English Education Association) Conference*. Economics and Technical University.

Aubusson, P., Fogwill, S., Barr, R., & Perkovic, L. (1997). What happens when students do simulation-role-play in science? *Research in Science Education*, *27*(4), 565–579. doi:10.1007/BF02461481

Aubusson, P. J., & Fogwill, S. (2006). Role-play as analogical modelling in science. In P. J. Aubusson, A. G. Harrison, & S. M. Ritchie (Eds.), *Metaphor and analogy in science education* (pp. 93–104). Springer. doi:10.1007/1-4020-3830-5_8

Bachelis, G. F., Maxim, B. R., James, D. A., & Stout, Q. F. (1994, April). *Bringing algorithms to life: Cooperative computing activities using students as processors*. Academic Press.

Bailey, S. D. (1993). *Wings to fly: Bringing theatre arts to students with special needs*. Rockville, MD: Woodbine House.

Bennett, J., & Hogarth, S. (2009). *Would YOU want to talk to a scientist at a party? Students' attitudes to school science and science*. University of York, Department of Educational Studies Research Paper 2005/08.

Berk, R., & Trieber, R. (2009). Whose classroom is it anyway? Improvisation as a teaching tool. *Journal on Excellence in College Teaching*, *20*(3), 29–60.

Boal, A. (2002). *Games for actors and non-actors*. New York: Routledge.

Bracha, B. A. (2007). *The integration of creative drama into science teaching* (Doctoral dissertation). Kansas State University. Retrieved from https://krex.k-state.edu/dspace/bitstream/handle/2097/531/BrachaArieli2007.pdf?sequence=1

Braund, M. (2015). Drama and learning science: An empty space? *British Educational Research Journal*, *41*(1), 102–121. doi:10.1002/berj.3130

Braund, M., Lubben, F., Scholtz, Z., Sadeck, M., & Hodges, M. (2007). Comparing the effect of scientific and socio-scientific argumentation tasks: Lessons from South African. *The School Science Review*, *88*(324), 67–76.

BSCS - Biological Science Curriculum Study. (2005). *BSCS Science and technology: Investigating physical systems.* Dubuque, IA: Kendall Hunt.

Buchanan, M. (2016). Why teach drama? *Child Drama.* Accessed June 3, 2016 from http://www.childdrama.com/the-essay.html

Burke, R. J., & Mattis, M. C. (2007). *Women and minorities in science, technology, engineering and mathematics: Upping the numbers.* Northhampton, NJ: Edward Elgar Publishing, Inc. doi:10.4337/9781847206879

Butler, J. (1989). Science learning and drama processes. *Science Education, 73*(5), 569–579. doi:10.1002ce.3730730505

Caldwell, K. (2011). Drama as a teaching tool. *Teaching Through the Arts.* Accessed June 3, 2016 from http://teachingthroughthearts.blogspot.com/2011/07/drama-as-teaching-tool.html

Cawthon, S., Dawson, K., & Ihorn, S. (2011). Activating student engagement through drama-based instruction. *Journal for Learning through the Arts, 7*(1). Retrieved from http://www.escholarship.org/uc/item/6qc4b7pt

Ceylan, E., Kerim Polat, R., Akpınar, M., & Kalender, S. (2015). Designing Science Instruction based on Creative Drama: The Effect on 6th Grade Students' Understanding and Elimination of Misconceptions. *International Online Journal of Educational Sciences, 7*(4), 1–10.

Close, H., & Scherr, R. (2015). Enacting Conceptual Metaphor through Blending: Learning activities embodying the substance metaphor for energy. *International Journal of Science Education, 37*(5-6), 5–6, 839–866. doi:10.1080/09500693.2015.1025307

Cobern, W. W. (1993). Contextual constructivism: The impact of culture on the learning and teaching of science. In K. Tobin (Ed.), *The Practice of constructivism in science education.* Washington, DC: American Association for the Advancement of Science.

Cornett, C. E., & Smithrim, K. E. (2001). *The arts as meaning makers: Integrating literature and the arts throughout the curriculum.* Don Mills, Canada: Pearson.

Cottrell, J. (1987). Creative drama in the classroom grades 4-6. Lincolnwood, IL: National Textbook Company.

Covell, K., & Howe, R. B. (2001). Moral education through the 3 Rs: Rights, respect and responsibility. *Journal of Moral Education, 30*(1), 29–41. doi:10.1080/03057240120033794

Covell, K., McNeil, J., & Howe, R. B. (2009). Reducing teacher burnout by increasing student engagement: A children's rights approach. *School Psychology International, 30*(3), 282–290. doi:10.1177/0143034309106496

Cristofi, C., & Davis, M. (1991, January). Science through drama. *Education in Science*, (141): 28–29.

Davidson, C. D., & Simms, W. (2017). Science theater as STEAM: A case study of "Save It Now". *e STEAM Journal, 3*(1), 14. Available at http://scholarship.claremont.edu/steam/vol3/iss1/14

Decoene, J., & DeCock, R. (1996). The children's rights project in the primary school "Devrijdagmarkt" in Bruges. In E. Verhellen (Ed.), *Monitoring children's rights*. The Hague: Martinus Nijhoff.

Dockett, S. (1995). *You make me alive! Developing understanding through play*. Deakin, ACT: Australian Early Childhood Association.

Dorion, K. R. (2009). Science through drama: A multiple case exploration of the characteristics of drama activities used in secondary science lessons. *International Journal of Science Education, 31*(16), 2247–2270. doi:10.1080/09500690802712699

Duckert, C. L., & DeStasio, E. A. (2016). Setting the stage for science communication: Improvisation in an undergraduate life science curriculum. *The Journal of American Drama and Theatre, 28*(2), 1–18. Retrieved from https://jadtjournal.org/2016/05/27/setting-the-stage-for-science-communication-improvisation-in-an-undergraduate-life-science-curriculum/

Duveen, J., & Solomon, J. (1994). The great evolution trial: Use of role-play in the classroom. *Journal of Research in Science Teaching, 31*(5), 575–582. doi:10.1002/tea.3660310510

Educational Theater Association. (2017). *2017 EdTA National Conference – The rise of STEAM. Nashville, TN September 14-20, 2017*. Retrieved from https://www.americansforthearts.org/events/artsmeet/2017-edta-national-conference-the-rise-of-steam-real-world-theatre-education-for-the-21st-century

Fels, L., & Meyer, K. (1997). On the edge of chaos: Co-evolving world(s) of drama and science. *Teaching Education, 9*(1), 75–81. doi:10.1080/1047621970090113

Freire, P. (2000). *Pedagogy of the oppressed* (M. B. Ramos, Trans.). New York: Continuum Publishing.

Gardner, H. (2011). *The unschooled mind: How children think and how schools should teach* (2nd ed.). New York, NY: Basic Books.

Gazzaninga, M. S. (1989). Organisation of the human brain. *Science, 245*(4921), 947–952. doi:10.1126cience.2672334 PMID:2672334

Gomes, L. F. A. M. (2016). *Teaching decision making analytical skills to engineers: A new paradigm.* Retrieved from http://citeseerx.ist.psu.edu/viewdoc/download?doi=10.1.1.201.3517&rep=rep1&type=pdf

Goodnough, K. (2001). Multiple intelligences theory: A framework for personalizing science curricula. *School Science and Mathematics, 101*(4), 180–193. doi:10.1111/j.1949-8594.2001.tb18021.x

Grady, S. (2000). *Drama and diversity: A pluralistic perspective for educational drama.* Portsmouth, NH: Heinemann.

Grocholski, E. (2018). *Research reveals boys interests in STEM careers declining; Girls interest unchanged.* Junior Achivement USA News. Retrieved from https://www.juniorachievement.org/web/ja-usa/press-releases/-/asset_publisher/UmcVLQOLGie9/content/research-reveals-boys'-interest-in-stem-careers-declining-girls'-interest-unchanged

Heathcoate, D., & Bolton, G. (1995). *Drama for learning: Dorothy Heathcote's mantle of the expert approach to education.* Portsmouth: Heinemann.

Heathcote, D. (1991). *Collected writings on education and drama.* Evanston, IL: Northwestern University Press.

Heaverlo, C. (2011). *STEM development: A study of 6th-12th grade girls' interest and confidence in mathematics and science.* Graduate Theses and Dissertations. Paper 10056. Retrieved from http://lib.dr.iastate.edu/cgi/viewcontent.cgi?article=1035&context=etd

Hendrix, R., Charles, E., & Shannon, D. (2012). The Integration of Creative Drama in an Inquiry-Based Elementary Program: The Effect on Student Attitude and Conceptual Learning. *Journal of Science Teacher Education, 23*(7), 823–846. doi:10.100710972-012-9292-1

HMI. (1989). Curriculum matters 17: Drama from 5-16. London: HMSO.

Hodson, D., & Reid, D. (1988). Changing priorities in science education: Part I. *School Science Review, 70,* 101-108.

Howe, R. B., & Covell, K. (2007). *Empowering children: Children's rights education as a pathway to citizenship.* Toronto: University of Toronto Press.

Inoa, R., Weltzek, G., & Tabone, C. (2014). A Study on the Relationship Between Theater Arts and Student Literacy and Mathematics Achievement. *Journal for Learning Through the Arts, 10*(1), 1–19. doi:10.21977/D910123495

Jablon, P. (1989). *A Documentation and impact analysis of a project-oriented, interdisciplinary, team-taught, peer-influence program for secondary schools and its effects upon science learning* (Doctoral Dissertation). New York University. (Dissertation Abstracts 9016410)

Jablon, P. (2010). *Analogous interrelationship role-plays: Engaging the previously disengaged in science*. Paper presented at The Association of Science Teacher Educators International Conference, Sacramento, CA.

Jablon, P. (2013). *Get real! Walking the walk to inform talking the talk. Full-time teaching in an urban high school. In Science teacher educators as K-12 science teachers: Practicing what we teach*. New York: Springer.

Jablon, P. (2014). *The synergy of inquiry: Engaging students in deep learning across the content areas*. Huntington Beach, CA: Shell Publications.

Jablon, P., & Born, T. (1988). STS works in the city. *SSTS Reporter., 4*(2), 5–8.

Kamen, M. (1991). Use of creative drama to evaluate elementary school students. In G. Kulm & S. Malcom (Eds.), *Science assessment in the service of reform*. Washington, DC: AAAS.

Kamen, M. (1991, August). *Creative drama and the enhancement of elementary school students' understanding of science concepts* (Doctoral Dissertation). The University of Texas at Austin.

Kase-Polisini, J., & Spector, B. (1992). Improvised drama: A tool for teaching science. *Youth Theatre Journal, 7*(1), 15–19.

Kelly, G. J., Carlsen, W. S., & Cunningham, C. M. (1993). Science education in sociocultural context: Perspectives from the sociology of science. *Science Education, 77*(2), 207–220. doi:10.1002ce.3730770208

Kelner, L. B. (1993). *The creative classroom: A guide for using creative drama in the classroom, PreK-6*. Portsmouth, NH: Heinemann.

Kentish, B. (1995). Hypotheticals: Deepening the understanding of environmental issues through ownership of learning. *Australian Science Teachers Journal, 41*(1), 21–25.

Kesar, S. (2018). *Closing the STEM gap: Why STEM classes and careers still lack girls and what we can do about it.* Digital Skills/Microsoft. Retrieved from https://query.prod.cms.rt.microsoft.com/cms/api/am/binary/RE1UMWz

Kinsella, K. (1996). Designing group work that supports and enhances diverse classroom work styles. *TESOL Journal*, *6*(1), 24–31.

Kuenzi, J., Matthews, C., & Mangan, B. (2006). *Science, technology, engineering, and mathematics (STEM) education issues and legislative options*. Washington, DC: Congressional Research Service, the Library of Congress.

Llewellyn, D. (2004). *Teaching high school science through inquiry: A Case study approach*. Corwin Press.

Mardirosian, G. H., & Lewis, Y. P. (2009, March 6). How to use theater to teach at-risk students. *The Chronicle of Higher Education*, *55*(26), 34–37.

Maslow, A. H. (1998). *Toward a psychology of being* (3rd ed.). New York, NY: Van Nostrand.

McCaslin, N. (1981). *Children and drama* (2nd ed.). New York: University Press.

McCaslin, N. (1986). *Children and drama*. New York: Longman.

McCaslin, N. (1996). *Creative drama in the classroom and beyond*. London: Longman Publishers.

McCaslin, N. (2006). *Creative drama in the classroom and beyond*. Boston, MA: Allyn and Bacon.

McGregor, D., & Precious, W. (2015). Dramatic science: Inspired ideas for teaching science using drama ages 5–11. New York: Routledge.

Metcalfe, R. J. A., Abbott, S., Bray, P., Exley, J., & Wisnia, D. (1984). Teaching science through drama: An empirical investigation. *Research in Science & Technological Education*, *2*(1), 77–81. doi:10.1080/0263514840020109

Miller, C., & Saxton, J. (2004). *Into the story: Language in action through drama*. Portsmouth, NH: Heinemann.

NGSS Lead States. (2013). *Next generation science standards: For states, by states*. Washington, DC: The National Academies Press.

O'Neill, C. (1995). *Drama worlds: A Framework for process drama*. Portsmouth: Heinemann.

Ødegaard, M. (2003). Dramatic science. A critical review of drama in science education. *Studies in Science Education, 39*(1), 75–101. doi:10.1080/03057260308560196

Onwukwe, E., Akuezuilo, E., & Ijioma, B. (2014). Science drama: Clue to underachievement in secondary school chemistry *STEMplanet, 11*. Retrieved from http://stemstates.org/journal/stemplanet-2014/2014-vol-11.html

Osborne, J., Simon, S., & Collins, S. (2003). Attitudes towards science: A review of the literature and its implications. *International Journal of Science Education, 25*(9), 1049–1079. doi:10.1080/0950069032000032199

Oxford, R. L., Hollaway, M. E., & Horton-Murillo, D. (1992). Language learning styles: Research and practical considerations for teaching the multicultural tertiary ESL/EFL classroom. *System, 20*(4), 439–456. doi:10.1016/0346-251X(92)90057-A

Papacosta, P. (2006). Humanities & Arts to the Rescue of Science. *Forum on Public Policy.* Retrieved from http://forumonpublicpolicy.com/archive07/papacosta.pdf

Peleg, R., & Baram-Tsabari, A. (2011). Atom Surprise: Using Theatre in Primary Science Education, Special Issue on Science Education in Preschools and Primary Schools: Classrooms, Teachers, and Children. *Journal of Science Education and Technology, 20*(5), 508–524. doi:10.100710956-011-9299-y

Pieczura, P. (2013). Decidedly dramatic! The power of creative drama in social studies. *Social Studies and the Young Learner, 25*(3), 9–12.

Pulkkinen, L. (1982). Self-control and continuity from childhood to adolescence. In Life-span development and behavior (vol. 4). New York, NY: Academic Press.

Reiff, J. (1992). *What research says to the teacher: Learning styles*. Washington, DC: National Education Association.

Roberts, D. (2012). *Fatal invention: How science, politics, and big business re-create race in the twenty-first century*. New York: The New Press.

Salisbury, B. T. (1996). *Theatre arts in the elementary classroom* (2nd ed.). New Orleans, LA: Anchorage Press, Inc.

Smilansky, S. (1968). *The effects of sociodramatic play on disadvantaged preschool children*. John Wiley.

Smilansky, S. (1990). Sociodramatic play: Its relevance to behavior and achievement in school. In E. Klugman & S. Smilansky (Eds.), *Children's play and learning: Perspectives and policy implications* (pp. 18–42). Teacher College Press.

Steinert, Y. (1993). Twelve tips for using role-playing in clinical teaching. *Medical Teacher, 15*(4), 283–291. doi:10.3109/01421599309006651 PMID:8139402

Stencel, J., & Barkoff, A. (1993). Protein Synthesis: Role-Playing in the Classroom. *The American Biology Teacher, 55*(2), 102–103. doi:10.2307/4449595

Strong, R., Silver, H., & Perini, M. (2001). *Teaching what matters most: Standards and strategies for raising student achievement*. Association for Supervision and Curriculum Development.

Stroupe, D., & Kramer, A. (2014, October). Students Modeling Molecule Movement Through Science Theater. *Science Scope, 38*(2), 70–77. doi:10.2505/4s14_038_02_70

Sturm, B. (2009). The drama of science. *Science in Schools, 13*, 29-33. Retrieved from http://www.scienceinschool.org/sites/default/files/teaserPdf/issue13_drama.pdf

Thompson, J. (2012). *Applied theatre: Bewilderment and beyond* (4th ed.). Oxford, UK: Peter Lang; doi:10.3726/978-3-0353-0309-4

United States Department of Education. (2006). *The Condition of education*. Washington, DC: National Center for Education Statistics, U.S. Government Printing Office.

United States Department of Education. (2017). *The Condition of education*. Washington, DC: National Center for Education Statistics, U.S. Government Printing Office.

Varelas, M., Pappas, C.C., Tucker-Raymond, E., Kane, J., Hankes, J., Ortiz, I., & Keblawe-Shama, N. (2010). Drama activities as ideational resources for primary-grade children in urban science classes. *Journal of Research in Science Teaching, 47*(3), 302–325.

Venkateswaran, T. V. (2008). *Communicating science through plays, games and simulations*. New Delhi: Scifun.

Wagner, B. J. (1979). *Dorothy Heathcote: Drama as a learning medium*. London: Hutchinson Education.

Weld, J. D. (2000). Fly Guys. *Teacher Magazine, 11*(8), 56–58.

Yager, R. E. (1993). The constructivist learning model. *Science Teacher (Normal, Ill.), 60*, 27–31.

KEY TERMS AND DEFINITIONS

Analogous Interrelationships: Fairly common interactions between persons that are familiar to a learner that are analogous to interactions between groups, societies, or nations that are not as yet understandable to learners. They can be used by educators to have their students better, and more easily, understand the larger, seemingly remote relationships.

Creative Dramatics: Creative drama is an improvisational, non-exhibitional, process-oriented form of drama in which participants are guided by a leader to imagine, enact, and reflect on experiences real and imagined. In education, used as a tool for better understanding by students of academic conceptual material.

Engaging Adolescents: The act of getting teens and pre-teens interested in some learning that they were previously not interested in.

Interdisciplinary Education: An educational approach in which two or more disciplines collaborate in the learning process with the goal of fostering interprofessional interactions that that enhance the practice of each discipline.

Science Education: The field concerned with sharing science content and processes with individuals not traditionally considered part of the scientific community.

STEAM: The use of the arts in engaging students and having them better understand the concepts and skills involved in the combination of science, technology, engineering, and mathematics (STEM) classes.

Theater: The activity of acting in, producing, direction, or writing plays.

Urban Education: The practice of having students learn who traditionally live in and attend school in urban cities.

Chapter 5
Exploring Simple Machines With Creative Movement

William Paul Lindquist
Hamline University, USA

Martha James-Hassan
Morgan State University, USA

Nathan C. Lindquist
Bethel University, USA

EXECUTIVE SUMMARY

This chapter explores the use of creative movement to extend meaning to inquiry-based science investigations. This process embraces the addition of A to STEM to realize the impact of STEAM. The chapter builds on the import of scientific and physical literacy, interdisciplinary learning, and the power of kinesthetic engagement. Students become active collaborative agents within a dynamic model using creative movement to bring meaning to the science of simple machines. The authors utilize working words into movement strategy to help students use their past experiences and motor memory to explore, interpret, and engage with as they seek understanding of simple machines. A Midwest urban elementary school provides the context for a unit plan culminating in a dance performance. The foundational ideas presented within this unit can be enacted within any classroom by creative movement (physical education or dance) specialists, science specialists, or classroom generalists. It follows with a presentation of science content on simple machines exploring the disciplinary core idea of force and motion.

DOI: 10.4018/978-1-5225-9631-8.ch005

INTRODUCTION

In the current climate of high-stakes testing and educational accountability extreme pressure is being placed on schools. The catchy binary of "Career and College Ready" serves adults' need to apply order to the messy process of educating young people. Those young people, however, may be better served by the less structured—and admittedly more challenging to evaluate—"life ready" educational framework. To be life ready learners must achieve alpha and numeric literacy as well as inquiry habits, history and systems knowledge, stewardship skills, aesthetic awareness, physical well-being, and most importantly, self-efficacy and critical thinking. To accomplish the ambitious goal of educating students rather than schooling them, teachers must build meaningful relationships with them and, from an instructional standpoint, use every minute twice. Interdisciplinary teaching and learning are critical tools to simultaneously deliver dynamic and engaging content to students as they support reflection, creative thinking, and critical problem solving.

In this chapter, the authors will explore the richness culminated when multiple and diverse scientific concepts are paired with creative movement in the classroom. First, a pedagogical background from both science and creative movement will be provided. This leads to a block plan for a sample unit in simple machines that can be carried out in the classroom, accompanied by suggestions about ways to integrate movement into any lesson. The authors' goal and hope are that the reader can walk away empowered by the thought of transforming their classroom into an interdisciplinary space, including the arts in STEM (science, technology, engineering, and mathematics education) and bringing to life the powerful potential of a $STEAM_2$-focused curriculum.

LITERATURE REVIEW

Scientific Literacy

Within an increasingly complex and technological world facing emergent global sustainability issues, it is critical for American students develop a foundational level of scientific literacy. Whether they become scientists, journalists, or refuse handlers, their participation in today's world requires the ability to think critically and act responsibly. It will be the abilities to think and act that allow our students to maintain and improve the quality of life for future generations. By kinesthetically teaching STEM concepts and teaching movement competency framed in STEM content, educators can empower their students to have the agency and confidence to become active participants rather than mere observers in the world.

To accomplish this, students need engagement in the active "doing" of science. This engagement helps them develop the proficiency to ask questions, design and carry out investigations, and reflect critically on the outcomes. Children learn science best by "doing" science (Martin, 2011). An engaged elementary classroom committed to the integrated needs of a $STEAM_2$ curriculum needs to begin with a robust commitment to providing hands-on use of science process skills in open-ended inquiries. With that foundation, learning is enriched and broadened throughout all aspects of $STEAM_2$. This includes the use of technology, the analysis of mathematics, meaningful interactions with text (in the broadest sense), and the integration of the arts – in this case, creative movement.

Physical Literacy

The purpose of education is to prepare young people with the knowledge, skills, and experiences they need to be successful citizens and productive members of society. In 2002, the United Nations General Assembly passed resolution 56/116: United Nations Literacy Decade: Education for All (United Nations, 2002). The foundation of this resolution was based on the collective conviction that literacy is crucial to the acquisition of essential life skills that enable every child, youth, and adult to address the challenges they face in life. It represents an essential step in basic education and an indispensable means for effective participation in the societies and economies of the twenty-first century. In their understanding of the complexities and arbitrary nature of the knowledge and experiences students will apply in the future, members of the General Assembly had the insight to recognize that authentic literacy is not restricted merely to reading and writing. According to the United Nations Education, Scientific, and Cultural Organization (UNESCO; 2003), literacy is concerned with how people communicate about knowledge, language, and culture within social practices and relationships.

This holistic, comprehensive understanding of literacy supports the development of interdisciplinary $STEAM_2$ instruction. merging scientific literacy with physical and dance literacy. As articulated by Mandigo, Francis, Lodewyk and Lopez (2009),

Physically literate individuals consistently develop the motivation and ability to understand, communicate, apply, and analyze different forms of movement. They are able to demonstrate a variety of movements confidently, competently, creatively, and strategically across a wide range of health-related physical activities. (p.7).

Physical literacy includes viewing physical activity, dance, and sport through a lens of creating, analyzing, interpreting, and evaluating movement as text. For an interdisciplinary theme-based instructional unit to be successful, standards within each of the individual content areas appropriate to that learning must have equal weight and representation. For physical literacy, instructors must focus not only on meeting dance and physical education standards, but they must also focus on acquiring and maintaining students' engagement in moderate-to-vigorous physical activity as much as possible. Traditionally, the concept of physical literacy focuses on physical education instruction. The authors have chosen to frame the artistic component of this chapter as creative movement for the intentional purpose of opening up space for movement both inside and outside of the gymnasium. Creative movement, as it is expressed in this writing, is an art form whose medium is the human body in motion (Dow, 2010). Aligned with both dance and physical literacy standards, teachers can plan—and, students can engage in—creative movement experiences in physical education, classroom, science lab, or any other learning space within the school community.

Interdisciplinary Teaching and Learning

Interdisciplinary teaching applies methods and language from several academic disciplines to examine a theme, issue, question, problem, topic, or experience (Coffey, n.d.). Interdisciplinary methods create connections between traditionally discrete disciplines in order to improve student engagement and development of rigorous mental habits around learning, understanding, and applying knowledge to novel situations. Well planned and executed interdisciplinary teaching enables students to see the links between theories, concepts, and subject areas. Heidi Hayes Jacobs (1989) defines interdisciplinary learning as a knowledge view and curriculum approach consciously applying the methodology and language from more than one discipline to examine a central theme, issue, problem, topic, or experience. Further, Barton and Smith (2000) explain that interdisciplinary teaching and learning enables teachers to use classroom time more efficiently while addressing content in depth, allowing students to see the relationship between content areas.

The importance of interdisciplinary teaching and learning is widely recognized as a critical element in educational experiences that are authentic and reflect the multi-faceted nature of real life, rather than compartmentalized subject-matter packages. Through repeated exposure to interdisciplinary thought, learners develop more advanced epistemological beliefs, enhanced critical thinking ability and metacognitive skills, and an understanding of the relations among perspectives derived from different disciplines (Ivanitskaya, Clark, Montgomery, & Primeau, 2002). Interdisciplinary teaching promotes natural and logical connections that reach across content areas organized around questions, themes, problems, or projects rather than traditional subject-matter boundaries. When students are enabled to recognize the greater value of authentic understanding, they will be less likely to wonder *when am I ever going to need to know this* or leave schooling with the stunted perception that science is only important from 9:00 am-9:50 am on weekdays.

It can be argued that the United States' industrialized education model was designed to indoctrinate students with just enough information to create a functional society. As education adapts to prepare students to live, work, and flourish in a global society, the impetus exists to move away from learning as recall and memorization towards a learning process that is on-going, respectful of diversity, and focuses on information-based communication. The goals of learning in the twenty-first century include the development of self-efficacy, cultural norms and expectations, and a fundamental understanding of academic subjects that support students in asking meaningful questions while becoming self-sustaining and lifelong learners (Bransford & National Research Council, 2000). Changing language included in the reauthorization of the Elementary and Secondary Education Act of 1965 under the title of the "Every Student Succeeds Act" recognizes the value of and need for students to experience various content areas and interdisciplinary study in order to be truly and wholly educated.

Interdisciplinary teaching and learning help students develop knowledge, insights, problem solving skills, self-confidence, self-efficacy, and a passion for learning. Interdisciplinary education can also be classified as a culturally relevant pedagogy because it values past experiences, interconnectivity, and context. Further, interdisciplinary teaching is culturally responsive because there is a focus on introducing content from a variety of perspectives that may challenge existing notions, while encouraging students to integrate concepts and ideas into a broader conceptual framework of analysis (Bransford & National Research Council, 2000).

It is the authors' belief that the "A" in $STEAM_2$ should be a driving force in the creation of highly engaging, relevant, interdisciplinary teaching and learning in STEM. Further, because of the role of movement in learning (Ratey & Hagerman, 2008) and the barriers to quality movement that exist in schools, we have squared the M to represent both math and *movement.* Our perspective is based on combining research illustrating that arts-based content integration may be critical to meeting the increased demand for creative thinkers in a global economy and that "art and the process of creation/making promote collaboration and help students synthesize and put into practice what they have learned" (Bedetti, 2015, p. 1). Further, in isolation, STEM is missing a key set of creativity-related components that are equally important in developing the creative-thinking and problem-solving that will be essential for success in future job markets (White, 2010). In addition to increasing students' meaning making ability through the creative process, there is a significant portion of the population whose dominant cognitive processing system is *not* language based and therefore is in need of alternate instructional and assessment opportunities such as those that arts— visual, dance, theater, and music—learning experiences provide in order to reach their full cognitive potential (Groff, 2013). Lastly, arts-based interdisciplinary learning fosters a positive community climate in schools and increases students' and teachers confidence and efficacy.

Emphasizing creative movement as a way to learn and express understanding of STEM concepts has several inherent benefits on health, neurological conditions, and social skill development that are worth acknowledging despite being beyond the scope of this chapter. This chapter will focus on the power of creative movement "as an authentic semiotic meaning-making tool that supports sophisticated levels of embodied thinking, aesthetically oriented expressive communication and empathetic social contagion" (Deans, 2016, p. 55).

Addressing Standards

The integration of arts-based interdisciplinary teaching and learning with authentic and effective $STEAM_2$ practices requires effort. Haring and Kelner (2015) describe three practices supporting cross- content collaborative teaching: (a) develop a common language, (b) be creative about combining classes, and (c) co-plan creatively. Developing a common language across all grade levels and content areas facilitates teacher collaboration and supports students in transferring and applying learning between classes. Creativity about connections in concepts, theories, ideas, and work products need not rest solely with the teachers. Interactions with community experts engaged in the $STEAM_2$ experience will contribute to making creative connections if the environment is supportive of critical thinking and reflection. Simultaneously encompassing and moving beyond these three practices offered by Haring and Kelner, Spitzer (2013) asserts that successful interdisciplinary education includes a clear and collective vision of interdisciplinary themes and didactics, as well as the insight and ability to exploit institutional opportunities. Exploitation of institutional opportunities may include leveraging local expertise, community partners, space, materials, schedules, and staffing. Supporting these additional contributions made by Spitzer, the authors add a fourth practice to support high quality $STEAM_2$ experiences for students: all instruction and assessment is standards-based.

Moving forward with a standards-based perspective to interdisciplinary $STEAM_2$ experiences in schools requires a few definitions:

1. ***Encore* Content Areas:** *Encore* content areas are those content areas often marginalized through the implementation of policies such as *No Child Left Behind*. They are the classes in schools that for the last several years have been considered to be outside the bounds of "core content" areas. Part of the importance of encore content areas is the high engagement levels often associated with them. Classes typically included as encore content areas are the

arts, science, social studies, world languages, health, and physical education. Encore content areas can serve either as a primary driver of interdisciplinary learning or as an enhancement for more traditional core content.

2. **Text:** In supporting authentic literacy development for students, Merriam-Webster offers this valuable supporting definition of text: "something considered as an object to be examined, explicated, or deconstructed." In this definition, text is seen as any symbolic representation of meaning. Utilizing this definition in relation to $STEAM_2$ teaching and learning is beneficial for two reasons:
    a) it calls to attention the academic value of observational, interpretational, and aesthetic texts in shaping meaning-making, and
    b) illustrates the literacy needs for comprehension beyond phonetic decoding and interpreting of the written word. In this broader interpretation, text includes but is not limited to charts, maps, graphs, movies, music, choreography, sculpture, posture, gesture, tapestry, painting, poems, books, short stories, and informational writings.

The expanded definition of text as symbolic representation of meaning becomes especially critical in the light of the literacy strand within the Common Core State Standards (CCSS) intended to guide alpha-literacy in science, social studies, history, and the technical subjects. In addition to supporting alpha-literacy development and usage, this robust definition of text supports the fundamental understanding that, while alpha and numeric literacy are critical, the primary driver of instruction in encore content areas should be the content standards for that area of study. As previously stated, relying on standards is a key factor in implementing an academically rigorous $STEAM_2$ unit. Therefore, the literacy stands included in the CCSS provide a common academic language and foundational structure for interdisciplinary teaching inclusive of encore content areas.

Quality interdisciplinary teaching should focus instruction and evaluation around content area standards. The "A" in this $STEAM_2$ project is a creative-movement based exploration and demonstration of student understanding of concepts associated with simple machines. From an instructional perspective, it is critical that educators embrace the relationship between concepts and the ways they can be metaphorically understood and represented through movement. The texts, or learning products, from this unit may include visual representations, science lab/observation reports, choreography, written or oral critical reflections, or other aesthetic representations of content, process, or application. Components of each of these texts can be introduced, developed, or assessed based on specific literacy strands in the CCSS (see Appendix A).

Models and Metaphors

NGSS Practice 2 focuses on the development and use of models as "explicit representations that are in some ways analogous to the phenomena they represent" (National Research Council, 2012, p. 56). The modeling of concepts is an integral component of contemporary scientific practice. Educators must engage children in using models as a way of exploring phenomena throughout their P-12 school years. Children possess a native desire for knowledge, a desire that is in place long before they arrive in school. Even very young children are able to utilize the practice of modeling in their quest to understand the world (DeLoache, 2000). Gopnik has shown there is a measurable degree of scientific thinking present in infants and toddlers (Gopnik, 2012). This provides a strong foundational cognitive structure to build on as students enter school.

The metaphors that models elicit can be used effectively to build understanding. Metaphors are able to tap into the pre-linguistic structures that developed our foundational cognitive schema (Lakoff & Johnson, 1980). Children are wired from an early age to develop these schemata of understanding through their basic bodily experience.

As children move through the early stages of cognitive development described by Piaget, they are reliant on physical engagement with the world as means to develop their basic understandings. The majority of elementary students reside in the concrete operational stage of development. During this stage students need continual access to physical and concrete experience as they more fully develop their understandings. To quote Thomas Edison (brainyquote.com/quotes/thomas_a_edison_131295), "Great ideas originate in the muscles."

The authors' approach in this chapter is to engage students as active agents within the conceptual model itself. Models are representations of the phenomena under investigation, and each model provides only a glimpse. The use of multiple models providing an aggregate of glimpses along the path toward greater understanding. No single model is perfect. Models are "deliberate simplifications of more complex systems" (National Research Council, p. 110). Within this framework of simplified glimpses, it is critical to be careful and deliberate about the modeling we consume and employ. Educators must be transparent with students about the explanatory power and limitations of each model utilized. Students must actively critique the model and focus on how it adds new understandings. To this end, the authors' framework here engages students in deliberate and recursive critique of kinesthetic models, discussion of effective and ineffective methods, and reflection upon how models may enhance understanding. It is critical to develop a well-worn path between the originating phenomena under investigation, the concepts in development, and the kinesthetic modeling activity. At each stage, students should be engaged in action, dialogue, writing, drawing, and further investigation. Using multiple representations can result in the understanding being sought after.

Invoking an expanded definition of text, inclusive of all symbolic representations of meaning, supports intentional reflection upon each product and step in the process. This expanded interpretation allows each participant in the thematic collaboration to focus on the text most relevant to addressing the academic standards of that content area. Further, this expanded definition of text allows students, with support of peers and the facilitator, to compare and contrast multiple texts on the same topic to verify and strengthen understandings.

Science Standards

Figure 1 identifies two third-grade performance standards drawn from the Next Generation Science Standards (Lead States, 2013) within Motion and Stability: Forces and Interactions. Captured as well is the Disciplinary Core Idea PS2.A: Forces and Motion.

CASE DESCRIPTION

The school is a typical elementary school in an upper Midwest urban school district. The gymnasium is bustling with teachers, parents, and students putting the finishing touches on tri-fold boards, suspended mobiles, and live experiments in various stages of composition, eruption, explosion, and decomposition. In the center of this gym-turned-science-fair stands an elevated stage where groups of students will soon share their artistic interpretations of scientific inquiry dance performance to a packed house theater-in-the-round.

In the midst of all this activity, Jason, a sixth-grade student, was anxiously waiting for the other members of his group to arrive. Jason's group had investigated and interpreted the cyclical stages of hurricanes. It quickly became apparent that his group members were not going to make it to the performance on time and could not be reached by phone. As his teacher, I spoke to Jason ensuring him his grade would not suffer if his colleagues did not arrive. Significantly overweight, much taller than his peers, an ethnic minority in this urban school, with a severe speech impediment, looked at me with pleading eyes and through stuttered speech asked, "If I can figure out how to do it by myself, can I please go on stage and dance by myself?" Jason did just that.

His mother met him afterwards with a beaming smile, presenting him with flowers picked up at the gas station on her way home from work and a congratulatory

Figure 1.Motion and stability: forces and interactions

Students who demonstrate understanding can:

3-PS2-1.

Plan and conduct an investigation to provide evidence of the effects of balanced and unbalanced forces on the motion of an object. [Clarification Statement: Examples could include an unbalanced force on one side of a ball can make it start moving; and, balanced forces pushing on a box from both sides will not produce any motion at all.] [*Assessment Boundary: Assessment is limited to one variable at a time: number, size, or direction of forces. Assessment does not include quantitative force size, only qualitative and relative. Assessment is limited to gravity being addressed as a force that pulls objects down.*]

3-PS2-2.

Make observations and/or measurements of an object's motion to provide evidence that a pattern can be used to predict future motion. [Clarification Statement: Examples of motion with a predictable pattern could include a child swinging in a swing, a ball rolling back and forth in a bowl, and two children on a see-saw.] [*Assessment Boundary: Assessment does not include technical terms such as period and frequency.*]

The performance expectation above was developed using the following elements from the NRC document *A Framework for K-12 Science Education*:

Science and Engineering Practices	Disciplinary Core Ideas	Crosscutting Concepts
Planning and Carrying Out Investigations Planning and carrying out investigations to answer questions or test solutions to problems in 3–5 builds on K–2 experiences and progresses to include investigations that control variables and provide evidence to support explanations or design solutions. • Plan and conduct an investigation collaboratively to produce data to serve as the basis for evidence, using fair tests in which variables are controlled and the number of trials considered. *Connections to Nature of Science* **Scientific Investigations Use a Variety of Methods** • Science investigations use a variety of methods, tools, and techniques. **Science Knowledge is Based on Empirical Evidence** • Science findings are based on recognizing patterns.	**PS2.A: Forces and Motion** • Each force acts on one particular object and has both strength and a direction. An object at rest typically has multiple forces acting on it, but they add to give zero net force on the object. Forces that do not sum to zero can cause changes in the object's speed or direction of motion. (Boundary: Qualitative and conceptual, but not quantitative addition of forces are used at this level.) **PS2.B: Types of Interactions** • Objects in contact exert forces on each other. **PS2.A: Forces and Motion** • The patterns of an object's motion in various situations can be observed and measured; when that past motion exhibits a regular pattern, future motion can be predicted from it. (Boundary: Technical terms, such as magnitude, velocity, momentum, and vector quantity, are not introduced at this level, but the concept that some quantities need both size and direction to be described is developed.)	**Cause and Effect** • Cause and effect relationships are routinely identified. **Patterns** • Patterns of change can be used to make predictions.

embrace. Jason's feeling of accomplishment and his mother's pride in him were readily visible for the world to see.

In strictly artistic terms, Jason's performance did not reach the high-end criteria described on the rubric. Yet the ownership of learning embodied by this young man, not a frequent visitor to school's typical markers of academic success, rose above all other indicators of success. Providing multiple and various means to explore scientific concepts allowed Jason agency over his learning, creating a deep, memorable learning experience. An effective integration of the arts, including the creative, kinesthetic qualities of dance, provides students like Jason an on-ramp to content.

The preceding example is one of many, many stories accrued over ten years of using creative movement as a tool to introduce, develop, and assess interdisciplinary content learning. Jason's story exemplifies the possibility of both content mastery and efficacy development. Following is an outline of another $STEAM_2$ unit carried out in a city-wide option urban elementary school in a Midwestern city of 300,000 people. The demographics of the district are >75% non-White, approximately 30% non-native English Speakers, and over 70% living in poverty as measured by those eligible for free or reduced meals. The unit brings together the best practices of $STEAM_2$'s interdisciplinary nature with a focus on science and the arts, particularly dance and physical education. The unit utilizes planned movement as a living model of scientific concepts related to the force and motion embodied in simple machines. Students build off concrete experiences with materials, engage in movement exploration activities, and culminate in a dance performance. Although students' mastery of physics, application of simple machines, and visual-oriented final products are the aim of this unit, the unit does not have to be confined to one physical space. This unit could be carried out in a classroom, laboratory, gymnasium, studio, or any learning space.

Educators implementing this unit should be comfortable with the Working Words into Movement Tool utilized in creative movement. The Working Words into Movement (WWM) Tool was developed to support the inclusion of movement as a pedagogical tool in introducing, developing, and assessing interdisciplinary content (Figure 2). WWM is based on the work of movement theorist Rudolf Laban (1879-1958) who developed the Laban Movement Analysis (LMA) taxonomy to generate and describe movement. LMA contains four major categories of movement: effort, body, shape, and space (Konie, 2011). For interdisciplinary use with young learners, the LMA has been translated into the WWM Tool (Figure 2). One purpose of the

Figure 2. Working words into movement (WWM) tool

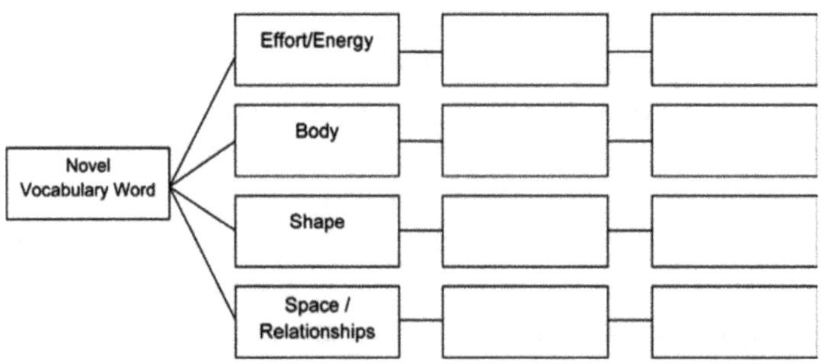

tool is to help students use their past experiences and motor memory to explore, interpret, engage with, and understand novel vocabulary. The second, and no less important purpose for the WWM Tool, is to serve as a catalyst for choreographing creative movement performances.

Prior to using the WWM Tool students must have ample time and experience exploring the four major categories of movement. It's important to ensure they have an understanding of the relationship to movement of these descriptive classifications.

Students begin their first phase with the Tool by brainstorming all possible ways the novel vocabulary word might "look" in each of these categories. For example, when brainstorming on the word "forces," in the Effort/Energy category, students may think of words such as push, pull, hard, light, constant, always, necessary, forward, or down. Encouraging students to include all of their thoughts, even if they think they are only guessing or are using parts of the word in an attempt to decode the whole, can be a very positive method. Educators should use student responses to the WWM Tool as a formative assessment to guide instruction based on abundance or absence of prior knowledge or misinterpretations. For example, if a student included "Yoda" as a word related to "force," they may be acknowledged for their humor and wide-range brainstorming but re-directed towards the scientific application sought to be understood through the WWM Tool.

The second phase of WWM Tool use is individual or peer evaluation of brainstorm lists. Evaluation takes place for two purposes. One purpose of this step is to eliminate any words or phrases that students think would not be easily translated to movement. The other purpose of this step is to discuss the vocabulary words being examined

Figure 3. Brief examples of some of the types or qualities of movements that may be included under each heading

Effort/Energy	Body	Shape	Space/Relationship
Weight: firm, light, strong **Flow:** Bound, free, jerky **Emotion/Intent:** Direct, indirect, forceful, angry, joyful **Speed:** Fast, slow, accelerating, sudden, sustained	**Actions:** curl, stretch, twist, support, lead, follow **Locomotion**: run, walk, gallop, skip, hop, jump, leap **Non-locomotor:** twisting, rising, spinning, pushing **Body Parts:** apart, asymmetrical, meeting, above, below	Straight, bent, curved, open, closed, tight, frozen, shifting	**Areas:** General/personal Directions: North, south, east **Facing:** Sideways, backward, forward Levels: Medium, high, low, changing **Pathways**: In air, on floor, direct, indirect, zig-zag, curved, straight **Planes:** Sagittal, frontal, horizontal **Other:** music, walls, props, literature, art, other dancers

further. Any words or phrases that are deemed irrelevant, inaccurate, or that do not readily and inherently relate to movement are simply crossed out. All remaining words are transferred to the next line on the flow chart. This is repeated for each of the four categories. In our example using the word "force," words such as "always" and "necessary" may be eliminated. While there are few "right" or "wrong" answers, it is the individual reflection or group discussion about which words to eliminate that holds power in helping all learners understand the novel vocabulary word (Figure 4).

Step three of the process can be student or teacher directed, based on instructional purposes and intended learning products. In this step the list of remaining words in each of the four categories: effort/energy, body, shape, and space are selected for application to a final movement problem. Step four, represented by the final area in the WWM Tool includes only the words from step three that the individual or group would like to use in their movement sequence. During step three (while preparing for step four), students should consider the parameters established either by the teacher, as a class, or by students for solving the *movement problem*. In this case a movement problem is the pre-determined criteria for a final movement performance. The final movement performance can range from a quick, unpolished informal performance in class to a very formal, highly rehearsed performance with an external audience. In the example of the word "force," students would select the words from each category(energy/effort, body, shape, space/relationship) that they believe best support their solution to the movement problem. Students must be able to justify, either orally or in writing, their choices.

It is anticipated this unit may be carried out with the help of the dance or physical education specialist, science specialist, classroom generalist, or a collaboration of

Figure 4. Working words into movement (WWM) tool: step one

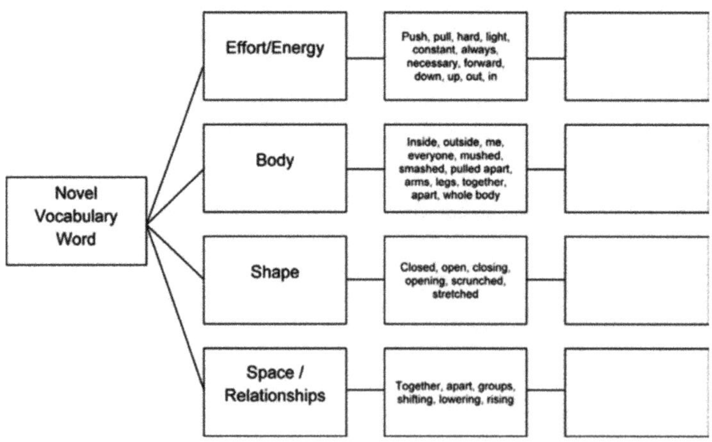

engaged educators. In this case study, our context was a collaborative effort between the creative movement specialist and a classroom generalist. The classroom generalist carried out an active 'doing-based' science program in her classroom, while students expanded their understanding through engagement in movement as text, metaphor, and modeling in physical education. Although implementing movement may be easier in open spaces that are bounded enough for teacher and peer engagement, having a gymnasium or dance studio is not essential. During implementations of our collaboration, we explored using classrooms, common areas, and courtyards. The spaces themselves had little impact on the students' comprehension or final products. In addition to finding effective work spaces throughout the school, we found this kind of collaboration between the generalist and specialist offered the ideal opportunity for students to fully engage in a robust exploration of content. The collaborative partnership allowed for alternative means of learning and expressing knowledge—for deep processing of the models and metaphors that emerged within the student performance.

An Explanation of Simple Machines Through Movement

Prerequisite knowledge, skills and experiences to ensure success in this unit:

- Clear and consistent rules, expectations, and behavior management in place in all collaborating classrooms
- Mastery of benchmarks and outcomes identified for English/Language Arts (Common Core), science, dance, and physical education through grade two
- Experience with Working Words into Movement (WWM) framework.

As in the use of any models, the critical reflection following the experience is able to make the transition from fun to rich meaning in science, arts, and self. Following are some kinds of questions that might be useful:

- What did you notice as you prepared and carried out your performance?
- What did you notice about your movements?
- How did your movement connect to force and motion?
- How was your movement like simple machines?
- How well did your performance 'model' force and motion?
- How did working with movement help you understand the principles of force and simple machines?
- What were some limitations of your model?
- How did your experience in modeling through movement help you better understand force and motion?

Table 1. An exploration of simple machines unit plan

Day One	Day Two	Day Three
INTRODUCTION/WARM-UP: moderate-to-vigorous physical activity (MVPA) in the form of a low organized tagging game suitable for the space **FOCUS:** Inquiry stations: Students rotate between 8-15 stations with a variety of common & novel simple machines for open exploration Science notebook: Describe how objects move or can be used Compare/Contrast notes to co-create classification of all materials into like- categories **CONCLUSION/COOL-DOWN:** Exit slip: At least one question generated during exploration or collaboration	**INTRODUCTION/WARM-UP:** How many categories did you co- create yesterday? **FOCUS:** Domain specific, rigorous vocabulary sorting race Naming: each group reviews vocabulary related to each category and determines a name and function for that particular simple machine **CONCLUSION/COOL-DOWN** 5-minute yoga w/intention on "where in your body do you find some similarities to simple machines"	**INTRODUCTION/WARM-UP:** 10-min walk/jog **FOCUS:** Direct instruction Review naming to ensure proper vocabulary Co-create definitions for each simple machine and generate examples list Movement Exploration Individual movement problem: How can you represent the actions of a simple machine through locomotor and non-locomotor actions? **CONCLUSION/COOL-DOWN** 1/3 of students at a time share their solutions, audience identifies which machines are represented, repeat
WORK PRODUCT/ ASSESSMENT:	**WORK PRODUCT/ ASSESSMENT:**	**WORK PRODUCT/ ASSESSMENT:** Visual recording of individuals presenting with group
Day Four	**Day Five**	**Day Six**
INTRODUCTION/WARM-UP: Repeat mvmt solution from Day Three, "trying on" mvmt observed in others Teacher loudly reinforce vocabulary through observation **FOCUS:** Forces Direct or inquiry-based instruction into forces, as related to simple machines	**INTRODUCTION/WARM-UP:** MVPA-low organized tagging game **FOCUS:** Working Words into Movement Groups of three (not partners from yesterday) Assigned vocabulary words to work into mvmt Create and rehearse one 30 second mvmt sequence to	**INTRODUCTION/WARM-UP:** Open space jog while playing video of Rube Goldberg machine **FOCUS:** Final Mvmt Problem introduced Co-determine a problem to be solved in school or community that a Rube Goldberg-like machine could tackle.
Mvmt exploration: Mvmt problem: (In pairs), ABA of simple machine o action (A), o acted upon by simple machine (B), o simple machine action (A) **CONCLUSION/COOL-DOWN** Presentation of mvmt problem solution Each pair evaluates another	reflect one of the words **CONCLUSION/COOL-DOWN** Presentation of mvmt problem solution Each group evaluates another	Review of group work norms/ expectations Each group is to create a repeatable 16 count mvmt sequence (approximately 10- 15 seconds) representing their assigned simple machine as a part of solving the community/school problem **CONCLUSION/COOL-DOWN** Fist-to-Five reflecting personal responsibility in the group
WORK PRODUCT/ ASSESSMENT: Peer evaluation exit slip	**WORK PRODUCT/ ASSESSMENT:** Peer evaluation exit slip	**WORK PRODUCT/ ASSESSMENT:** None
Day Seven	**Day Eight**	**Day Nine**
INTRODUCTION/WARM-UP: High cardio/low organized tagging activity with assigned group member teams **FOCUS:** Continued group creation/rehearsal of 16 count sequence As groups complete rehearsal, send "spies" to give feedback to each other's groups Groups each determine how they need to be acted on to initiate and to pass along action **CONCLUSION/COOL-DOWN** Put it all together, one 16 count sequence after another	**INTRODUCTION/WARM-UP:** 5-10 min walk/jog while watching prior day's video What can make your sequence more accurate & more aesthetically interesting **FOCUS:** Refining individual 16 count sequences Put it all together Structure: Initial resting shape (held by each member of each group until it is time for that group to begin their 16 count sequence for the first time and between executions) **First verse: Each group** 16 count, one group at a time, following the previous group, holding resting shapes when complete) Second verse: The second verse begins when the final group concludes the first execution of their 16 count sequence and the dance starts over with the first group. This verse is an ABAB add-on structure: A=16 count movement sequence, B=16 count rest,	**INTRODUCTION/WARM-UP:** Socratic Jog Topic: What makes for a quality dance performance? **FOCUS:** Rehearse and refine performance (Performance should have an audience and be recorded if, that is 'in class' a tenth day should be added for drafting of the blueprint prior to performance so that they may also be displayed. If limited time, collaboration with visual arts or classroom teachers valuable here. **CONCLUSION/COOL-DOWN:** **Oral reflection on learning, process, and product**
	repeat. The first group begins verse two with their 16 counts, group two then begins as group one rests. Then groups one and three execute their A while all others hold resting shapes, then groups two and four move, and so on until every group is participating in the 16 count move, 16 count rest. Third verse: Continuous repetition of 16 count sequence, culminating in an 'overheating' or 'explosion' action leading to the... Final resting shape **CONCLUSION/COOL- DOWN:** Yoga sequence with intention of appreciating group contributions	
WORK PRODUCT/ ASSESSMENT: Video recording of initial attempt	**WORK PRODUCT/ ASSESSMENT:** None	**WORK PRODUCT/ ASSESSMENT:** Video recording of performance Rube Goldberg illustration/ blueprint of Simple Machine Dance solution to school or community problem

SCIENCE CONTENT

This section is meant to provide the physics content for the educator to understand

and explain forces, motion, and simple machines to students. Examples are provided to link these scientific terms and ideas to the WWM tool and the teaching unit.

Forces and Motion

All the motion we observe can be explained with Newton's Three Laws. Newton's first two laws of motion may be summarized as follows: Motion that is constant requires no explanation, and all changes in motion are explainable by forces (that is to say, a "push" or a "pull" as discussed in the WWM tool above). Here "constant" motion means either no motion or motion in a straight line without changes in speed or direction. Importantly, Newton's First Law identifies the concept of inertia, or the property of all mass to resist changes in motion. Newton's Second Law describes how changes in motion do occur, namely, that there must be an external force applied to an object. This can cause it to change its speed or direction. Finally, Newton's Third Law, while often more mysterious, is another way to say that you can't lift yourself up by your bootstraps, or, that there are no "free" forces.

Unfortunately, the persistent presence of friction can lead to some misconceptions. While it may seem that a *constant* force must be exerted on an object for it to stay moving at a *constant* speed, this is not true since friction is *also* a force that must be taken into account. Consider, for example, pushing a desk across the floor. The force of the push and the force of friction are in opposite directions. If the forces are balanced, the desk moves at a constant speed. If the force of the push is greater than the force of friction, the desk will speed up; if it is less, the desk will slow down accordingly. This speeding up or slowing down is called "acceleration." Newton's Second Law provides a quantitative summary of that change via an equation:

Acceleration = Force ÷ Mass

What this means is that if you apply more force, the object's motion changes more abruptly and the motion is no longer constant. Students could also engage with the concept of "constant motion" by considering whether or not forces are at work when a book is simply sitting on a desk, e.g. gravity pulling down and the force from the desk holding the book up. Since the forces balance, the zero motion remains constant at zero motion.

Finally, Newton's Third Law provides a summary about how objects exert forces on each other. In effect, there are no "free" forces. By pushing on something, I exert a force on it and cause it to accelerate. However, that object also pushes back equally hard on me. I have to push against the ground to gain more force and get the object moving, but the ground pushes back equally hard against me. What does the ground push against? Well, it is the Earth itself, but because the Earth has nothing to push against in empty space, it changes its motion ever so slightly in an equal and opposite manner. This change is imperceptible because the Earth is so large. Taken together, Newton's three laws tell us that: (1) mass has a property called inertia that resists changes in motion; (2) to get a massive object to move, a force must be applied on it; (3) there are no "free" forces, and all forces of interaction are equal and opposite.

Forces and Rotation

Forces can also cause objects to rotate. Consider the situation shown below in Figure 5. The two boxes are being pushed from opposite sides with two equal forces. Since the forces balance, the two boxes will not accelerate to the left or to the right. However, when these balanced forces are *offset* from each other, they can cause the object to *rotate*. Offset forces produce what is called "torque," a force that causes rotation. If students are asked to brainstorm words for how they get things to rotate, perhaps terms like "twist" or "spin" may arise. It is also likely that the same terms "push" or "pull" would get discussed, providing a direct link between forces and torques. Indeed, rotation is also a form of motion, so forces *must* be involved. Similarly to the way unbalanced forces cause an object to accelerate to the left or right, an unbalanced set of *torques* can cause "rotational acceleration" in either a clockwise or counterclockwise direction. The amount that the force or forces are offset from the point of rotation tells us how much torque, and therefore how much rotational acceleration, those forces will provide. Again, in Figure 5, the different placements of the two pushing forces will either cause no rotation, some rotation, or quite a bit of rotation. Mathematically, the amount of torque provided by a force is proportional to both the force and to the "lever arm" or distance from the point of rotation. This is described as:

Torque = Force × Lever Arm

Figure 5. Forces and Rotation. Even when two forces are balanced, so that the boxes don't accelerate to the left or to the right, an offset force can cause an object to start to rotate.

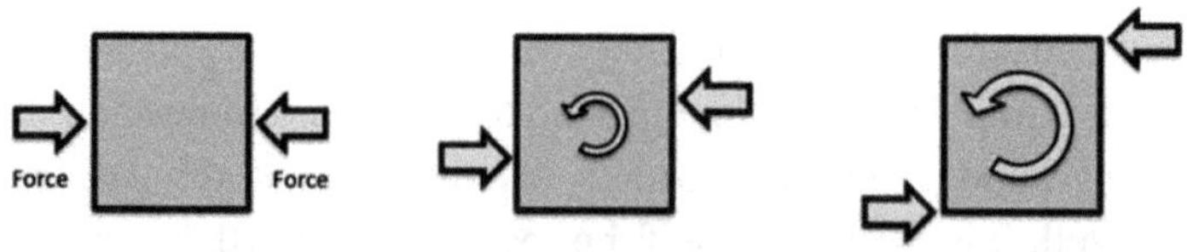

As with forces pushing objects along a path, we can now calculate the torque produced by each separate force, add them up, and therefore predict whether or not there will be any rotation. If the total net torque is non-zero, the object will rotate. In terms of creative movement, the instructor might connect why your arms rotate downward from your shoulders when held straight out: they are rotating due to the torque produced by the force of gravity. The muscles in our shoulders are providing the balancing torque to keep the arm held straight out. If there is another force, say, someone placing a weight in the outstretched hand, there is now another torque wanting to rotate the arm downwards and the shoulder muscles need to work even harder. If the weight is placed instead at the elbow, the "Lever Arm" is less, and the torque provided by the shoulder doesn't need to be as strong.

Forces and Work

When we exert a force on an object and the object starts to accelerate, it will move a certain distance. The relationship between the *distance* the object moves and the *force* applied to that object is described with the concept of *work* (or energy spent). I have to do work on an object (expend energy or exert effort) to move it a certain distance by applying a force. You can understand this concept by considering the effort of pushing a desk across a rug. To describe this, students would likely think of words like work, effort, hard, try, power, strong, or difficult. But when considering how to move the desk from here to there, students would also naturally understand that you need to either "push" or "pull" the desk, e.g. apply a force. The instructor might then ask if it takes more effort to push the desk a farther distance, and, clearly, it does. Mathematically, the concept of work is expressed as follows:

$$\text{Work} = \text{Force} \times \text{Distance}$$

To move a massive object, we require more force, and therefore exert more effort. To push an object a farther distance, we also need to exert more effort or energy.

Work and Simple Machines

With this physics background, it is now clear how simple machines do what they do. A simple machine is a mechanical arrangement that allows the operator to do work on an object — e.g., move it a certain distance by applying a certain force. However, keeping in mind how work (or energy or effort) is related to the force and the distance, the mechanical arrangement can be made so that —*for the same amount of total work* — the force applied is much less. This can only happen if the distance over which the force operates is much longer. For example, students could consider two scenarios: pushing the desk across the rug or putting it on wheels. While they will understand that the wheels (a simple machine) make movement a lot easier, the instructor could ask them if eventually, after pushing for a long time over a much longer distance, they might still get just as tired as if they were pushing it across the rug. For the same amount of work or energy spent, we can either choose to apply much more force (hard) over a small distance or much less force (easy) over a larger distance. Of course, in using a simple machine, the amount of effort we put in (small force/ large distance) must equal the amount of work done by the machine (large force/ small distance). This is a statement of "Conservation of Energy." Ignoring any negative effects of friction, the amount of input work will equal the amount of output work, or "Work In = Work Out."

The amplification of the output force relative to the input force provided by a simple machine is called the "Mechanical Advantage" or MA. Depending on the particular mechanical arrangement, simple machines can be characterized by one or more of the following:

1. Amplify the magnitude of an input force.
2. Amplify the distance (or the speed) of an input force.
3. Change the direction of an input force.
4. Transfer an input force from one location to another.

Classically, there are six simple machines: (1) the lever; (2) the pulley; (3) the inclined plane; (4) the wheel and axle; (5) the screw; and (6) the wedge. Many of these are related (e.g. the screw is a twisted ramp and a pulley is similar to a wheel and axle but with ropes). Therefore, the following sections will provide examples and the physics background for only three of these.

1. The Lever

Consider the situation illustrated below in Figure 6: lifting a heavy box by using a lever. The point of rotation is called the fulcrum. In this case, a small force down on the lever over a larger distance is able to lift with a large force over a smaller distance. The product of each of the "Force × Distance" on both sides of the lever is the same, but the force exerted by the lifter is much less than it would be if the box was lifted directly. Unfortunately, the box isn't lifted nearly as high. That is the trade-off with all simple machines using this principle; we can lift things as heavy as a car with our own hands and a long lever, but only a few inches. Notice the characteristics of this simple machine: (a) the magnitude of the force that the lifter exerts is magnified; (b) the speed at which the box lifted was slower than the speed that the lifter was pushing down (over the same amount of time, the box went

Figure 6. The Lever. Using the concept of the forces acting over a distance allows us to design a lever that requires less force, but only lists the box a small distance. The same amount of work is done in two cases: using the lever (small force x large distance = large force x small distance) or lifting the box up that same small distance directly (large force x small distance) without the lever.

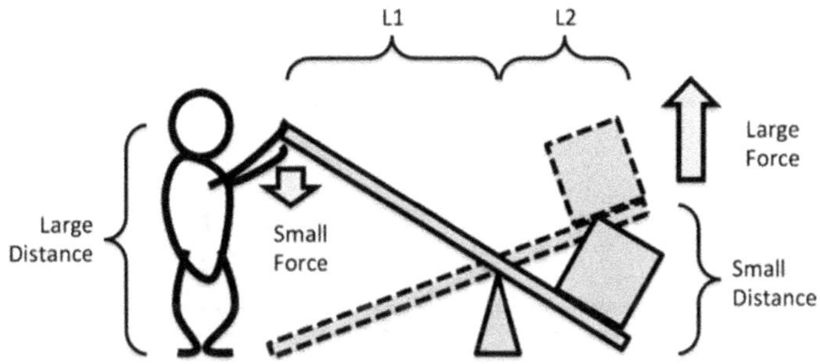

a shorter distance up than the hands pushing down, thereby moving more slowly); (c) the direction of the force was flipped from pushing down on the lever to pushing up on the box; (d) the force was transferred from one end of the lever to the other.

Considering the geometry of the lever, we can see that the MA of the lever is: *L1 / L2,* or the ratio of the lengths from the fulcrum. This is labeled in the figure. We can also understand the mechanism of the lever by using and balancing *torques.* Recall that to determine whether something will rotate (i.e. whether or not the lever will lift by rotating around the fulcrum) each force and how much torque that force is producing must be considered. This is diagrammed in Figure 7. By using the equation stating that torque is

proportional to the force and the lever arm, the clockwise torques and the counterclockwise torques can be tallied to determine the net torque. If the net torque is balanced, the lever will not rotate. If somehow there can be a harder push on one end, the balance of torques becomes upset, and the lever will rotate, lifting the object up. In terms of body movement, students can consider the simple machine of their arm as a lever. With an elbow on a table, a student can try to lift up a heavy book. The instructor would ask the students where the fulcrum is (the elbow) and what is providing the input force (the tendon and bicep muscle). But by considering how much the tendon actually moves (short distance) compared with the book (long distance), we see that the elbow simple machine actually has a MA less than one. The force that the tendon is applying to the bone is actually quite a bit greater than the weight of the book.

Figure 7. The balanced lever. By considering each force and the torque produced by those forces, it is possible to determine whether or not an object, in the case a lever, will rotate or remain balanced. Recall that torque is force times lever arm. Therefore, a large force with a small lever arm can be balanced (or overcome, if you are trying to lift something heavy!) by a small force and a large lever arm.

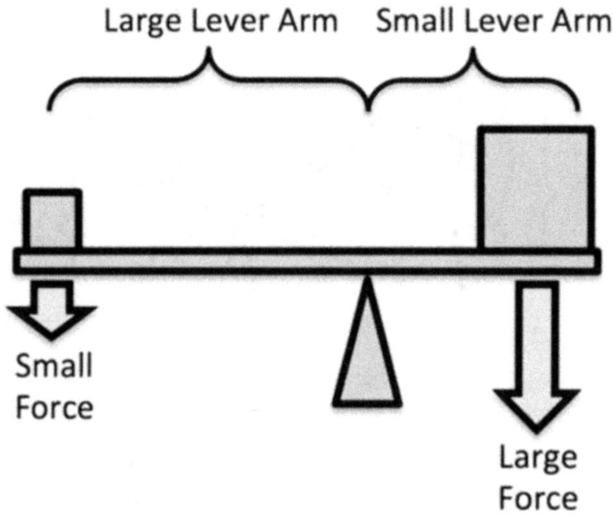

2. The Pulley

A second type of simple machine is the pulley. Here, the concern again is with reducing the amount of force required to perform some work (i.e. move) some heavy object. Pulleys also change the direction of a force, depending on how many pulleys are used. A single pulley offers no mechanical advantage, as shown below in Figure 8, because the force is only changing direction (not acting) over a different distance. If, however, two pulleys are used, the rope must be pulled over a greater distance than beforeto lift the object, the same amount as shown on the right. This means that the force of the pull is less than before since the same amount of work (force lifting the weight up the same distance) is performed. For example, a double pulley offers a mechanical advantage of "2" because twice as much rope has to be pulled out to move the far end of the rope (the object being moved) by the same distance as before.

One rope can also be tied around a different sized pulley wheel, meaning that more rope has to be pulled out over a longer distance to move the heavy object over a shorter distance, thereby again gaining mechanical advantage. This is similar to

Figure 8. The pulley. The single pulley system on the left only changes the direction and location of an input force. There is no mechanical advantage because the force acts over the same distance. A double pulley however does offer mechanical advantage because to lift the heavy object a small distance, you must pull the rope (e.g. exert a force) over a longer distance. The mechanical advantage is equal to how many portions of the rope are pulling vertically on the hanging weight. The more pulleys, the higher the mechanical advantage, but the farther the rope needs to be pulled.

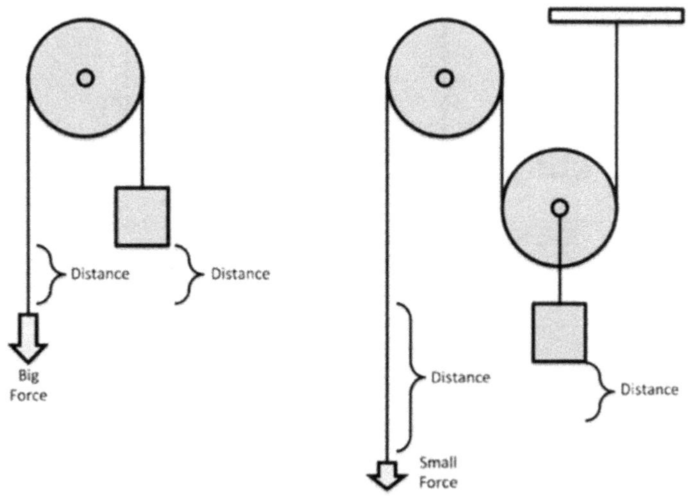

how a "wheel and axle" gives mechanical advantage. Since the inner pulley and the outer pulley each rotate the same amount, say a full 360-degree turn, the rope attached to the outer pulley will unwrap a much longer amount compared to the rope wound around the inner pulley. This is shown in Figure 9. This is also understood in the sense that the rope being pulled on now has a larger *lever arm*, so for the same amount of force, more torque is produced, and the pulley will now begin to rotate. Another example is trying to open a door with a doorknob that wasonly a thin rod, not a bulging handle. The mechanical advantage is the size of the door handle to the size of the actual axle that releases the lock.

3. The Inclined Plane

To lift an object to a certain height, the lifter must do work against the force of gravity. Therefore, the amount of work that is performed is equal to the "Force of Gravity (the weight)" *times* the "Distance Lifted." If done directly, i.e. straight up, a force equivalent to the weight of the object for the full height must be exerted.

Figure 9. The puller as wheel and axle. Wrapping a rope around a pulley/wheel with a different diameter offers mechanical advantage because the force both (1) has a larger lever arm and can produce more torque and (2) the force must act over a much longer distance to turn the pulley/wheel one full revolution. The mechanical advantage here is the ratio of the two diameters of the two wheels.

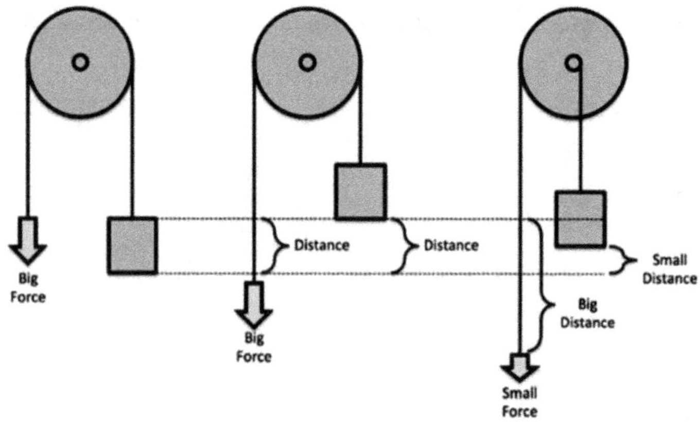

This is difficult. If, however, the lifter can spread the force of the push over a longer distance while still lifting the weight to the same height, the same amount of total work has been performed, but with much less force. This is the advantage of the inclined plane, shown in Figure 10. Since the weight is lifted upwards, "Work Out" for the ramp is set to be "Work Out = Weight × Height." The amount of work put in is "Work In = Force of Push × Ramp Length." Since "Work in = Work Out" we find that the mechanical advantage of the ramp is "Ramp Length ÷ Height." Therefore, the longer the ramp, the greater the mechanical advantage less pushing force is required. However, it will have to be pushed for a longer distance. Connecting creative movement or dance in the classroom could explore jumping up in the air. Why can't we jump to the second floor of a building but can easily climb stairs or a ramp to get there? The amount of force required to jump straight up is more than our legs can give us, so the tradeoff in taking the stairs is that we'll need to walk a much longer distance to get there.

CONCLUSION

Combining the scientific concepts of force and motion with human movement is an organic collaboration that simply makes sense for students. Deep and lasting understandings on the theory, concepts, and skills develop in the students. Further,

Figure 10. The inclined plane. You can perform the same amount of work by either: (1) lifting a weight directly against the force of gravity; or (2) pushing with less force over a longer distance, but still getting the weight to the same height. The mechanical advantage of a simple ramp is the ramp distance divided by the vertical distance. The longer the ram, the greater the mechanical advantage, but you have to push over a much longer distance to lift the object up to the same final height.

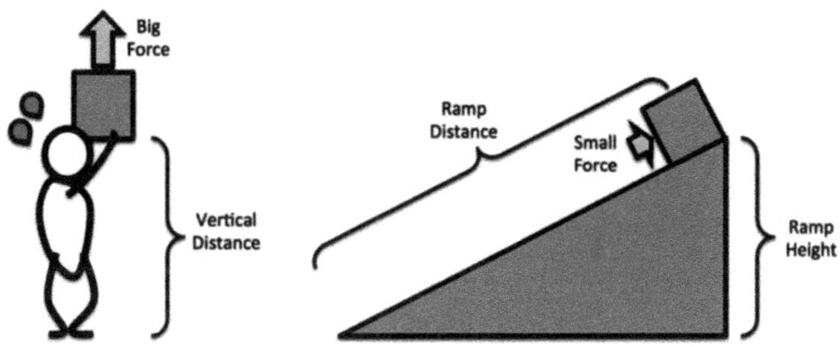

students routinely share that, not only do they understand simple machines, they also get along better as a community and feel pride in themselves for being able to express their understandings through a movement performance. For the teachers involved, each time we have co-facilitated the unit, we learn something new about our strengths and needs as teachers. We have learned to take ownership and release control. Over the years, we have invited other content area teachers to join in the project, adding their content area expertise and learning from our process. While the unit structure always starts the same, each experience is different and perhaps better, than the one before. From each experience we have learned how to make the collaboration and the learning experience better, stronger, more effective, and more fun because life—and certainly life teaching in urban schools—is way too much work to not also be fun!

REFERENCES

Barton, K., & Smith, L. (2000). Themes or motifs? Aiming for coherence through interdisciplinary outlines. *The Reading Teacher*, *54*(1), 54–63.

Bedetti, G. (2015). Collaborative college playwriting and performance: A core course "Trespassing" onto the dramatic arts. *International Journal of Education & the Arts*, *16*(19).

Bransford, J. (2000). *How people learn: Brain, mind, experience, and school.* Washington, DC: National Academies Press.

Coffey, H. (n.d.). *Interdisciplinary teaching*. Retrieved October 3, 2016 from http://www.learnnc.org/lp/pages/5196

Deans, J. (2016). Thinking, feeling and relating: Young children learning through dance. *Australian Journal of Early Childhood*, *41*(3), 46–57. doi:10.1177/183693911604100307

DeLoache, J. S. (2000). Dual representation and young children's use of scale models. *Child Development*, *71*(2), 329–338. doi:10.1111/1467-8624.00148 PMID:10834468

Dow, C. B. (2010). Young children and movement: The power of creative dance. *Young Children*, *65*(2), 30–35.

Gopnik, A. (2012). Scientific thinking in young children: Theoretical advances, empirical research, and policy implications. *Science*, *337*(28), 1623–1627. doi:10.1126cience.1223416 PMID:23019643

Groff, J. S. (2013). Expanding our frames of mind for education and the arts. *Harvard Educational Review*, *83*(1), 15–39. doi:10.17763/haer.83.1.kk34802147665819

Haring, D., & Kelner, T. (2015). Why we got serious about interdisciplinary teaching. *Educational Leadership*, *73*(4), 68.

Ivanitskaya, L., Clark, D., & Montgomery, G. (2002). Article. *Innovative Higher Education*, *27*(95). doi:10.1023/A:1021105309984

Jacobs, H. H. (1989). The growing need for interdisciplinary curriculum content. In H. H. Jacobs (Ed.), *Interdisciplinary Curriculum: Design and Implementation* (pp. 1–11). Alexandria, VA: ASCD.

Konie, R. (2011). *A brief overview of laban movement analysis*. Retrieved April 27, 2016 from http://www.movementhasmeaning.com/wp-content/uploads/2010/09/LMA-Workshop-Sheet.pdf

Lakoff, G., & Johnson, M. (2003). *Metaphors we live by*. Chicago, IL: University of Chicago Press. doi:10.7208/chicago/9780226470993.001.0001

Lead States. (2013). *Next generation science standards: For states, by states*. Washington, DC: National Academies Press.

Mandigo, J., Francis, N., Lodewyk, K., & Lopez, R. (2009). *Position paper: Physical literacy for educators*. St. Catherines, ON, Canada: Brock University. Retrieved May 10, 2016 from https://www.phecanada.ca/sites/default/files/pl_position_paper.pdf

Martin, D. J. (2011). *Elementary science methods: A constructivist approach* (6th ed.). Belmont, CA: Wadsworth Publishing.

Merriam-Webster. (n.d.). *Definition of text*. Retrieved Oct 4, 2016 from http://www.merriam-webster.com/dictionary/text

National Research Council. (2012). *A framework for K-12 science education: Practices, crosscutting concepts, and core ideas. Committee on a Conceptual Framework for New K-12 Science Education Standards. Board on Science Education, Division of Behavioral and Social Sciences and Education*. Washington, DC: National Academies Press.

Ratey, J. J., & Hagerman, E. (Collaborator). (2008). Spark: The revolutionary new science of exercise and the brain. New York, NY: Little, Brown and Co.

Spitzer, H. (2013). Introduction of interdisciplinary teaching: Two case studies: Commentary on Teaching Science, Technology, and Society to Engineering Students: A Sixteen Year Journey. *Science and Engineering Ethics*, *19*(4), 1451–1454. doi:10.100711948-013-9475-y PMID:24085355

The Common Core State Standards Initiative. (2018). Retrieved September 1, 2016 from http://www.corestandards.org/ELA-Literacy/introduction/key-design-consideration/

United Nations. (2002). *56/116. United Nations Literacy Decade: education for all. Resolution adopted by the General Assembly, Fifty-Sixth Session*. Retrieved May 10, 2009 from http://www.undemocracy.com/A-RES-56-116.pdf

United Nations Educational, Scientific and Cultural Organization. (2003). *Literacy, a UNESCO perspective*. Retrieved May 10, 2016 from http://unesdoc.unesco. org/images/0013/001318/131817eo.pdf

White, H. P. (2010). *STEAM not STEM whitepaper: An agreement on what drives the US economy in the future*. Retrieved from http://steam-notstem.com

KEY TERMS AND DEFINITIONS

Creative Movement: Objective driven activities through which learners explore movement for the purpose of developing skills, building efficacy, or stimulating imagination with the intent of communicating knowledge of facts, concepts, principles, ideas, and/or feelings.

Encore Content Areas: Encore content areas are those content areas taught in schools that are often marginalized through the implementation of policies such as No Child Left Behind that focus nearly exclusively on content and direct instruction in reading and math courses. They are the classes in schools that for the last several years have been considered to be outside or adjacent to the bounds of "core content" areas. Examples of encore content areas are social studies, the Arts, health, science, physical education, career/technical education and world languages.

Force and Motion: A push or a pull that can accelerate, rotate, or otherwise alter the motion of an object. The relationship between applied forces and the resulting motion is the subject of Newton's Laws of Motion.

Physical Literacy: The ability to move with competence and confidence in a variety of physical activities in multiple environments that benefit the healthy development of the whole person.

Scientific Literacy: The knowledge and skills necessary to participate in scientific decisions impacting personal, local, and global affairs.

Simple Machines: A mechanical arrangement used to apply a force. Typically used to amplify a small input force into a larger output force. Examples include a lever, pulley, or an inclined plane.

STEAM$_2$: Science, technology, engineering, the arts, mathematics, and movement.

Text: Symbolic representation of meaning. Text includes but is not limited to charts, maps, graphs, movies, music, choreography, sculpture, posture, gesture, tapestry, painting, poems, books, short stories, graphic organizers and procedural manuals.

Working Words Into Movement: A process and support tool to bring to light the organic movement embedded in vocabulary words.

APPENDIX

CCSS Standard Benchmarks That May Apply To A Creative Movement/ Simple Machines Unit For Grade 3

See Table 2.

Steam$_2$ Lesson Plan Template

See Table 3.

Table 2.

3NCAS-Dance Standards that *may* Apply to Simple Machines Unit Plan, Grade 3	
Strand Identification	Standard
Artistic Process: Creating. Generate and conceptualize artistic ideas and work.	
DA: Cr1.1.3	b. Explore a given movement problem. Select and demonstrate a solution.
Artistic Process: Creating. Organize and develop artistic ideas and work	
DA: Cr2.1.3	b. Develop a dance phrase that expresses and communicates an idea of feeling. Discuss the effect of the movement choices.
Artistic Process: Creating. Refine and complete artistic work	
DA: Cr3.1.3	a. Revise movement choices in response to feedback to improve a short dance study. Describe the differences and changes made in the movement.
Artistic Process: Performing. Select, analyze, and interpret artistic work for presentation	
DA: Pr4.1.3	a. Judge spaces as distance traveled and use space three dimensionally. Demonstrates shapes with positive and negative spaces. Perform movement sequences in and through space with intentionality and focus.
DA: Pr4.1.3	c. Change use of energy and dynamics by modifying movements and applying specific characteristics to heighten the effect of their intent.
Artistic Process: Performing. Develop and refine artistic technique and work for presentation	
DA: Pr5.1.3	a. Replicate body shapes, movement characteristics, and movement patterns in a dance sequence with awareness of body alignment and core support.
DA: Pr5.1.3	b. Adjust body-use to coordinate with a partner or other dancers to safely change levels, directions, and pathway designs.
DA: Pr5.1.3	c. Recall movement sequences with a partner or in group dance activities. Apply constructive feedback from teacher and self-check to improve dance skills.
Artistic Process: Responding, interpreting intent and meaning in artistic work.	
DA: Re8.1.3	a. Select specific context cues from movement. Explain how they relate to the main idea of the dance using basic dance terminology.
Artistic Process: Connecting, synthesize, and relate knowledge and personal experiences to make art.	
DA: Cn10.1.3	b. Ask and research a question about a key aspect of a dance that communicates a perspective about an issue or event. Explore the key aspect through movement. Share movements and describe how the movements help to remember or discover new qualities in these key aspects. Communicate the new learning in oral, written, or movement form.

SHAPE America Physical Education Standards that *may* Apply to Simple Machines Unit Plan, Grade		
Strand Identification	Standard/Outcome	Application to Simple Machines Unit Notes
Standard 1. The physically literate individual demonstrates competency in a variety of motor skills and movement patterns.		
S1.E6.3	Performs a sequence of locomotor skills, transitioning from one skill to another smoothly and with- out hesitation.	
S1.E7.3	Balances on different bases of support, demonstrating muscular tension and extensions of free body parts.	
S1.E8.3	Transfers weight from feet to hands for momentary weight support	
S1.E11.3	Combines locomotor skills and movement concepts (levels, shapes, extensions, pathways, force, time, flow) to create and perform a dance.	
S1.E12.3	Combines balance and weight transfers with movement concepts to create and perform a dance.	
Standard 2. The physically literate individual applies knowledge of concepts, principles, strategies and tactics related to movement and performance		
S2.E3.3	Combines movement concepts (direction, levels, force, time) with skills as directed by the teacher.	
Standard 3. The physically literate individual demonstrates the knowledge and skills to achieve and maintain a health-enhancing level of physical activity and fitness.		
S3.E2.3	Engages in activities of physical education without teacher prompting.	
Standard 4. The physically literate individual exhibits responsible personal and social behavior that respects self and others.		
S4.E4.3a	Works cooperatively with others.	
S4.E4.3b	Praises others for their success in movement performance.	

CCSS ELA Standards that *may* Apply to Simple Machines Unit Plan, Grade 3	
Strand Identification	Standard
Key Ideas and Details	
CCSS.ELA- LITERACY.RL.3.1	Ask and answer questions to demonstrate understanding of a text, referring explicitly to the text as the basis for the answers.
CCSS.ELA- LITERACY.RI.3.3	Describe the relationship between a series of historical events, scientific ideas or concepts, or steps in technical procedures in a text, using language that pertains to time, sequence, and cause/effect.
Craft and Structure	
CCSS.ELA- LITERACY.RI.3.4	Determine the meaning of general academic and domain-specific words and phrases in a text relevant to a grade 3 topic or subject area
Integration of Knowledge and Ideas	
CCSS.ELA- LITERACY.RI.3.9	Compare and contrast the most important points and key details presented in two texts on the same topic.
Writing	
CCSS.ELA- LITERACY.W.3.2.B	Develop the topic with facts, definitions, and details.
CCSS.ELA- LITERACY.W.3.7	Conduct short research projects that build knowledge about a topic.
CCSS.ELA- LITERACY.W.3.10	Write routinely over extended time frames (time for research, reflection, and revision) and shorter time frames (a single sitting or a day or two) for a range of discipline-specific tasks, purposes, and audiences
Speaking and Listening: Comprehension and Collaboration & Presentation of Knowledge and Ideas	
CCSS.ELA- LITERACY. SL.3.1.B	Follow agreed-upon rules for discussions (e.g., gaining the floor in respectful ways, listening to others with care, speaking one at a time about the topics and texts under discussion).
LITERACY. SL.3.1.C	Ask questions to check understanding of information presented, stay on topic, and link their comments to the remarks of others.
CCSS.ELA- LITERACY. SL.3.1.D	Explain their own ideas and understanding in light of the discussion.
CCSS.ELA- LITERACY.SL.3.2	Determine the main ideas and supporting details of a text read aloud or information presented in diverse media and formats, including visually, quantitatively, and orally.
CCSS.ELA- LITERACY.SL.3.6	Speak in complete sentences when appropriate to task and situation in order to provide requested detail or clarification. (See grade 3 Language standards 1 and 3 here for specific expectations.)
Language: Conventions of Standard English	
CCSS.ELA- LITERACY.L.3.3.B	Recognize and observe differences between the conventions of spoken and written standard English.

Table 3.

Lesson Title:			Sequence in unit:
Scientific Big Idea(s):		Encore Big Idea(s):	
Target Grade level:		Lesson Date:	Duration & # Ss:
Academic/Scientific Standards Addressed			
Introduced	Developed	Assessed for Mastery	Included for repetition/extension/ transference
Academic/Encore Standards Addressed			
Introduced	Developed	Assessed for Mastery	Included for repetition/extension/ transference
Academic/CCSS Addressed			
Introduced	Developed	Assessed for Mastery	Included for repetition/extension/ transference
Standards-based Performance Objectives and/or I can Statements (cognitive, affective & psychomotor)			
Key Vocabulary			
Space, equipment, materials:			
Instruction	What	Who is responsible?	Organization of students
Engaging message/ Instant Activity			
Lesson Focus			
Assessment			
Culmination/closing/ wrap-up/cool down/ processing			

Chapter 6
Getting to "Know" STEAM

Merrie Koester
University of South Carolina, USA

EXECUTIVE SUMMARY

This chapter describes the evolution of an arts-integrated approach to science curriculum inquiry which has been evolving since the 1990s—before the national science standards, the acronym STEM, much less STEAM, appeared across educational horizons. It reads as ethnography and has been performed in community, in association with the most caring of souls, with the goal of achieving a more inclusive/empowering, aesthetic science education, and a deep appreciation of the importance of the creative arts in the science learning process. It presents two research-based iterations of STEAM education in practice: 1) the creation of arts-integrated middle school ocean science curricula and 2) the development of a pedagogical tool called the "Know"tation as a way for teachers and students to make learning visible and integrate the languages of science throughout the process of inquiry. The cases described here apply many features of the STEAM model developed in Chapter 2 of this book.

INTRODUCTION

How can educators engage those students who struggle to read and for whom science informational text seems particularly threatening? I encountered this same challenge myself with my very first teaching job, which tasked me with teaching physics to thirty-five older ninth graders, who had previously failed the class. The football coach, their former teacher, dubbed them his "dummy class." "Don't expect too much from them," he cautioned. I was astonished. Certainly, no one in that class was stupid! Yet, they had most definitely been made to *feel* that way through tracking,

DOI: 10.4018/978-1-5225-9631-8.ch006

neglect, and marginalization. The real kicker was that the more affluent students up the hall in the "gifted science class" had a nice lab with equipment and space to do experiments. My class and I, on the other hand, had thirty-five desks crammed into rows in a small corner classroom and an old textbook none of my students could read. I was so mad I could spit, but I was also lucky. I had been trained in the arts, and I knew how to tell a story through drawing. I used that as a way to connect, with language we could all understand. Together, we drew and played with funny little cars, which we rolled down those crummy old textbooks. These kids started passing, and I started to get a sense of what it meant to *design* a science curriculum as an artist might.

What I did not realize then, as I do now from study of Elliot Eisner, John Dewey, Johann Pestalozzi, Maxine Greene, Richard Siegesmund, Liora Bresler, Bruce Uhrmacher, and other arts education researchers is that what I had observed in my students and myself was an *aesthetic* transformation. Uhrmacher (2010) noted that a student who acquires "aesthetic capital" may feel or act differently in a good and positive way. As a brand-new teacher, I was so transformed by the way an artistic pedagogy turned around previously failing students that I set out to change science education. Ah, to be 25 again! I applied to MEd program at the University of Hawaii, where I took graduate level classes in both the visual and performing arts as well as art and science curriculum theory. I became determined to identify the common language that science and art share and developed an experimental curriculum called Teaching Science through the Arts (TSTA) at a time before the acronym STEM had become popularized (Koester, 1989). I researched, developed, and then field-tested curricula for teaching science through the creative arts of drawing, poetry, music, dance, creative drama, and fictional literature. In my master's thesis, I tentatively proposed that virtually any area of K-12 science could be taught through the arts. The lesson planning objective, I reasoned, was to match the art form to the dynamics of the science content being taught. Right away I recognized that a science teacher who would implement the TSTA model would need to either personally acquire basic skills in multiple art forms and/or collaborate with an art teacher specialist. Because I had significant arts training, I managed to create and field-test about fifteen TSTA lessons as part of my final thesis. However, I did not have near enough data to propose a *theory* about best practices for teaching science through the arts. That process has taken nearly two more decades.

What follows is a true story that narrates the transformation of a science teacher into a researcher and the evolution of an idea into a curriculum model that is still growing and developing.

BACKGROUND

Creating As-If Worlds

A few months after completing my masters at UH, I returned to my home state of South Carolina with my hundred-page thesis carefully packed in a special box. I had accepted a middle school science position and was eager to field-test my arts integration techniques in a sixth-grade physical science and eighth-grade Earth science class. After only one week, however, my principal said that she had forgotten to tell me that I would be required to take my sixth-graders through their first ever science fair projects. Oh, no! Which art form would possibly match the teaching of THE scientific method, something our curriculum dictated *all* students were to learn? I was full of dismay and fretted for nearly a week, when suddenly, the answer came to me as clearly as I imagine Archimedes' Eureka Moment occurred to him. Late one Sunday evening after I put my two boys to bed, I began writing a fictional story about a twelve-year-old girl who unwittingly let her father complete her entire science project – an impressive, meter tall electromagnet crane. Even worse, the girl won first prize for work that was not her own. Her best friend, who *had* done her own project, came in second. The next morning, I shared with my sixth-graders the beginning of a story that later became the first of four science education novels written over the course of two decades and recently released in their third editions (Koester, 2015 b-e). Almost instantly, my students entered into the shame my young protagonist felt for cheating her best friend out of first place. We reflected at length on the importance of doing independent, honest work. They learned the steps of the experimental process by performing the "Science Rap" which I had written for them. To my amazement, nearly all of them eagerly dug in to their own projects, determined to do their own work. That same year, one of my sixth-graders, so shy that she rarely spoke, won first prize in the regional science fair. She, like the character in my story, later grew up to be a science teacher.

Who among us did not dread doing their first ever science fair project? Indeed, I was that twelve-year-old girl *and* the teacher who was now all grown up, encouraging her students to actually embrace this arduous task. In my creative drama classes at UH, I had discovered the power of creating *as-if* worlds in the science classroom (more on that in just a bit). The effect seemed extra-ordinary if that world came from a fictional trade book. I developed this hypothetical learning equation: Imaginative story (in a trade book featuring science content) + creative drama (with movement) → Learning + Engagement. To test my claim, I developed for my then fourth-graders a lesson on digestion. I began by reading the trade book, *Gregory, the Terrible Eater*, by Mitchell Sharmat with them. This allegorical tale is about a goat named Gregory, whose parents think he is a TERRIBLE eater because he wants to eat foods like fruits, vegetables, eggs, and fish instead of the *normal* GOAT diet of shoes, paper, tin cans, coats, pants, etc. My students eagerly responded when I asked them to get on all fours and pretend to eat a pile of shoes like good little goats. They produced Oscar-worthy performances when I asked them to refuse to eat as their parents demanded. The pawing and stomping and mournful bleating were hilarious. Some rolled on their backs and kicked their feet in the air in defiance. To add a musical performance component, we improvised an operatic, call and response exchange between a group of students, pretending to be goat kids (refusing to eat perfectly yummy shoes), and their worried goat parents. The "tune" we used was Beethoven's Fifth—da, da, da, daaaaaaa (Koester, 1989):

Kids: *We will not eat!*
Parents: *Oh, yes you will!*
Kids: *Shoes make me sick!*
Parents: *Don't talk like that!*
Kids: *We're feeling weak.*
Parents: *'Cause you don't eat.*
Kids: *It tastes like junk!*
Parents: *What will we do?*

Anderson (2004) has written that "the methods of drama in education have the potential to create 'as-if' worlds within the classroom that can foster *situated* learning" (p. 284). Anderson was referring to the cognitive psychology studies of Brown, Collins, and Duguid (1989) on the benefits of learning in *authentic* contexts, *situated* within a context with which one could personally relate. Anderson argued that through creative drama and the creation of *as-if* worlds, students could enter into simulations of what *felt* like authentic experiences. Through the vehicle of our imaginations, stories take us places where we might never, pragmatically speaking, be able to go—in spaceships to Mars; in deep sea submersibles to deep ocean hydrothermal vents, where giant tube worms with bright red tips twist and turn in front of the front window; and into science research laboratories, where people like we race to discover a vaccine against *Ebola*. We enter into deep, empathic relationships with a story's characters.

Neuroscientists have determined that there is a scientific basis for the empathy we feel in *as-if* situations. In 1992, a team of researchers at the University of Parma, Italy, working with macaque monkeys discovered that the same brain cells fired when the monkeys performed a task (like picking up a peanut) as they did when the monkeys simply watched a person do the same task. They named these brain cells *mirror neurons* (Ehrenfeld, 2011). Neuroscientists became very excited about this discovery, and soon found that human beings also had mirror neuron systems. Were mirror neurons the scientific explanation for empathy? Gallese (2005), an Italian neuroscientist, certainly thought so. He took a lead in the mirror neuron research and discovered that indeed, mirror neurons fire in our brains even when we just *imagine* doing a specific task. For example, we can look at a picture of a hammer, and the motor neurons that we would use to pick up a real hammer are excited. When we see another person smile, the muscles in our own face that control smiling light up on functional MRI. Gallese (2005) had determined that our brain can "model the acting body of other individuals." He called this phenomenon "embodied simulation" and conceptualized what was happening as a kind of *resonance* between bodies, which he theorized could be at the heart of all social cognition (p. 23-48, in Koester, 2017).

Rollo May (1985) explained that artists portray their reality through *images*—poetic, aural, plastic, dramatic, and embodied. The processes described by science are a veritable gold mine for *imag*inative, empathic visualization. Their very names conjure wonderful images and possible as-if situations. Scientific words like evolution, metamorphosis, growth, decay, reproduction, birth, and symbiosis are just a few examples of the exciting events described in a life science class. The physical and Earth sciences provide just as many opportunities for artistic, creative expression and interpretation (Koester, 1989). I still take every kind of creative arts class I can. The process is invigorating and enlightening as long as I check my ego at the door. Just recently, I bumbled my way through an improvisational theatre class, and came away understanding the power of "Yes, *and*" messages, as opposed to "Yes, *but,*" which kills a scene. Every time I take a new art class, I come away with renewed appreciation of the work ethic of the performing artist and the important roles that trying, failing, and being critiqued each have in improving. The process is scary, too. There is no podium behind which to hide. A supportive, caring teacher makes all the difference between whether you will risk ditching your cover and discovering your inner artist (It's there somewhere!) or quit the class for fear of embarrassing yourself.

I am very fortunate to live and work in a state where I have received enthusiastic support for my experimental pedagogy by professional science and art education organizations, as well as many colleges and universities. Beginning in 1993, I began giving workshops on teaching science through the arts at the South Carolina Science Council (SC2) annual meetings. In the first 1993 session, I shared videos of my students, who had choreographed their version of "The Tectonic Drag." The beginning verse went like this:

On the floor of the seas
are rifts and ridges,
that circle the world
like gigantic zippers.

I had written "The Tectonic Drag" as a poem to be choreographed. It was my way of trying to bring my eighth-grade students into the deep, to the ocean floor, to the beginning. Here's the chorus:

Let your arms go up and out and down;
they're convection currents going round and round.
Convection currents are the driving force that
push the Earth's plates along their course.
The plates are doin' the TECTONIC DRAG –
spreading, crashing, diving – making buildings sag.

Can we ever know the oceans' effects on us, or more importantly, our effects on them? We should give thanks to marine algae every single day for the oxygen we breathe! David Ballentine, a noted marine phycologist, believes that if all marine algae were to die, we would have less than one week to live (personal interview, 2007). These primitive life forms have been making oxygen for the planet for millions of years. Why do sponges produce natural antibiotics? We don't yet know. How many fishes can be caught before their populations collapse completely and entire populations of all sorts of living things collapse? Do we dare find out? In Lloyd Alexander's novel, *Taran Wanderer*, the wizard Morda sneeringly observes: "I have seen enough of human kind and have judged them for what they are—lower than beasts, blind and witless, quarrelsome, caught up in their own selfishness" (in Koester, 2011). Alexander's novel creates an *as-if* world, inside of which students develop understanding of human impact on the environment and each other. Through the elements of narrative— challenge, conflict, and struggle—eventually resolution is achieved. When teachers learn to develop lessons that move through periods of dissonance and struggle, for which no one right solution is apparent, students will make creating some kind of resolution their own responsibility. They key words here are *making, struggle,* and *responsibility*. In real life, resolution is not always achieved. In real *science*, the struggle is to discover what we don't know. This theme of persistence in the face of struggle was most definitely the clarion call of Chapter 2 of this book.

DRAWING FROM THE WATER

On June 7, 2006, I was invited to attend the closing session of the first ever Conference on Ocean Literacy (C.o.O.L.), where a consensus of mostly marine scientists and government policy makers had gathered to review the draft of a publication called the *Seven Essential Principles of Ocean Literacy and Detailed Fundamental Concepts* (http://oceanliteracy.wp2.coexploration.org/). I was very excited about attending this event, as my students and I had been participating with the field-testing of NOAA's extensive, virtual, and interactive *Ocean Explorer* marine science curriculum being developed by the education department of their office of ocean exploration (http://oceanexplorer.noaa.gov/welcome.html). My eighth-grade students sent real-time questions to NOAA research scientists, as they whisked about in deep sea submersibles. Images of giant tube worms beamed into my classroom. This educational technology was unlike any I had ever experienced as a science teacher, *and* it was free!

At the C.o.O.L gathering, the legendary oceanographer and first female aquanaut, Sylvia Earle, spoke. During our break, I got to meet this great and gracious scientist and pioneering explorer. John Michel Cousteau, son of Jacques, reminded us that his father had always said that we will only protect what we love and understand. My heart and mind were stirring in the way they always do, right before I get an idea or make an important decision. I have learned to pay close attention when this happens, to be alert for the clue of what to do next. I didn't have to wait long. Suddenly, the lights went out in the conference hall, just as I thought the meeting was about to close. Our attention was directed to a large screen on a side wall. Flickers of light played across the screen, and then, Woods Hole Oceanographic Institute scientist and deep ocean explorer Robert Ballard appeared. He was speaking with us via telepresence from a submersible on the bottom of the Aegean Sea! Ballard declared that the battle for ocean literacy (and hence the future of our planet) was to be won at the *middle* school, because that's when we started "losing" our nation's science students. Wait! *I* was a middle school teacher! Was Ballard talking to me? Exhorting me to become some kind of ambassador for ocean literacy? I was certain he was! In spite of the fact that I had said I did not have another science education novel in me (I had been writing them for nearly eighteen years by then), I vowed to write one *last* book, this time about the dire need for ocean literacy.

At about the same time, in my hometown of Charleston, SC, the SC Maritime Foundation had just launched a tall ship, the *Spirit of South Carolina*. She was a full-scale replica of the nineteenth century pilot schooner, the *Frances Elizabeth*. She was built as sail training vessel to provide experiential education to South Carolina youth. While onboard, students would learn about the math, science, and natural history of the surrounding waters, as well as how to work as a cooperative crew. The schooner was outfitted with the most modern ocean science equipment and technology. I was fortunate enough to join multiple school groups on several sailing experiences in the Charleston Harbor. Many of these children had never been aboard any kind of boat, much less a nearly 100 foot, three-masted sailing ship! I determined to prominently feature this proud vessel in my new book.

I next applied for and received a research and development grant from the National Marine Sanctuary Foundation (NMSF). With these funds, I set out to research and have vetted the ocean science content for this book, a journey that took me into some of the finest aquaria and ocean science centers in our country—The Shedd (Chicago), The Baltimore Aquarium, The Monterey Bay Aquarium, The Sant Ocean Hall (Smithsonian Natural History Museum), and The South Carolina Aquarium. I traveled to Puerto Rico, where Ruperto Serrano Chaparro, director of the Sea Grant Puerto Rico, introduced me to prominent marine ecologists and researchers, experts all in coral reef diseases and the phenomenon of dead zones. I snorkeled in the waters off the coast of Puerto Rico and also the island of St. John and saw with my own eyes the devastation beneath the waves—the gray "boneyards" where once great, colorful "gardens" of coral once thrived. I spent three days and nights sailing on the *Spirit of South Carolina*, keeping midnight watches, mustering with the crew, and even practicing man overboard drills. I don't think I slept the entire time for fear of missing something!

By now, I was a very active member of both the South Carolina and National Marine Educators Association. I had learned that the breeding ground of the North Atlantic Right Whale, the most endangered of all marine mammals, was very near me at Gray's Reef National Marine Sanctuary, whose staff sent me classroom sets of information packets on this gentle giant.

In 1995, Daniel Pauly, principal investigator at the Sea Around Us Project, coined the term "shifting baselines" to describe changes which occur slowly enough that we barely, if at all, notice them occurring. As a result, our "baseline" of what is acceptable shifts. For example, tomorrow's coral reef divers will come to see "boneyards" as the norm (https://www.ted.com/talks/daniel_pauly_the_ocean_s_shifting_baseline?language=en). I was fascinated with this concept. Could it be that science, technology, engineering, and mathematics had somehow played a role in the ocean's downfall? I suspected the answer was yes. Before I knew what was happening, my writing and my message took a critical turn. I started looking for what had been left *out* of the history books. Settlage and Southerland (2007) have written that teachers must be vigilant about the contents of the *null curriculum*. Foucalt (1977) declared that critical researchers in science education have a responsibility to show how dominant classes have manipulated truth to their advantage. I felt I surely must do the same with this new work. Indeed, my book would be fiction, but my motivation for writing it had become deeply personal. It was rooted in my own repugnance of the racism I observed growing up in the deep South and my passion for preserving and protecting the Ocean, upon which all life forms, from diatoms to despots, depend for their very existence.

Bill Bigelow wrote a chapter for Wayne Au's book, *Rethinking Multicultural Education* (2009) called "Once Upon a Genocide." He cited a need for books for young students which 1) portray what "the Indians might have been thinking about the arrival of the Spaniards" (p. 78) and 2) asks children to "think about how today's world was shaped by the events of 1492" (p. 84). I sought to answer both appeals. In my novel, I lay bare the crimes of imperialism, exploitation, murder, and racism of Columbus and the Spanish *conquistadores* who followed him. My long-range action plan was to communicate at the school, community, and professional levels the significance of understanding that entire chapters of human history have been rewritten by people bent on obfuscating the truth about the human and environmental exploitation and destruction set into motion in 1492. Through art, using the rhythm of djembe drums and storytelling to painting scenes in the mind's eye, where they would not likely be forgotten, I would attempt to tell the story of what really happened on and after that first "Columbus Day." I wrote and wrote and wrote and wrote. When I slept, the plot twisted and turned like sharks twirling inside the barrel of incoming wave. A translation of Christopher Columbus' 1492 journal sat on my bedside table. His scribblings about the Tainos *naïveté* read like a horror story:

They are like the other peoples I have found with the same beliefs, and they believed that we had come from heaven and they gave what they have for whatever they are given without saying that it is too little and I believe that they would do the same with spices and gold if they had any.

Finally, I was ready for the professional development phase of what I called L.A.S.T. Book Project—a literacy-based, arts-infused, science-centered, technology-driven ocean science curriculum for middle grade learners (and, the completion of my *last* book). I had just enough funding left to provide training and art supplies to two middle school teams. I created an application and selected two schools, whose faculty seemed most committed to entering in a year-long, collaborative, interdisciplinary effort centered on advancing ocean literacy. Both these schools' principals were fully on board, a factor shown to be very important to the success of any professional development project (Loucks-Horsley, 1996). A year later, the South Carolina Science Council invited me to give a keynote about what happened during the project. What an exciting November day in 2011 that was! Onstage with me was Tiffany, a middle school science teacher from one of the lowest income districts in my state. During our study, she had collaborated with Annie, the school's art teacher. Behind us, filling nearly the entire stage and the wall behind it was the arts installation their students had created. It featured a gigantic ocean creature, with the body of a fish and a human head. In the first chapter of my book, entitled "The End and the Beginning," a twelve-year old girl walking on the beach against an oppressive, northeast wind and freezing rain, comes across a dead North Atlantic Right Whale, its gills ensnared with fishing lines and its dorsal fin jaggedly slashed by marks only a ship's propeller could have made. Spelunking into the whale's cavernous rib cage, the girl had discovered a small, beautifully crafted wooden box, inside of which she found a carved and polished stone amulet, with the head of a human and whale's body. It was a 500-year-old carved amulet, a symbol of our human responsibility to preserve and protect ocean resources, but the girl did not know that yet. (Figure 1)

As I was told later, as soon as she read this part of the story, Annie, the art teacher, sprang into action. "What if," she imagined out loud, "what if our students design and construct a giant replica of this human fish, constructed from the kind of debris that has harmed or killed so much of marine life?" What if, indeed! (Figure 2)

In a joyful act of collaborative creation, Tiffany and Annie's students designed and then constructed their "trash" creature. Through their art, they communicated the ever-growing threat of marine debris on ocean life. Students transformed other plastic bottles into schools of cheerfully colored fish. (Figure 3)

Tiffany, Annie, and I were overjoyed with the results! The school principal arranged for there to be a special school assembly for the unveiling. (Figure 4)

As a result of participating in this kind of arts-infused exploration of ocean science, students came to understand that humans have a responsibility for protecting and preserving the ocean. They also realized that not all scientific discoveries result in good, and that the discovery of the Gulf Stream marked the beginning of an era of human conquest of the ocean and of their fellow human brothers and sisters. They engaged in critical thinking about critical matters by examining how their own history intersected with the history of ocean science and exploration. (Figure 5)

This ocean literacy action research project was my first sustained effort to bring science teachers, arts teachers, and their shared students *together* in the process of curriculum inquiry. Acting as role models and mentors, these teachers showed their students that deep learning can happen when balanced art/science teaching partnerships are formed. They cross-trained one another in the knowledge, skills, and

Figure 1. The human fish, a metaphor for our responsibility to preserve and protect ocean resources

Figure 2. Students constructing a "fish" artifact from potential marine debris

Figure 2. Students constructing a "fish" artifact from potential marine debris

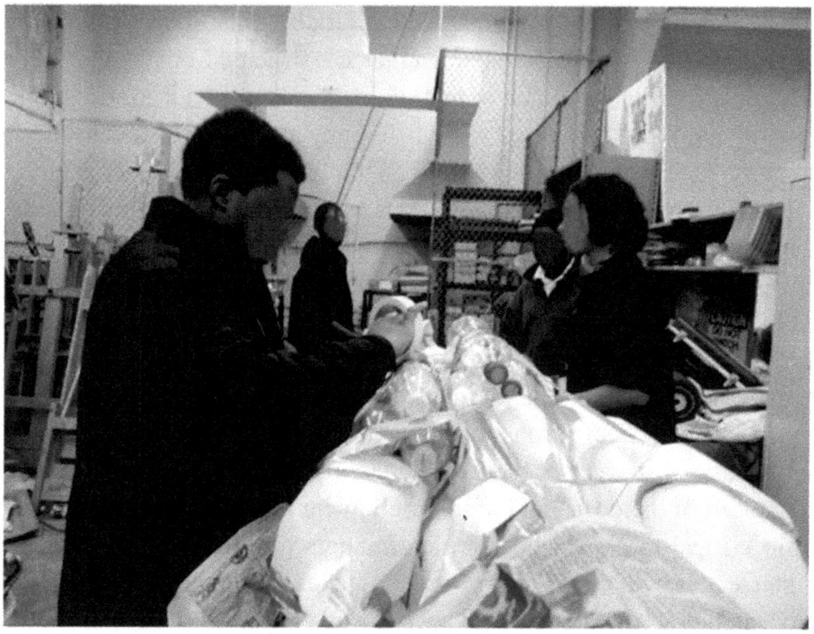

Figure 3. Arts installation of "human fish" and colorful reef "creatures"

Figure 4. Art teacher, university researcher, and science teacher celebrate installation of STEAM marine education project

Figure 5.

Figure 5. Atlantic Ocean currents associated with "New World" exploration, colonization, slave trade, and commerce by Europeans

practices of each other's teaching specialty, and each reported being transformed by the collaborative effort. What was happening at this school was a deeply *aesthetic* inquiry of human impact on the environment—a joyful, participatory, sensory and arts-based form of learning described by scholars like Pestalozzi (1894/1973), Dewey (1938), Greene (2001), Eisner (1991, 2002), and Siegesmund (2010) as promoting inclusivity, democracy, and freedom.

A similarly inspiring and highly synergistic art/science education collaboration was simultaneously taking place at the second school enrolled in The L.A.S.T. Book Project. Sadie, the eighth-grade science teacher, and Shannon, the school visual arts teacher, had received permission from their principal to create a special projects art/science class. Over coffee one afternoon, Sadie, Shannon, and I hatched the idea to literally *draw* their students into the plight of the nearly extinct North Atlantic Right Whale. Shannon would teach them the elements of design and the art of batik so that they might eventually work together to create a large mural in celebration of this endangered and majestic animal. Meanwhile, in her science class, Sadie would teach about human impact on the ocean, especially our impact on the lives of once thriving ocean animals. The topics in the science curriculum included overfishing, shifting baselines, coral reef bleaching, global warming, harmful algal blooms, dead zones, and the power of youth. On my first day meeting with this combined class, we talked about the potential impact of the mural as an art form to communicate profound and important messages. I asked students to contemplate a projected image of Picasso's *Guernica* (1937). This mural displays the tragedies of war through its graphic display of the suffering of innocents, killed by Nazi bombs on the town of Guernica during the Spanish Civil War. As such, Picasso's mural also became a symbolic call for peace (http://www.pablopicasso.org/guernica.jsp). We talked about how artists need to develop deep knowledge of their subjects before they can make powerful art about them. They agreed that Picasso clearly must have witnessed and known a lot of suffering to have so strongly portrayed the devastation of war.

I then showed them an image of a beached North Atlantic Right Whale, brought to an ignoble death by becoming ensnared in commercial fishing tackle and with propeller gash wounds across its dorsal surface. Many exclaimed, "It's the whale in the book we're reading! They're going extinct!" Sadie, who was also in the room with us, jumped up and said, "That's right! And now it's our turn to try to do something to save this whale."

Next, I shared a video which was similar in effect to this newer NOAA release: http://oceantoday.noaa.gov/whalecall/. The students were thoughtful and quiet. Shannon and Sadie stood up and explained that we were all going to work together on a special art/science project whose goal was to learn as much as possible about North Atlantic Right Whale and then to create art that symbolized its *right* to live (pun intended). Next, I told them about Leonardo da Vinci, an artist/scientist who said that to draw something is to know it. Clearly, we could not draw from a live whale, but the staff at Gray's Reef National Marine Sanctuaries (http://www.graysreefnmsf.org/), a breeding ground for these whales, generously sent us a set of huge posters depicting the anatomy of these magnificent mammals. I placed a poster on each art table, and Shannon distributed brand new sketchbooks and drawing pencils, paid for by the NMSF grant. In addition, Gray's Reef specialists made available to us a wide array of educational materials, now accessible at http://graysreef.noaa.gov/education/materials/welcome.html.

Over the next week, in both science and art class, students were challenged to design their own sketchbook journal entries, depicting their growing knowledge about the North Atlantic Right Whale. (Figure 6)

We then scheduled a class during which all students rotated around the art room to offer appreciative critique of each other's sketchbook entries. Shannon and Sadie announced to the students that they were now ready to make art about and *for* this whale—art as an act of care. Shannon tasked each student with drawing a mural design in their sketchbooks, which they would soon draw again with white crayon on a 40 x 40 cm^2 of canvas. The paper sketchbook would be their "experimenting" space, where they would practice batik techniques before moving on to the more expensive canvas. Shannon was engaging her students in the art studio habit of mind involved with *developing craft* through practice (Hetland, Winner, Veenema, & Sheridan, 2003). In this stage of inquiry, students engaged in trial-and-error reasoning as they experimented with cause and effect. This stage of "messing around" or "playing with the material" is no different than the tinkering an engineer does as he or she works through the elements of a design process. It's an essential part of the work-in-progress. In this case, the amount of water added to the acrylic paint was a very important independent variable. The kind of brush and how tightly the fabric was stretched were important factors to consider as well. Students *had* to know it was okay to make a big mess at first. Mess is as much a part of the nature of art practice as it is of science. Insights evolve as mistakes get made.

Before they started, Shannon modeled all the techniques she would expect her students to practice and learn. She held up a canvas square for them to see and then drew an abstract design on the cloth with her own crayon. Then she demonstrated several different kinds of mistakes that might occur in the process, and said that they would likely make them, too. She emphasized that artists need to *engage and*

Figure 6. Student sketchbook entries representing the characteristics of the North Atlantic Right Whale

persist at the task at hand (Hetland et. al, 2003). Prior to this stage in the project, we had engaged in many other art studio habits of mind, characterized by such evocative verbs as *observe, envision, integrate, express, differentiate,* and *connect worlds* (Hetland et. al, 2003). I thought, *if only more science classes functioned like art studios!* Shannon's students then entered the *work* stage. I noticed, too, that everywhere were signs reminding the students to take care of their studio and their materials. Truly, this was a space suffused with caring messages.

One of the most significant ways that Shannon and Sadie made their curriculum inclusive and demonstrated care for their students' feelings was the decision to make a quilt, pieced together from *every* student's mural design. (Figure 7)

The class voted unanimously on one of their classmate's designs, a gifted artist named Rose (pseudonym), who later earned acceptance to the public magnet high school of the arts. (Figure 8)

Rose was also appointed to oversee the final mural's completion as general "contractor" of an artifact that was to measure 3.5 meters in length and 0.75 meters in width. Bringing the project to completion would require students to work collaboratively through issues of scale and proportion and practice making precise measurements.

Rose was born leader. She expertly organized her classmates into work crews, each with its own sub-contractor. Pictured here are members of the measurement and layout team. (Figure 9).

Rose made sure that every single classmate participated in painting this mural. Here she is, putting on the finishing touches, just before we presented it at the National Marine Educators Association (NAEA) meeting in Savnnah, Georgia. (Figure 10)

Figure 7. Student's paper and canvas mural designs (left) and assemblage of class quilt (right)

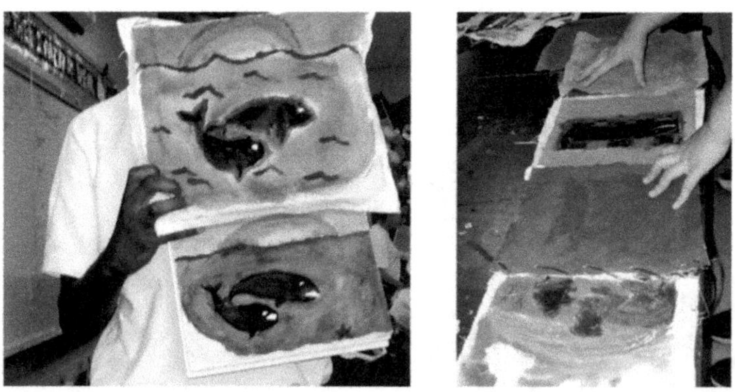

Figure 8. Winning North Atlantic Right Whale mural design on paper and cloth

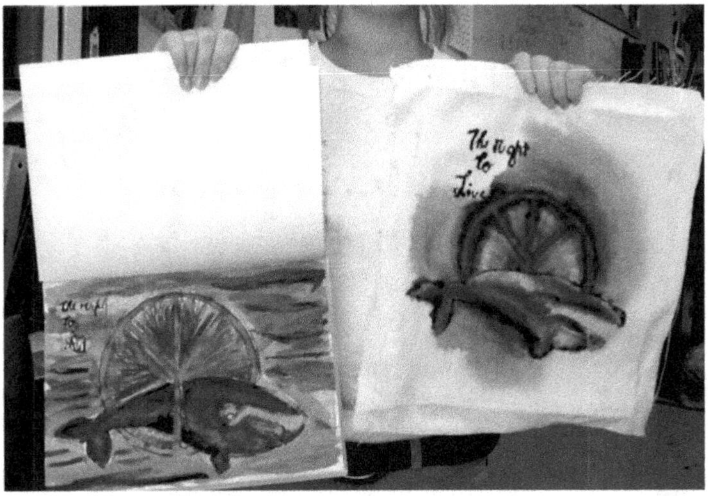

THE MANY FACES OF STEAM

What Is STEAM Anyway?

In 2011 or so, while I was working on this ocean literacy project, I discovered that a pedagogy called STEAM had been "invented" (though I would argue that Leonardo da Vinci had prior rights to that patent). Regardless, I regarded this "newcomer" on the educational landscape as a fantastic harbinger of promising reform. When I began a literature review, however, I grew increasingly concerned that there seemed to be no consensus on just what STEAM was and what it was not. My list of questions

Figure 9. Mural measurement and layout team

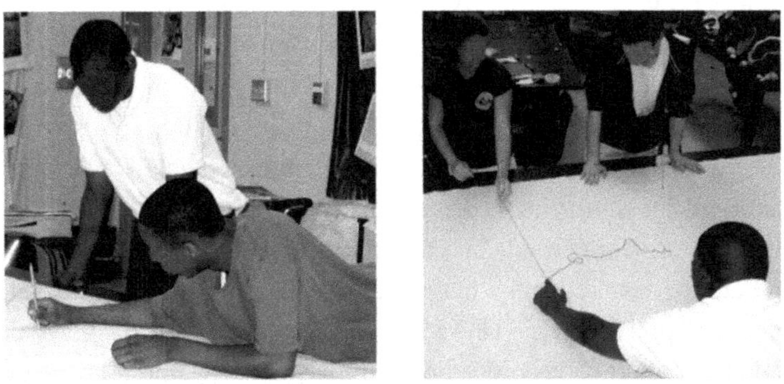

Figure 10. Rose puts the finishing touches on the mural, entitled "The Right to Live!"

was long: Which teachers *did* STEAM—art, science, ELA, social studies, all, some? Did it "happen" after school, or during? If both, how were the versions different from one another? Were specific art forms or science/STEM content areas being privileged over others? What did STEAM look like when it was "working"? What steps were being taken to make sure that STEAM was not just one more task being added to the already over-taxed art teacher's planning book? Were science, STEM, and arts teachers co-constructing project-based lessons? What kind of academic rigor did STEAM pedagogy offer? Further, how were the engineering, technology, and mathematics components being addressed? What kinds of STEAM might be funded? Would competing interests jockey to brand their own versions of STEAM as profit earning, educational commodities, thus achieving, at best, an isolated impact (for the developers) versus a collective one (for all stakeholders)? This competitive market strategy seemed counter-intuitive to me. Dewey (1938) challenged us all to consider what conditions have to be satisfied so that education become a "reality and not just a name or slogan" (p. 116).

Full disclosure: I have never been much for semantics, especially where educational acronyms supposedly signifying some collection of agreed upon skills, knowledge sets, and practices are concerned. The name readers choose to assign to this idea is trivial. I am here to serve readers whether they represent an organization that labels itself as doing science, STEM, STEAM, TSTA, or something similarly catchy. We are all together in this business of educating. I have seen that a collaborative and collective STEAM education supports teachers and students so that both science/ STEM and the arts actually matter to all stakeholders (Koester, in press). I further believe that STEAM in which the arts are used as "handmaidens" in the service of science/STEM is *not* STEAM. Likewise, when science/STEM understanding is not an integral part of a more arts-based curriculum, this is *not* STEAM either. It's something else. The most glorious explosions of STEAM learning I have witnessed emerged when art and science teachers worked together, each celebrating the gifts of the other and joining in the collective goal of effecting deep learning and appreciation of the importance of their respective areas of human endeavor.

Project Draw for Science

Just before I began The L.A.S.T. Book ocean literacy project, I accepted a position as an adjunct professor at the College of Charleston teaching science methods for preservice teachers, grades 2-8. I very quickly discovered with a pre-test that most of the elementary pre-service and many of the middle grade candidates lacked deep science content knowledge, especially in the physical sciences. I told them about my first ever teaching job, how I had used drawing to first deepen my own knowledge of physics, and how teaching through drawing had resulted in positive academic turn-arounds. I shared Leonardo da Vinci's claim: *To draw something is to know it.* We studied the maestro's detailed drawings of anatomy, machines, and the movement of water. I then asked these pre-service science teachers to consider the corollary claim: To *know* something is to be able to draw a simple picture explaining it. Immediately came cries of protest, "Not if you can't draw!"

"A valid point," I acknowledged. Let's test my claim *and* your counterclaim. I reached for a bag I brought with me and withdrew its contents. "Here is a loaf of bread and a jar of peanut butter," I announced. "Using them as a visual reference, can you draw a simple picture that explains how to make a peanut butter sandwich? The drawing only has to be *readable*—not a da Vinci masterpiece," I added. They could all readily complete this task, each in his or her own way. I continued, "Could you argue that you have now used drawing as a kind of *language* to communicate your understanding of how to make a PB & J sandwich? Could I (as your teacher) readily determine from your drawing whether your understanding was well developed (or not)?" They agreed with both propositions and quickly surmised that that if they did indeed have knowledge about a concept or procedure *and* they had a visual reference, they could produce a simple explanatory diagram even if they were not gifted artists.

I explained to these science-teachers-to-be that few science teachers whom had ever trained me in the graphic art of visually organizing text and images on a single page were very organized themselves. Often their notes are splattered across their classroom white boards like symbolic vomitus. (Figure 11)

Figure 11. Teacher science notes that are visually chaotic in their presentation

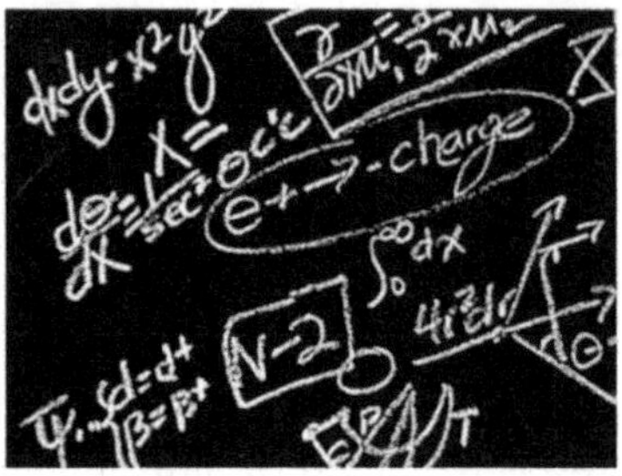

Student versions of these notes are often even more confusing and certainly do not invite later study because they have no story—nothing to make the scribbles (strange symbols written in a strange language) *mean* something. The result is complete semiotic breakdown (Koester, 2015).

Finson and Pederson (2011) have argued for the importance of visual data in science and posit that a "comprehensive understanding of a problem and potential solutions to it are possible only when one knows not only particular skills in visualization, but also in knowing the right questions to ask" (p. 71). I regard this claim as a call to action (in Koester, 2015a). I ask, "Why is drawing science concepts not afforded at *least* equal status with speaking and writing, especially in teacher education programs?" Michael Polanyi (1966) has said that we all know more than we can tell. I challenged my pre-service science students to find evidence to support Polanyi's claim throughout their student teaching experience. Over time, I began to imagine an action research project that asked the following question: What if drawing were conceptualized as science language *and* as a legitimate means of formatively assessing science "knowing"? I placed quotations around the prefix "knowing" to remind one that the very nature of science involves a constant re-testing and revision of what we *think* we "know" based on the evidence we may have at a given time (Koester, 2015a).

In 2012, with funding and support by the University of South Carolina Center for Science Education, I began the Project Draw for Science research with five teachers from the poorest to the most affluent districts in our state, all of whom self-identified as not being able to reach their struggling readers and wanted to learn how to teach through drawing. They exhibited deep levels of care for their students and took part in this study with no remuneration at all. After seven months, the students who made the most significant improvement were either struggling readers and/or special needs students. When asked to explain how it was that learning science through drawing affected them, these same students replied with responses like, "It makes me feel like I'm getting help. It helps me think way better than just listening does. It makes me feel joyful and energetic!"

I knew from the years leading up to this study that the science teachers I trained felt empowered when they learned the art of visually organizing content for and with their students. They also confessed that creating drawing repertoires as part of their lesson preparation deepened their own content knowledge. The process of teaching science through drawing which I had developed to this point began with learning basic drawing skills and the elements of graphic design. This is followed by identifying only the most important science vocabulary words for each concept. Struggling readers, especially those with dyslexia—some 20% of learners (Shaywitz, 2003)—benefit greatly from having new words syllabicated and pronounced out loud as they are written. Inviting students to create illustrated flash cards for new terms is very helpful for English Language Learners. At the advanced level of teaching science through drawing, I teach educators how to combine creative drama techniques, scriptwriting, and drawing to effect what I call the Performative Narrative Drawing (Koester, 2015a).

At the beginning of the Project Draw for Science study (http://merriekoester.wix.com/project-dfs), I began an earnest quest to deepen my own understanding of science *language*. My readings took me straight into the important work of Jay Lemke (1999, 2002). Never had I come across a scholar who had had so keenly studied the qualities of science classroom discourse practices and the many ways in which a student could become completely overwhelmed by the multiple and simultaneous de-coding processes required in a science classroom setting. Lemke determined that at any one time in the teaching and learning of science, students were bombarded by not one, but four different languages: WORDS, IMAGES, SYMBOLS, and ACTIONS, all part of a single science *sign system,* which must be integrated if any kind of significant sense making (other than just low level memorizing) are to be achieved (2002).

I was thrilled by Lemke's semiotic portrayal of a science teaching/learning *system*, as it was an important missing ingredient in my own model, which was, at that time, more about "what's" and "how's," but not so much about how those dimensions were integrated. Two epiphanies immediately occurred to me: 1) Lemke's emphasis on *contextualizing* the languages of science was essential practice, and 2) we science teachers use too many words in isolation from the mathematical symbols, images, and actions associated with the rest of the science inquiry.

The Art and Design of the "Know"tation

I set out, as part of the Project Draw for Science research, to construct and field-test a graphic visualization tool for the teaching and formative assessment of science content learning. I visualized a one-page composition which integrated all of Lemke's four science languages: 1) science terminology (the WORDS); 2) explanatory drawings (the IMAGES); 3) the equipment and procedures used during an investigation (the ACTIONS); 4) and mathematical equations, icons, graphs, and tables (the SYMBOLS). I would call this pedagogical tool a "know"tation. Its purpose was to tell the *story* of situated, contextual science inquiry as it unfolded in real-time in the classroom. While there would be no one *right* way to design the "know"tation, the relationship between the four science languages should be readily apparent and the meaning capable of being "read" (Koester, 2015). Depending on the stage of inquiry during which it was created, the "know"tation might have more words than symbols, but all designs would feature a prominent explanatory and narrative image. Further, an image might serve multiple purposes, such as depicting an experimental procedure (an action) and/or explaining a phenomenon (such as why the sky is blue). Criteria for formative evaluation could be assessed using the developmental terms like "emerging," "proficient," and "mature" along criteria that would include scoring the following: misconception quotient (out of 5); graphic design; creativity; language integration; and clear meaning conveyed. I am still developing the evaluative component of the "know"tation and enthusiastically welcome all feedback and suggestions!

The readability of a "know"tation or any graphic design comes from the arrangement of its elements. My favorite book on the subject of visual note design is Mike Rhode's *The Sketchnote Handbook* (2013). I discovered Rhode's work when I enrolled in an online course called "Rockstar Scribe" (https://become-a-rockstar-scribe-at-school-or-work.teachery.co/register) as a way of improving my own graphic visualization skills (before trying to teach it myself). In the Project Draw for Science capacity building training, we began by considering four essential elements of page design: 1) typography, 2) images and their placement, 3) white space, and 4) hierarchy and flow between the elements. We practiced making different kinds of directional arrows and icons for bullets. (Figure 12)

We also reflected on the importance of creating a clear visual path through the composition. Rhode (2013) presented several readable "pathways" which I have represented here. (Figure 13)

Even teachers with very deep content knowledge readily recognize that unless they employ the elements of design in their graphic representations of content, the result can be rather chaotic. (Figure 14)

In very short order, teacher participants progressed to making "know"tations during each stage of 5E science inquiry (Bybee,1989). One teacher, "Ms. Maya," most of whose students were reading well below grade level, created *Explain* "know"tations with her students on large post-it note sheets, which they later used to review for their end-of-year standardized tests. (Figure 15)

Figure 12. Elements of graphic design, which enhance visual "flow"

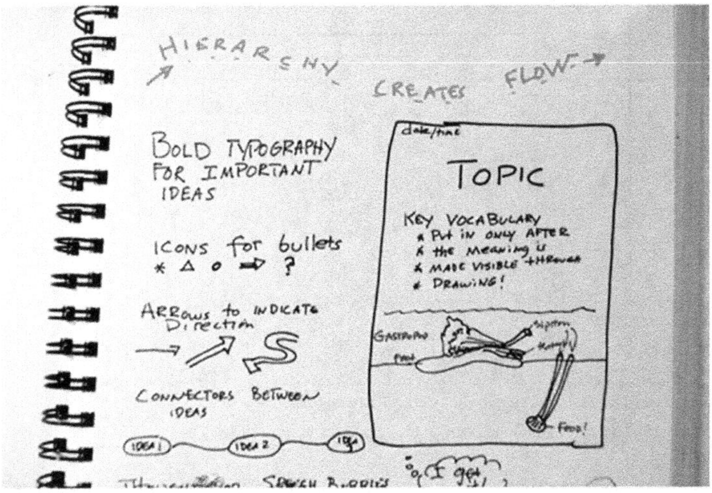

Unlike *Explain* "know"tations, which are mostly words and images, with perhaps some symbols, *Explore* "know"tations feature words, images, symbols, and actions, thoughtfully integrated through the applied use of graphic design elements to tell the story of an actual experiment in progress. (Figure 16)

Figure 13. Visually "readable" pathways through a graphic representation

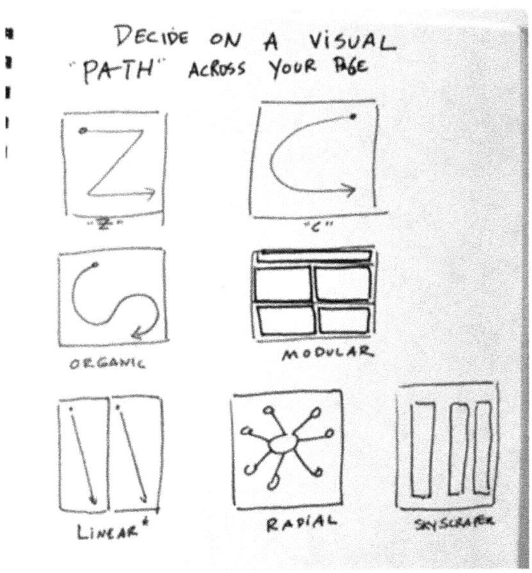

Figure 14. Pre-training attempt by teacher to create a visual explanatory model of climate change

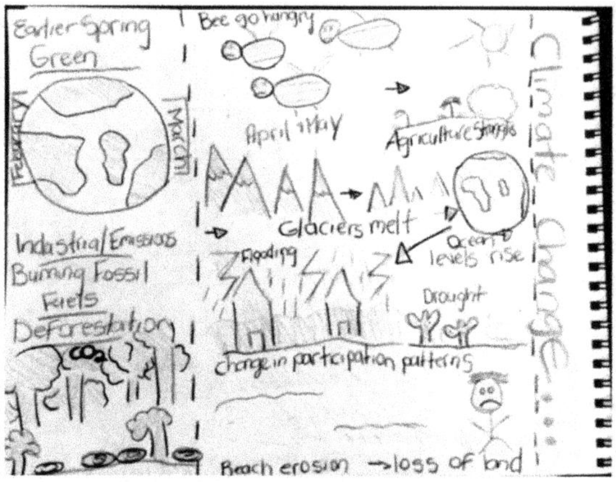

In every way, we came to regard the design of "know"tations as acts of collaborative critical making— as public artifacts to be shared and revised as needed and through whose *construction* students learned to integrate the languages of science.

Figure 15. Large poster-sized explain "know"tation with a large prominent image and only key words

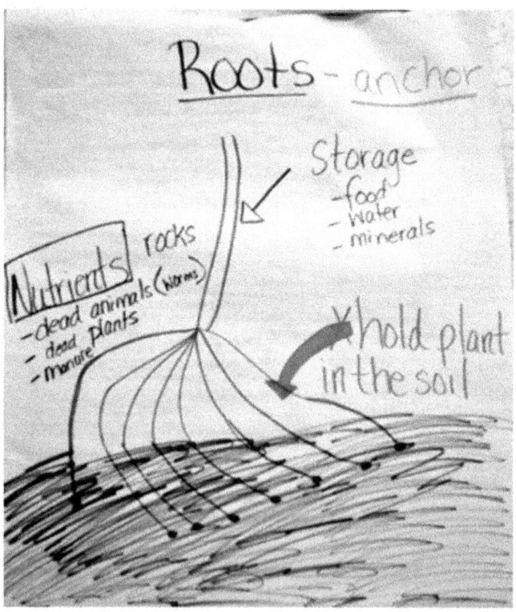

Figure 16. "Know"tation created during the Explore phase of inquiry. Note that all four languages of science are featured here.

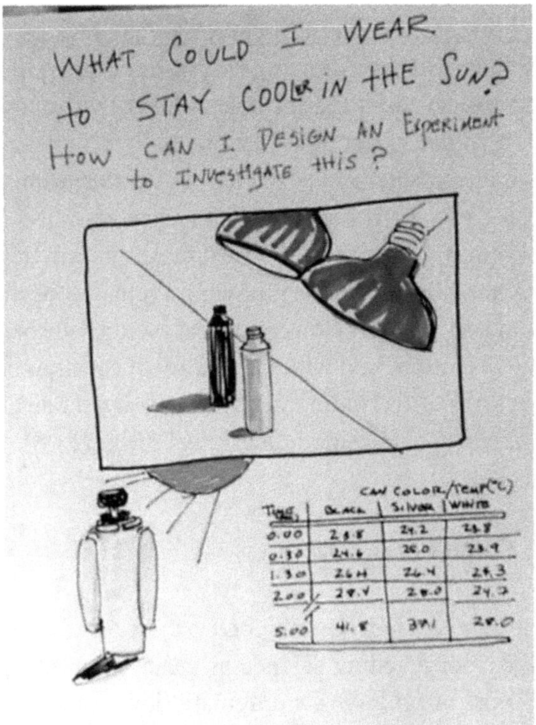

The Problem of Teacher Content Knowledge

The process of teaching science through drawing and visual thinking has its drawbacks. Hashweh (1987) has shown that poor content knowledge significantly impacts teaching and is a source of misconceptions directly communicated to students (in Cochran,1991, p. 12). I have discovered that one of the best ways for teachers to identify weaknesses in their content and procedural knowledge is what I call the "back-of-the-napkin-test." This can be a liminal experience for many teachers. What they quickly discover is that even if they say they have zero drawing *talent*, they are more than capable of making a simple explanatory drawing of a subject they know well and about which they care. However, when tasked to draw about a science subject which they understand poorly, the napkins quite often remain blank. Only a few moments later, they more or less confess that they do a terrible job of teaching about their weak areas. Right away, they also realize that if they are to actually *teach* science through drawing, they must draw their way into a deeper relationship with their most dreaded content area. As stated earlier, many middle school science teachers have quite under-developed content knowledge about the physical sciences. (Figure 17)

Part of the work I do through long-term professional development is advocating for the creation of standards-based drawing repertoires. Along the way, I work very hard to uncover and reflect upon the misconceptions that many science teachers commonly believe about the content they are teaching (Koester, in press). Critique of student drawings is also essential. The evidence of both learning and *mis*-learning is right there but can easily be missed by the teacher who becomes distracted by the aesthetic appeal of a student's work. For example, an artistically gifted student produced a visually appealing explanatory drawing about photosynthesis, incorrectly depicting oxygen being given off by flowers and carbon dioxide being taken up by the roots. Her teacher was so impressed with the quality of the drawing that he failed to notice the misconceptions. This finding was not an isolated one at all. Accordingly, we are developing a theory of "aesthetic masking." (Figure 18)

CONCLUSION: C'ING STEAM

In the conclusion of *Science Teachers Who Draw: The Red Is Always There*, I set forth a tentative model for teaching science as caring, non-prescriptive, aesthetic inquiry, one whose goal is achieving a pragmatic level of science literacy for all students regardless of their reading ability. I believe that an aesthetic paradigm in general and drawing instruction in particular should not only be included in science

Figure 17. The Back-of-the-Napkin test. Teacher draws a "blank" for physical science concept

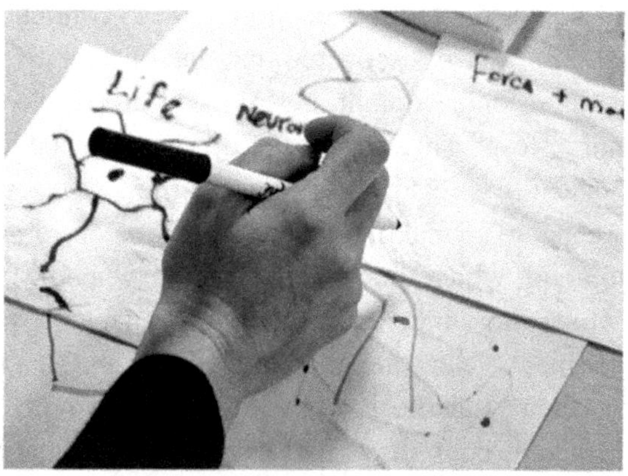

Figure 18.

Figure 18. Aesthetic masking of student misconceptions about photosynthesis

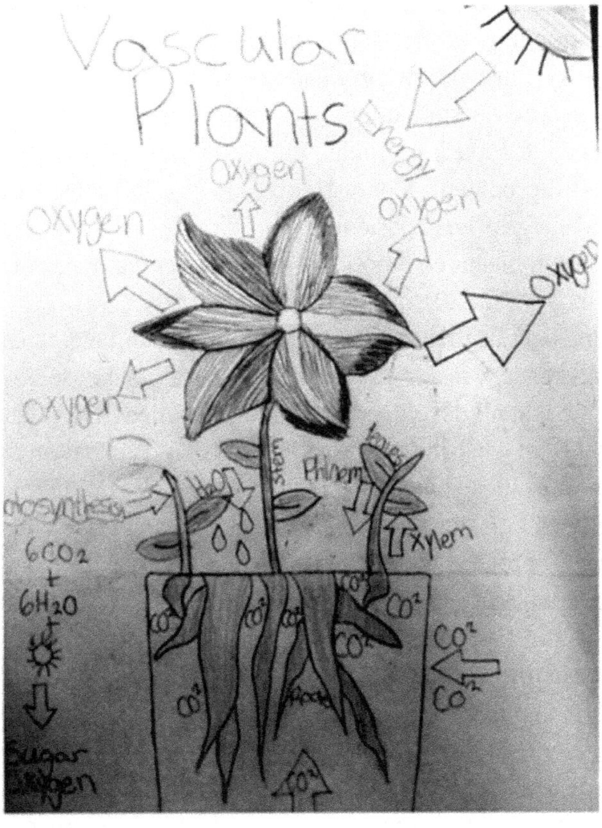

teacher education programs but become the subject of further participatory action research and professional development efforts. Shulman (1987) concluded that that the more we learn about teaching, the more "we will come to recognize new categories of performance and understanding that are characteristics of good teachers, and will have to reconsider and redefine other domains" (p. 13). To this end, I created a Facebook professional community page called "Science Teachers Who Draw" so that we can keep this conversation going and learn from each other. I believe that drawing to teach and to learn can improve heuristic problem-solving skills and result in deep learning of science, providing a platform to reveal misconceptions *and* epiphanies. The time required to visually work out a problem creates a context for dialogue, reflection, and the refinement of ideas that may not be possible in a heavily algorithmic, screen-delivered curriculum.

Lederman's (1992, 2007) reviews of the research on the natural science teaching ultimately concluded that the "instructional approach, style, rapport, and personality of the teacher are all important variables in effective science teaching" (2007, p. 845). An aesthetic orientation to teaching implies that teachers can't just script and *direct* student learning, they must jointly *produce* learning *performances*, at the end of which they must also assess student emotions and feelings about what just happened (or didn't). Indeed, feedback is crucial, but it can't just be about whether students have memorized the content. Conquergood (1998), has stated that the performance paradigm privileges "an experiential, participatory epistemology" (p. 27), and wrote that only through shared emotional experiences can a performance achieve transformation (in Koester, 2015, p. 186).

I have come a very long way from that twenty-something young woman, whom a class of very bright non-readers inspired to change her way of being a science teacher. They were with me when I developed a pedagogy I called TSTA. They are with me now as I tentatively call what I am doing STEAM. I could not have made this journey without the support of those who believed that what I was doing mattered. All along the way, I have been building tribes. I seek to develop practice-embedded educational research (PEER) projects that have been called for by Catherine Snow in her 2014 Wallace Foundation Distinguished Lecture for AERA, published in 2015. This PEER work would be characterized by a constellation of C's: *Caring Collaboration* among *Content* specialists, *Cultural relevance*, *Cross-training, Curating*, *Creative process*, *Community*, and the achievement of a *Collective* impact. Participants in this project would collectively build a professional STEAM teaching/learning community characterized by the sharing of lessons with which they have found success, as well as obstacles they may have encountered in the process. I am optimistic about the possibilities!

A democratic and inclusive science education must work to decrease inequalities and achievement gaps. To date, non-dominant culture and low SES students are often tracked into the lowest level classes with the least qualified teachers and with consistently low expectations made of them (Oakes, 2005, in Koester, 2015). The result can be complete semiotic breakdown and loss of human capital. Who knows how many of these students might otherwise have actually *enjoyed* the study of science had they also been assigned to experienced educators who established *ways in* to meaning-making that did not depend on the written or verbal expression of a "foreign" language. Indeed, I imagine such abandoned students shaking their heads in disgust, declaring, "Forget science! How can I be in your community if I can't speak your language or understand your sign systems?" Who can blame them? The Lakota had much to teach us all: "*Mitakuye oyasin.* We are all related."

REFERENCES

Anderson, C. (2004). Learning in "as-if" worlds: Cognition in drama in education. *Theory into Practice*, *43*(4), 281–286. doi:10.120715430421tip4304_6

Bigelow, B. (2009). Once upon a genocide: Columbus in children's literature. In W. Au (Ed.), *Rethinking multicultural education* (pp. 73–85). Milwaukee, WI: Rethinking Schools.

Bresler, L. (2006). Toward connectness: Aesthetically based research. *Studies in Art Education*, *48*(1), 52–69. doi:10.1080/00393541.2006.11650499

Brown, J. S., Collins, A., & Duguid, P. (1989). Situated cognition and the culture of learning. *Educational Researcher*, *18*(1), 32–42. doi:10.3102/0013189X018001032

Bybee, R. W. (1989). *Science and technology education for the elementary years: Frameworks for curriculum and instruction*. Washington, DC: The National Center for Improving Instruction.

Carnine, L., & Carnine, D. (2004). The interaction of reading skills and science content knowledge when teaching struggling secondary students. *Reading & Writing Quarterly*, *20*(2), 203–218. doi:10.1080/10573560490264134

Columbus, C. *(1492). Journal of the first voyage (Diario del primer viaje) 1492* (B. W. Ife, Trans. & Ed.). Warminster, UK: Aris and Phillips Ltd.

Conquergood, D. (1998). Beyond the text: Toward a performative cultural politics. In S. Dailey (Ed.), *The future of performance studies: Visions and revisions* (pp. 25–36). Annandale, VA: National Communication Association.

Dewey, J. (1938). *Experience and education*. New York, NY: Kappa Delta Pi.

Ehrenfeld, T. (2011). Reflections on mirror neurons. *Observer, 24*(3). Retrieved from http://www.psychologicalscience.org/index.php/publications/observer/2011/march-11/reflections-on-mirror-neurons.html

Eisner, E. (1991). *The enlightened eye: Qualitative inquiry and the enhancement of educational practice*. New York: Macmillan.

Eisner, E. (2002). *The arts and the creation of mind*. New Haven, CT: Yale University Press.

Finson, K., & Pederson, J. (2011). What are visual data and what utility do they have for science education? *Journal of Visual Literacy*, *30*(1), 66–85. doi:10.1080/23796529.2011.11674685

Gallese, V. (2005). Embodied simulation: From neurons to phenomenal experience. *Phenomenology and the Cognitive Sciences*, *4*(23), 23–48. doi:10.100711097-005-4737-z

Greene, M. (2001). *Variations on a blue guitar: The Lincoln Center Institute lectures on aesthetic education.* New York: Teachers College Press, Columbia University.

Hashweh, M. Z. (1987). Effects of subject-matter knowledge in the teaching of biology and physics. *Teaching and Teacher Education*, *3*(2), 109–120. doi:10.1016/0742-051X(87)90012-6

Hetland, L., Winner, E., Veenema, S., & Sheridan, K. (2013). *Studio thinking 2: The real benefits of visual arts education.* Academic Press.

Koester, M. (1989). *Science and art together again* (Unpublished master's thesis). University of Hawaii Manoa.

Koester, M. (2011). What's your story? How can we bring students to science through reading and the arts? Keynote address delivered at the South Carolina Science Council meeting, Myrtle Beach, SC.

Koester, M. (2014). *Project Draw for Science.* Retrieved from http://merriekoester.wix.com/project-dfs

Koester, M. (2015a). *Science teachers who draw: The red is always there.* Blue Mounds, WI: Deep University Press.

Koester, M. (2015b). *Agnes pflumm and the stonecreek science fair* (3rd ed.). Charleston, SC: Read for Science Publishing.

Koester, M. (2015c). *No place like periwinkle* (3rd ed.). Charleston, SC: Read for Science Publishing.

Koester, M. (2015d). *Pond scum and agnes pflumm* (3rd ed.). Charleston, SC: Read for Science Publishing.

Koester, M. (2015e). *Agnes pflumm and the secret of the seven* (2nd ed.). Charleston, SC: Read for Science Publishing.

Koester, M. (2017). Deep learning in science: An argument for an aesthetic paradigm of science education. In F. V. Tochon & D. Busciglio (Eds.), Deep education in the disciplines. Blue Mounds, WI: Deep University Press.

Lederman, N. (2007). Nature of science: Past, present, and future. In S. Abell & N. Lederman (Eds.), *Handbook of research on science education*. Abingdon, UK: Taylor and Francis.

Lederman, N. G. (1992). Students' and teachers' conceptions of the nature of science: A review of the research. *Journal of Research in Science Teaching*, *29*(4), 331–359. doi:10.1002/tea.3660290404

Lee, O., Fradd, S. H., & Sutman, F. X. (1995). Science knowledge and cognitive strategy use among culturally and linguistically diverse students. *Journal of Research in Science Teaching*, *32*(8), 797–816. doi:10.1002/tea.3660320804

Lemke, J. (2002). Enseñar todos los lenguajes de la ciencia: palabras, símbolos, imágenes y acciones. In M. Benlloch (Ed.), *La educación en ciencias: ideas para mejorar su práctica* (pp. 159–186). Barcelona, Spain: Paidos.

Lemke, J. L. (1999). Typological and Topological Meaning in Diagnostic Discourse. *Discourse Processes*, *27*(2), 173–185. doi:10.1080/01638539909545057

Litts, B. (2015). *Making learning: Makerspaces as learning environments* (Doctoral dissertation). Retrieved from www.informalscience.org/sites/.../Litts_2015_Dissertation_Published.pdf

Loucks-Horsley, S. (1996). Principles of effective professional development for mathematics and science education: A synthesis of standards. *NISE Brief, 1*(1). Retrieved from http://eric.ed.gov/?id=ED409201

May, R. (1985). *The courage to create*. Toronto, Canada: Bantam.

NGSS Lead States. (2013). *Next Generation Science Standards: For States, By States*. Washington, DC: The National Academies Press.

NOAA Office of Ocean Exploration. (2016). *Ocean Explorer*. Retrieved from http://oceanexplorer.noaa.gov/

Oakes, J. (2005). *Keeping track: How schools structure inequality*. New Haven, CT: Yale University.

Ocean Literacy Framework. (2005-20015). Retrieved from http://oceanliteracy.wp2.coexploration.org/

Papert, S. (1991). Situating constructionism. In I. Harel & S. Papert (Eds.), *Constructionism* (pp. 1–11). Norwood, NJ: Ablex.

Pestalozzi, J. (1973). *How Gertrude teaches her children*. New York, NY: Gordon Press. (Original work published 1894)

Polanyi, M. (1966). *The tacit dimension*. New York, NY: Doubleday.

Rhode, M. (2013). *The Sketchnote handbook: The illustrated guide to visual notetaking*. Berkeley, CA: Peachpit Press.

Settlage, J., & Southerland, S. A. (2007). *Teaching science to every child: Using culture as a starting point*. Abingdon, OX: Taylor & Francis.

Shaywitz, S. (2003). *Overcoming dyslexia: A new and complete science-based program for reading problems at any level*. New York, NY: Vintage Books.

Shulman, L. S. (1987). Knowledge and teaching: Foundations of the new reform. *Harvard Educational Review*, *57*(1), 1–23. doi:10.17763/haer.57.1.j463w79r56455411

Siegesmund, R. (2010). Aesthetics as a curriculum of care and responsible choice. In T. Constantion & B. White (Eds.), *Essays on aesthetic education for the 21st century* (pp. 81–92). Rotterdam: Sense Publishers.

Snow, C. (2015). Rigor and realism: Doing educational science in the real world. 2014 Wallace Foundation distinguished lecture. *Educational Researcher*, *44*(9), 460–466. doi:10.3102/0013189X15619166

Tochon, F. V. (2013). *Educational semiotics: Signs and symbols in education*. Blue Mounds, WI: Deep University Press.

Trefil, J. (2008). *Why Science?* New York: Columbia University, Teacher's College Press.

Uhrmacher, P. B. (2010). The power to transform: Implementation as aesthetic awakening. In T. Constantion & B. White (Eds.), *Essays on aesthetic education for the 21st century* (pp. 183–203). Rotterdam: Sense Publishers.

Section 3
Theme-Based

Chapter 7
Tower Design, Build and Test as a STEAM Project:
Tower Design, Build, and Test

Judith Bazler
Monmouth University, USA

EXECUTIVE SUMMARY

The next generation science standards promote the teaching of engineering skills including the designing, testing, and building of models. Tower building can yield real world experience that not only provides the student with physics and mathematics through motion and stability but also through the explanation of the use of models and the engineering practice of design, redesign, and testing of these models. Tia Pliskow used the project of building a tower with her middle school students in order to provide a cooperative team long-term project. She focused first on the design, using background information on existing towers. She required each team to design their tower first using graph paper and scale. This process stressed the need for science, technology, engineering, art, and mathematics. The case included in this article expands her process by including a cost analysis attempting to promote real world engineering, links to more content, and final project photos. In addition, by building a shake platform, a test for the tower is added.

INTRODUCTION

Designing and building a tower project requires that students utilize and explore each component of STEAM (science, technology, engineering, art, and mathematics). The physics of force and motion combine with the effects of earth movement, wind,

DOI: 10.4018/978-1-5225-9631-8.ch007

and other geological factors. Also, the mathematical calculations affect the design and structural integrity of the tower. The history of towers including the newest and largest structure may motivate students to embark on a fascinating journey that perhaps winds back to the pyramids and ends at the Burj Khalifa. This journey also provides students with historical and modern engineering role models with information concerning their interests and their passion for this art form. Also suggested is research on disasters such as the collapse of the bridge in Minneapolis, Minnesota in 1961 with information on whether the collapses are caused by design flaws, material problems, and/or lack of maintenance.

Researching past and present structures accompanied by foundational scientific and mathematical research before students design and carry out the activity is the basis for the STEAM approach to learning. All disciplines combine to develop the problem and to form a new solution. The building, testing, and redesigning not only strengthens the process but emphasizes the importance of the research process.

This approach varies from past practice because it embraces the research process and the STEAM philosophy. This chapter provides the practitioner with a sampling of the history of a number of towers with suggestions for further research. The history intertwined with science, mathematics, and engineering provides the foundation needed to proceed to design and redesign. The activity itself leads students into the building and testing which naturally leads them back to the design. The students will:

- Practice structural design, model development, and engineering to be assessed by their designs, models, model tests, and redesigns.
- Focus on motion and stability, assessed by observation questions and applied design.
- Use mathematics and computational thinking, assessed by observation questions and applied design.
- Explain the use of models in writing, assessed by observation questions.
- Research towers globally, assessed by observation questions.
- Interact in teams, assessed by teacher informally and by observation questions.
- Research structural disasters.

STANDARDS

Next Generation Science Standards

Dimension I: Scientific and Engineering Practices (1,2,3,4,5,6,7,8) including asking questions and defining problems, developing and using models, planning and carrying out investigations, analyzing and interpreting data, using mathematics and computational thinking, constructing explanations and designing solutions, engaging in argument from evidence, and obtaining, evaluating, and communicating information.

Dimension II: Crosscutting Concepts (2, 3, 4, 6, 7) including cause and effect: mechanism and explanation, scale, proportion, and quantity, systems and system models, energy and matter: flows, cycles and conservation, structure and function, and stability and change.

Dimension III: Disciplinary Core Ideas/Physical Science (PS1, PS2, PS3) Matter and its interactions, Motion and stability: forces and interactions, Energy /Earth and Space Science (ESS2, ESS3) the Earth's systems, Earth and human activity/ Engineering, Technology and Application (ETS1, ETS2) Engineering design, Links among engineering, technology, science, and society.

Standards for Mathematical Practice

MP1: Make sense of problems and persevere in solving them.

MP2: Reason abstractly and quantitatively.

MP5: Use appropriate tools strategically.

National Social Studies Thematic Standards.

8: Science, Technology, and Society.

9: Global Connections.

CONTENT INFORMATION

The Physics of Building a Tower

The foundation of a tower has a number of forces acting upon it. There are the compressive/download forces caused by the weight of the tower and there are the forces of the wind causing an uplift effect. Three major items also need to be addressed including the load, the wind speed and seismic forces.

From a mathematical standpoint, the equation for the tower is:

$$f(x) = xw(x) - \sqrt{x^2 w^2(x) + f_0^2}$$

where $f(x)$ is the half-width of the tower at height x, f_0 is the half-width of the tower at the ground and $w(x)$ is the maximum wind pressure the tower can withstand at a height x.

Figure 1. Graphic representation of mathematics involved in towers

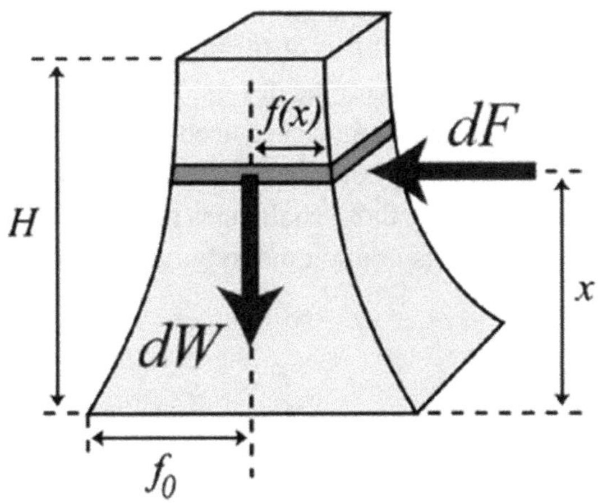

This is a root of the quadratic equation describing a parabola resting against the right side of the tower, so the curve of the tower's right side is described by the negative arm of the parabola. For example, the fastest winds recorded at the tower reached a speed of 214 km/h in 1999 and would have produced pressures of just 2.28 kN/m^2. The diagram below provides a graphic representation of the mathematics involved in towers. The force of the wind (dF) produces a torque around the bottom left corner of the tower which is countered by the force of the tower's weight (dW).

CASE DESCRIPTION

Lesson Plan Concerns

This is a STEAM project; therefore, this is not a one-day activity. The time for part A and part B is necessary for the student to reflect on how much time is given to the planning and design components of building a tower. This project may take months to complete. The materials are easy to find and not costly. The most difficult portion of this project is refraining from jumping to the building and spending an adequate amount of time on the design. A great show on LIFEtime television is Project Runway and Project Runway Kids. This show creates a challenge, perhaps to create an outfit for women/girls out of material from a hardware store. The participants first sketch an example of what they might create. Then most of the participants get their materials but first create a pattern/model. Then and only then do they construct the garment. The viewer senses that both the pattern and the actual finished garment is a model for future distribution because in both phases the designer faces many challenges that they must resolve. In the tower project, building the design may present additional challenges that the designer did not focus on during the design process but that need to be resolved in order to proceed with the design. The research prior to the design should provide the incentive, but the teacher needs to elaborate on the structural science and mathematics including, but not limited to, force and motion. Even though there is not a section for redesign, those challenges that were not initially identified will arise, and redesign becomes implicit in the building process.

PROJECT PLAN

Tower Design and Building Challenge

How do engineers design and build towers? Engineers need to develop many solutions by asking questions and defining constraints. Designs are then reviewed for cost, improvements, time, and function. Models are then developed. The engineering design process includes defining the problem, gathering pertinent information, generating multiple solutions, analyzing and selecting a solution, and testing and implementing the solution.

You will first research other existing towers, design a tower, and then develop a test to determine its stability and strength. You will then build the designed tower out of straws, paper clips, and tape. You will be assessed on the height of the tower, its stability and its cost.

Safety

Be careful using scissors and the sharpened pencil; both can cause damage if used improperly. Do not drink or open the soda.

- *Materials for Tower*50 straight plastic drinking straws
- Scissors
- Two feet of masking tape
- 30 paper clips
- One sheet of graph paper
- Two sheets of lined paper
- One sharpened pencil
- Metric ruler
- Can of soda

Figure 2. Eiffel Tower

- Cost of Materials Data Sheet.
- Integration of Science and Engineering Practices Form

Materials for Shake Platform

- Stiff board, small piece of plywood
- Hot glue or masking tape to secure the shake platform

Procedure

Design Part A

1. **Keep track of time spent on Part A and B!**

2. **Read** all materials and explore on the internet the tallest buildings in the world and how they were designed and constructed. Look up the Burj Khalifa, the Eiffel tower, and La Torre di Pisa.
3. **Graph** the design that you will try to make in class, making sure that your graph uses metric measurements.
4. **Determine** the height of the tower that you will build.
5. **Design** a test for the tower.

Building Tower Part B

1. **Gather** all the materials that you will use.
2. **Build** the tower using your design graph.
3. **Measure** the height of your tower, record the height on the data sheet.
4. **Test** the tower using your original test design.
5. **Determine** the cost of building your tower using the Cost of Materials Data Sheet.
6. **Answer** the Questions on Observations list below on a separate piece of paper.
7. **Clean** up area.

Building Shake Platform Part C

1. **Gather** the platform and put it on a table.
2. **Test** the platform so that it can be controlled by hand.
3. **Download** the free app “Smart Toolkit” onto a smart phone to test the vibration using the Scale.
4. **Place** the smart phone onto the shake table and observe the vibration on the Scale.

Testing Tower Part D

1. **Secure** your tower to the shake platform.
2. **Place** your smart phone on the shake platform with the Mercalli Scale app open.
3. **Shake** your shake platform to test your tower at Mercalli V. Your tower should survive Mercalli V.
4. **Shake** your shake platform to test your tower at Mercalli VIII. Your tower should not survive Mercalli VIII.

Observations

1. What was the height of your tower?
2. What was the cost of building your tower? How did design compare to cost?
3. How much time was spent on both the design and the building of the tower?
4. If you had a chance to do this project again, what would your team have done differently?
5. Do you think that this activity was more rewarding to do as a team, or would you have preferred to work alone on it? Why?
6. If you could have used one additional material (ex: glue, wood sticks, etc.) which would you choose and why?
7. If you used the Smart Toolkit, what was the reading before your tower collapsed? What was it after the tower collapsed?
8. Do you think that once a building is designed and approved for construction that many aspects are changed during the building process? Why or why not? How does the design consider any imposed state or federal regulations, as well as public safety, health, and/or intellectual property? Describe any constraints the design may pose relative to health or safety hazards (noise, radiation, allergies, etc.)
9. How long do you think it will take before a building is constructed that surpasses the height of the Burj Khalifa? Where do you think it will be built? Why?
10. What environmental constraints exist for engineers to take into account when designing a building?
11. Pick one of the towers you researched in #1 Design Part A and write a short one-page report on the tower. Include in your report the location, the height and dimensions, any constraints in building the tower, the cost of building the tower and important information about the tower.
12. Research structural disasters caused by human error or neglect. What was the cause of the disaster? What was the name of the structure? What was the date of the disaster? Where did this happen?
13. Research safety and precautions must engineers consider when designing a tower? How often are structures checked for safety and by whom? How do you guarantee a structure is safe? What should engineers be looking for when trying to keep a structure safe?

Cost of Tower Materials Data Sheet

See Table 1.

Table 1. Cost of materials data sheet

Materials/Labor	Cost/Unit/Item	How Many Used	Total Cost
50 straws	$1.50/box 100		
Scissors	$1.89 each		
2 feet masking tape	$.06/meter (1m=3.28ft)		
30 paper clips	$0.01/each		
One sheet of graph paper	$1.34/ream of 500		
Two sheets of lined paper	$1.34/ream of 500		
One sharpened pencil	$2.41/pack of 12		
Metric ruler	$19.34/24 rulers		
One 12 ounce soda	$3.56/6 cans		
Civil engineer each one	$40.00/hour/engineer		
Total			

CURRENT CHALLENGES FACING TEACHERS

Integrating the design with a hands-on component, including the building of a tower with design, enables students to focus first on the design, art, and science of a tower, then on the practicality of their design. Without the design, students rarely look at the serious problems with force on any structure and on the aesthetics of their product. Included in the activity is an economic component again examining the economic feasibility of the design. Student responses to this case included both the positive motivation factor and also the understanding of the science, mathematics, engineering, and art design involved in the process of creating towers. The time factor for any project is greater than in the traditional forty-five minute or double class activity. Usually any STEAM project will be spread over a longer time schedule.

SOLUTIONS AND RECOMMENDATIONS

This project can be completed as a competition as evidenced on a number of spaghetti tower challenges on YouTube. One could substitute spaghetti if desired, but the project building is more efficient using straws due to the spaghetti's fragility. Some of the challenges use spaghetti and marshmallows for structural integrity instead of tape or glue. The marshmallows and glue seem to be messier. One could choose to change the goal of the project from being a tower that is tall and structurally sound to just tall. Examples of tall towers that were built for this project is in the Appendix.

Finally, this project lends itself to group teaming. There seems no need to assign tasks to each member due to the motivating factor implicit in the process itself. Over time, the motivation of achieving success with the structural strength and sturdiness seems to be incentive enough. Perhaps that alone is the challenge.

REFERENCES

Ashley, S. (2001). When the twin towers fell. *Scientific American*. Retrieved from https://www.scientificamerican.com/article/when-the-twin-towers-fell/

Burj Khalifa. (2011). *Civil Engg. Seminar*. Retrieved from http://civilenggseminar.blogspot.com/2011/09/burj-khalifa.html

Case study: The Leaning tower of Pisa. (n.d.). *Madrid Engineering*. Retrieved from http://madridengineering.com/case-study-the-leaning-tower-of-pisa/

Craven, J. (n.d.). The World Trade Center twin towers, 1973-2001. *About.com*. Retrieved from http://architecture.about.com/od/worldtradecenter/ss/worldtrade.htm#step1

Eiffel Tower. (2006). *Engineering.com*. Retrieved from http://www.engineering.com/Library/ArticlesPage/tabid/85/ArticleID/62/Eiffel-tower.aspx

Gallant, J. (2002). The shape of the Eiffel tower. *American Journal of Physics*, *70*(2), 160–162. doi:10.1119/1.1417530

Leaning tower of Pisa. (n.d.). In *Wikipedia*. Retrieved from https://en.wikipedia.org/wiki/Leaning_tower_of_Pisa

Pliskow, T. (2008). *Reaching new heights in middle school science: Straw tower competition builds*. Academic Press.

Pratap, T. (2015). What were the engineering challenges in building the Burj Kahalifa? [Msg 3]. Message posted to https://www.quora.com/What-were-the-engineering-challenges-in-building-the-Burj-Khalifa

Willis Tower. (n.d.). Retrieved from http://www.willistower.com

KEY TERMS AND DEFINITIONS

Engineering Design: A mindset that expects students to learn about a problem because of failures.

Force: A quantitative description of the interaction between two physical bodies. Force is proportional to acceleration.

Load: The forces to which a structure is subjected due to superposed weight or to wind pressure on the vertical surfaces; *broadly*: the forces to which a given object is subjected.

Mercalli Scale: Is a seismic intensity scale used for measuring the qualitative intensity of an earthquake. It measures the effects of an earthquake. It was invented by Giuseppe Mercalli in 1902.

Torque: A force that has a twist and can cause a rotation around an axis.

Tower: A building or structure typically higher than its diameter and high relative to its surroundings that may stand apart or be attached to a larger structure.

APPENDIX

Examples of Towers

The Eiffel Tower

Eiffel's tower was completed in a matter of months with a small labor force. His results started a revolution in civil engineering and architectural design. With the completion of the tower, Eiffel earned the nickname "magician of iron." Over time the tower faced demolition but then became an important radio and television transmission tower.

The tower's dimensions are a height of 1069 feet (320.75 meters), which is about the equivalent of a 105-story building. The public has access to heights of 189 feet (57.63 meters), or 19 stories; 380 feet (115.73 meters), or 38 stories; and 896 feet (273 meters), or 89 stories. For an image of the Eiffel Tower go to https://www.toureiffel.paris/en.

The base of the tower covers a square area of 328 feet (100 meters) on a side. Visitors can stand in the center of the area at the base and look directly up at the floor of the second level, 38 stories above.

The tower is built of iron and weighs 7300 metric tons. It is extremely light. The four pillars supporting the tower are aligned to the points of the compass. Visitors can use the tower as a reference to find their way in Paris. Another unique feature is the tower's base. The four semi-circular arches required elevators to ascend on a curve. The glass-cage machines were designed by the Otis Elevator Company of the United States. These elevators provide visitors a fantastic view as they ride to the top of the tower. This tower moves in the wind. On days with high, gusting winds, the wind can reach speeds in excess of 100 mph at the summit of the tower. Visitors can feel the tower swaying gently at the summit. The magnitude of the sway in the tower, under worst-case conditions, is about six inches (15 cm). There is no danger of the tower being damaged by wind-induced movement since it is designed to withstand movements easily five times beyond those produced by the highest winds ever recorded. Today, the movements are monitored by a laser alignment system. The tower also leans very slightly in bright sunlight, as one side is heated by the sun and expands slightly.

Other statistics include:

- 2.5 million rivets.
- 300 steel workers, and 2 years (1887-1889) to construct it.
- Sway of at most 12 cm in high winds.

- Height varies up to 15 cm depending on temperature.
- 15,000 iron pieces (excluding rivets).
- 40 tons of paint.
- 1652 steps to the top.

Eiffel, Alexandre Gustave 1832-1923, French engineer

A noted constructor of bridges and viaducts, he also designed the Eiffel Tower and the internal structure of the Statue of Liberty. He was initially charged with corruption in the 1888 scandal of Ferdinand de Lesseps's failed Panama Canal project. Lesseps (1805-1894), the successful promoter and construction supervisor for the Suez Canal (1859-69), later assumed presidency of the French company building the Panama Canal starting in 1881. Eiffel was hired to provide engineering services for the project. Seven year later the project was forced into bankruptcy amid charges of corruption. Lesseps was sentenced to prison for misappropriation of funds, but the sentence was not carried out. Eiffel was cleared of all wrongdoing by a French appeals court in 1893. Nonetheless, he withdrew from commercial life and spent the rest of his years studying aerodynamics" ("Eiffel tower," 2006)

The Burj Khalifa

Burj Khalifa is a skyscraper in Dubai, United Arab Emirates and is currently the tallest structure in the world at 828 m (2,717 ft). Construction began on 21 September 2004, with the exterior of the structure completed on 1 October 2009. The building officially opened on 4 January 2010.

For an image of the Burg Khalifa go to HTTPS://en.m.wikipedia.org/wiki/Burj_Khalifa.

The tower's architecture and engineering were performed by Skidmore, Owings and Merrill of Chicago, with Adrian Smith as chief architect and Bill Baker as chief structural engineer. The primary contractor was Samsung C&T of South Korea. The total cost for the project was about US $1.5 billion.

Burj Khalifa has more than 160 stories. It is the tallest building in the world. It has a spiraling "Y" shape. (see diagram below)

The top section of the tower consists of a structural steel spire utilizing a diagonally braced lateral system. The structural steel spire was designed for gravity, wind, seism, and fatigue in accordance with the requirements of AISC Load and Resistance Factor Design Specification for Structural Steel Buildings. The exterior exposed steel is protected with a flame-applied aluminum finish.

Figure 3. Typical floor plan

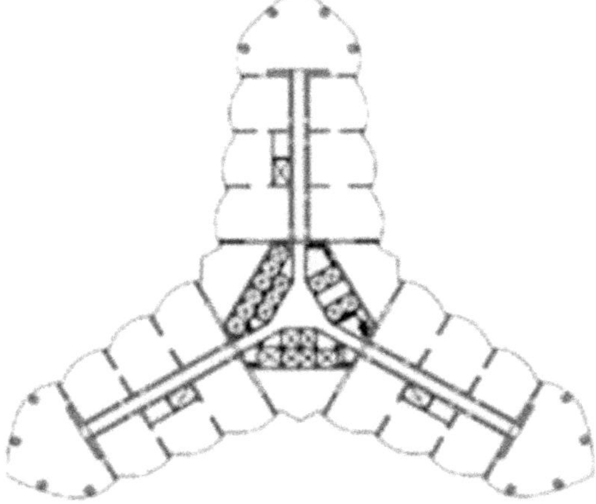

Willis (Sears) Tower in Chicago

"One of the tallest buildings in the world and the tallest building in America. In 1969, Sears Roebuck and Company was the largest retailer in the world with about 350,000 employees. Deciding it needed a central office space for its many employees, the company hired architects Skidmore, Owings, and Merrill to design what would become one of the largest office buildings in the world. After breaking ground in 1970, it took three years to complete and used enough concrete to make an eight-lane, five-mile-long highway. In 1988 Sears moved out, and it was renamed Willis tower in 2009." (https://www.flickr.com/photos/meezersqueezer/41379548562).

Other facts include:

- 1,450 feet high (443 meters);
- 1,730 feet high (520 meters) including twin antennae
- Eighth-tallest building in the world; tallest in Western Hemisphere
- Cost more than $175 million to build" (Willis tower).

For an image of the Willis tower go to https://www.willistower.com.

Pisa in Italy

For an image of the Leaning tower of Pisa go to http://www.towerofpisa.org.

"The Leaning tower of Pisa is a freestanding tower of the city of Pisa. It is known for its unintended tilt. The tower's tilt began during construction, caused by an inadequate foundation on ground too soft on one side to properly support the structure's weight. The tilt increased in the decades before the structure was completed, and gradually increased until the structure was stabilized (and the tilt partially corrected) by efforts in the late 20th and early 21st centuries. The height of the tower is 55.86 metres (183.27 feet) from the ground on the low side and 56.67 metres (185.93 feet) on the high side.

There has been controversy about the real identity of the architect of the Leaning tower of Pisa. For many years, the design was attributed to Guglielmo and Bonanno Pisano. However, recent studies, seem to indicate Diotisalvi as the original architect due to the time of construction and affinity with other Diotisalvi work.

Galileo Galilei is said to have dropped two cannonballs of different masses from the tower to demonstrate that their speed of descent was independent of their mass. During WW II, the Germans were using the tower as an observation post.

Numerous efforts have been made to restore the tower to a vertical orientation or at least keep it from falling over. Most of these efforts failed; some worsened the tilt. On February 27, 1964, the government of Italy requested aid in preventing the tower from toppling.

On January 7, 1990, the tower was closed to the public After a decade of corrective reconstruction and stabilization efforts, the tower was reopened to the public on December 15, 2001, and was declared stable for at least another 300 years. In May 2008, engineers announced that the tower had been stabilized." (Leaning tower of Pisa).

Twin Towers in NYC

Designed by American architect Minoru Yamasaki (1912-1986), the original World Trade Center consisted of two 110-story buildings (known as the "Twin towers") and five smaller buildings. The North tower (1 WTC) was finished in 1970 and the South tower (2 WTC) was completed in 1972. tower One was 1,368 feet (414 meters) tall, 110 stories high. tower Two was 1,362 feet (412 meters) tall, 110 stories high. Both towers weighed about 1,500,000 tons together. On August 7, 1974 Philippe Petit used a bow and arrow to assemble a steel cable between the two towers and then he walked across the tightrope. Other stunts included parachuting from the top and climbing from the bottom of the towers. (Craven, n.d.)

For an image of the Twin towers in NYC go to https://wpclipart.com/AmericanHistory/otherevents/Twintowersc2000.jpg.html.

After first describing the highly redundant structural system that kept the 110-story twin towers standing for decades despite hurricane-force winds and a terrorist truck bomb, the engineers then delineated how that system was breached and finally overcome on that fateful day when America was attacked. The main culprits in bringing the famously lofty buildings down, they concluded, were the two intensely hot infernos that erupted when tens of thousands of gallons of aviation fuel spilled from the doomed airliners. Once high temperatures weakened the towers' supporting steel structures, it was only a matter of time until the mass of the stories above initiated a rapid-sequence "pancaking" phenomena in which floor after floor was instantly crushed and then sent into near free fall to the ground below. Significantly, the panel stated that any mitigating reinforcements and redundancies added to these buildings could have only delayed the inevitable failure, though they would have bought more time for the evacuation of the occupants. No existing or foreseeable economically viable skyscraper structure, they agreed, could have withstood this kind of cruel onslaught. Clearly, prevention is the best defense against this kind of assault. (Ashley, 2001).

Chapter 8
The Great Race:
Using Air to Move Paper Airplanes and Balloon Rockets

Jena Valdiviezo
Long Branch School District, USA

Letitia Graybill
Monmouth University, USA

EXECUTIVE SUMMARY

This is a case in which students build paper airplanes from templates provided by the instructor as well as those which they can design themselves. They extend their ideas on flight in by using the principles developed in the paper airplane race to power simulated rockets made out of balloons. They consider such variables as materials, mass, and design to see which combination of material design and mass are most effective in constructing an airplane or a rocket that flies the fastest and the furthest in a competition. Contestants are rated on team consideration of variables needing to be controlled in order to have a fair assessment of the designs. When the designs are agreed upon and constructed, a race is conducted. The ideas developed in the paper airplane competition are then used to design a rocket carrying a paper airplane capable of flying across the classroom in the fastest time with the most direct route. This is a simulation of the space shuttle flights. The parameters of the races are developed by the participants.

DOI: 10.4018/978-1-5225-9631-8.ch008

LITERATURE REVIEW

Humans have yearned to fly like birds for eons. Many designs were created through the ages that had some disastrous results. However, because of the desire, inventive attitudes, and almost fanatic devotion to the creation of new knowledge, the pursuit of flight went on and on and on. It took until 1903 in Kitty Hawk, North Carolina (Shapell & Willen, 2012) for Wilbur and Orville Wright to produce the first sustained flight. Engineering took over after that and resulted in a revolution in travel by air. The history of the saga has been well documented in literature. We have learned through experiment and engineering principles to develop crafts that can fly supersonically and transport people and material around the earth in increasingly shorter times. We have revolutionized travel to the point that travel beyond the Earth is within possibility and may even be accessible for the average human. Those who were alive in 1969 to see Neil Armstrong walk on the moon were enthralled by the vision. What an exciting moment that was! This initiated a series of flights in which more and more powerful rockets were used to launch these spacecrafts beyond the Earth's gravitational field. More extensive interplanetary flight has already been accomplished by non-manned space flights. Two Voyager crafts were launched in September of 1977. These Voyagers spent eleven years exploring Jupiter, Saturn, Uranus and Neptune before escaping the gravitational pull of the sun in 2012 to begin interstellar explorations (NASA). The Voyager space crafts are the third and fourth human creations to fly beyond all the planets in our solar system. They were preceded by Pioneers 10 and 11 in escaping the gravitational pull of the sun. This was accomplished in 1998. Voyager 1 passed Pioneer 10 to become, as far as we know, the most distant human-made object in space. As of September 2013, Voyager 1 was at a distance of 18.7 billion kilometers from the Sun while Voyager 2 was at a distance of 15.3 billion kilometers. All of this has been achieved in a mere one hundred and eleven years. Imagine what is possible in the future. This review of the literature is sufficient for a basic understanding of flight mechanics. There are many other books that can give a deeper understanding. The Completer History by R. G. Grant is a well-received source that gives additional technical detail. This could serve as a reference for teachers.

What a wonderful history of flight we can discuss in our classrooms! Students can become excited by the prospect of future developments. How many of them might become space travelers of the future? How many of them might become part of interplanetary missions that are now being planned? If teachers can "touch the future," as the first teacher in space, Christa McAuliffe, said (brainyquote.com/quotes/christa_mcauliffe_134582), how many science teachers can actually affect the future of space flight through the students that they teach? It is an awesome responsibility that most teachers, especially science teachers, should gladly accept. It could be

Figure 1. Titan rocket with voyager spacecraft mounted on it

inspiring to have students develop their own ideas of designing for flight by using simple materials that are inexpensive and yet effective in considering questions of design, speed, and flight.

CONTENT INFORMATION

The lessons included in this case are interdisciplinary and can be used in various grade levels. Cross curriculum concepts in mathematics, science, social studies, and language arts are integrated with standards in technology and design. The Core Curriculum Mathematics Standards, the Next Generation Science Standards, and the latest standards in technology and social studies education are supported by the activities described within this paper. These standards include:

A. Next Generation Science Standards: http://www.state.nj.us/education/aps/cccs/science/
   1. Standard 1: 3-PS2-1
      Plan and conduct an investigation to provide evidence of the effects of balanced and unbalanced forces of the motion of an object.
   2. Standard 2: 3-PS2-2
      Make observations and/or measurements of an object's motion to provide evidence that a pattern can be used to predict future motion.
   3. Standard 3: 5-PS2-1
      Support an argument that the gravitational force exerted by Earth on objects is directed down.

4. Standard 4: MS PS 2-2
   Plan an investigation to provide evidenced that the change in an object's motion depends on the sum of the forces acting on the object and the mass of the object. Designers will take into account Newton's Laws (e.g. balanced and unbalanced forces and consequent changes in motion). They must also consider specification of units and changes in variables.
5. Standard 5: MS-PS2-4
   Construct and present arguments to support the claims that gravitational forces are attractive and depend on the masses of the interacting object.
6. Standard 6: 3-5-ETS1-1
   Define a simple design problem reflecting a need or a want that includes specified criteria for success and constraints on materials, time, or cost.
7. Standard 7: 3-5-ETS1-2
   Generate and compare multiple possible solutions to a problem based on how well each is likely to meet the criteria and constraints of the problem.
8. Standard 8: 3-5-ETS1-3
   Plan and carry out fair tests in which variables are controlled and failure points are considered to identify aspects of a model or prototype that can be improved.
9. Standard 9: MS-ETS1-1
   Define the criteria and constraints of a design problem with sufficient precision to ensure a successful solution, taking into account relevant scientific principles and potential impacts on people and the natural environment that may limit possible solutions.
10. Standard 10: MS-ETS1-2
    Evaluate competing design solutions using a systematic process to determine how well they meet the criteria and constraints of the problem.
11. Standard 11: MS-ETS1-3
    Analyze data from tests to determine similarities and differences among several design solutions to identify the best characteristics of each that can be combined into a new solution to better meet the criteria for success.
12. Standard 12: MS-ETS1-4

Develop a model to generate data for iterative testing and modification of a proposed object, tool, or process such that an optimal design can be achieved.

B. New Jersey Student Learning Standards Mathematics: http://www.state.nj.us/education/aps/cccs/math/
   - Organize, represent, and interpret data with up to three categories; ask and answer questions about the total number of data points, how many in each category, and how many more or less are in one category than in another.
   - This standard is found in all grades from Grade 1 through high school.

Specific Math Standards linked the Science and Engineering Standards

- MP.2: Reason abstractly and quantitatively. *(3-PS2-1)*, (MS-PS2-1), *(*MS-PS2-2*) (3-5-ETS1-1), (3-5-ETS1-2),(3-5-ETS1-3)*
- MP.4: Model with mathematics. *(3-5-ETS1-1),(3-5-ETS1-2),(3-5-ETS1-3)*
- MP.5: Use appropriate tools strategically. *(3-PS2-1), (3-5-ETS1-1),(3-5-ETS1-2),(3-5-ETS1-3)*
- 3.MD.A.2: Measure and estimate liquid volumes and masses of objects using standard units of grams (g), kilograms (kg), and liters (l). Add, subtract, multiply, or divide to solve one-step word problems involving masses or volumes that are given in the same units, e.g., by using drawings (such as a beaker with a measurement scale) to represent the problem. *(3-PS2-1)*
- 3-5.OA: Operations and Algebraic Thinking (3-ETS1-1),(3-ETS1-2)
- 6.NS.C.5: Understand that positive and negative numbers are used together to describe quantities having opposite directions or values; use positive and negative numbers to represent quantities in real-world contexts, explaining the meaning of 0 in each situation. (MS-PS2-1) (MS-PS2-2)
- 6.EE.A.2: Write, read, and evaluate expressions in which letters stand for numbers. *(MS-PS2-1)* (MS-PS2-2)
- 7.EE.B.3: Solve multi-step real-life and mathematical problems posed with positive and negative rational numbers in any form, using tools strategically. Apply properties of operations to calculate with numbers in any form, convert between forms as appropriate, and assess the reasonableness of answers using mental computation and estimation strategies. *(MS-PS2-1)* (MS-PS2-2)
- 7.EE.B.4 Use variables to represent quantities in a real-world or mathematical problem, and construct simple equations and inequalities to solve problems by reasoning about the quantities. *(MS-PS2-1)* (MS-PS2-2)

C. New Jersey English Language Arts Standards: http://www.state.nj.us/education/aps/cccs/lal/

Reading, writing, and speaking grounded in evidence from texts, both literary and informational

- RI.3.1: Ask and answer questions to demonstrate understanding of a text, referring explicitly to the text as the basis for the answers. *(3-PS2-1)*
- RI.5.1: Quote accurately from a text when explaining what the text says explicitly and when drawing inferences from the text. (5-PS2-1), *(3-5-ETS1-2)*
- RI.5.9: Integrate information from several texts on the same topic in order to write or speak about the subject knowledgeably. (5-PS2-1), *(3-5-ETS1-2)*
- W.3.7: Conduct short research projects that build knowledge about a topic. (3-PS2-1), 3-PS2-2)
- W.5.7: Conduct short research projects that use several sources to build knowledge through investigation of different aspects of a topic. (3-5-ETS1-1), (3-5-ETS1-3)
- W.3.8: Recall information from experiences or gather information from print and digital sources; take brief notes on sources and sort evidence into provided categories. (3-PS2-1), 3-PS2-2)
- W.5.8: Recall relevant information from experiences or gather relevant information from print and digital sources. Summarize or paraphrase information in notes and finished work, and provide a list of sources. *(3-5-ETS1-1), (3-5-ETS1-3)*
- W.5.1: Write opinion pieces on topics or texts, supporting a point of view with reasons and information. (5-PS2-1)
- W.5.9: Draw evidence from literary or informational texts to support analysis, reflection, and research. *(3-5-ETS1-1), 3-5-ETS1-3)*
- RST.6-8.1: Cite specific textual evidence to support analysis of science and technical texts, attending to the precise details of explanations or descriptions. *(MS-PS2-1), (MS-PS2-2)*
- RST.6-8.3: Precisely follow a multistep procedure when carrying out experiments, taking measurements, or performing technical tasks. *(MS-PS2-1), (MS-PS2-2)*
- WHST.6-8.7: Conduct short research projects to answer a question (including a self-generated question), drawing on several sources and generating additional related, focused questions that allow for multiple avenues of exploration. *(MS-PS2-2)*

D. Social Studies Standards: http://www.state.nj.us/education/aps/cccs/ss/

- Social studies education provides learners with the knowledge, skills, and perspectives needed to become active citizens in local national and global communities in the digital age.

E. Technology Education: http://www.state.nj.us/education/aps/cccs/tech/

Students must be given the opportunity to study technology which provides an opportunity for them to learn about the technology-related processes and knowledge that are needed to solve problems and extend human capabilities. (ITEA)/ They must have an opportunity to:

1. Design, develop, and utilize technological systems such as communication, transportation, manufacturing, and construction technologies
2. Perform open-ended, problem-based design activities
3. Learn cognitive, manipulative, and effective learning strategies
4. Apply technological knowledge and processes to real world experiences using up-to-date resources
5. Work individually as well as collaboratively to solve problems (ITEA)

    F. National Core Arts Standards: http://nccas.wikispaces.com/Overarching+Anchor+Standards

    This site gives an overview of the current thinking of arts education in the nation. They were adopted by New Jersey in 2014.

Air is what makes airplanes fly. Aerodynamic forces operating on the airplane must be considered in designing crafts. Four forces (drag, thrust, lift, and gravity (mass)) must be balanced in order to have the plane fly any appreciable distance (Ross-Nazzal, 2004). Drag is the force that opposes an aircraft's motion through the air. Drag is caused by the resistance of the air to the plane flying through it. This can easily be felt by running a hand through the air. Students can feel the resistance of the air as their hands pass through it. Designers try to minimize drag by designing a streamlined plane. Students can see a reduction in drag on their hands by streamlining the shape of their hand as it passes through the air. Thrust is the force that opposes drag. This is the force that moves the plane forward. Lift must oppose gravity. Gravity is the force due to the Earth's gravitational field and is well recognized by students. Lift is related to the mass of the plane and is accomplished when the pressure below the wings is greater than the pressure above them. Design of the wing is an important factor in minimizing mass and maximizing lift. These opposing forces keep the plane in the air. The plane's mass must be kept to a minimum in order to maximize lift. The effect of air on a balloon can be demonstrated by filling a balloon with air and then letting go. What a topsy-turvy flight is initiated and what child has not experienced this for him or herself? If only that random flight can be controlled, then what? With this background, which has hopefully engaged the students' interest, the teacher and students can design and build an aircraft that will fly far and true. Perhaps the teacher and students can increase the aircraft's flight by taking the jet effects of air escaping from a balloon into account, increasing thrust, and considering streamlining design. These ideas will incorporate mathematics, engineering principles, social studies topics, science concepts, and aesthetics.

The world record for a paper airplane flight is held by Joe Ayoob and aircraft designer John M. Collins (both USA), at McClellan Air Force Base on February 26, 2012, in North Highlands, California (Guinness). The plane was constructed from a single sheet of uncut A4 paper. A4 paper is a twenty-pound copy paper that is readily available. The aircraft was designed by John M. Collins and flown by his colleague, Ayoob. The record stands since 2012. The plane flew 69.14 meters, or 226 feet 10 inches (Guinness). The highest altitude paper plane launch was 35,043 meters (37,963 yards) and was achieved by David Green (UK) in Ellsworth, Cambridgeshire, UK, on 24 June 2015 (Guinness).

Aerodynamic forces operating on the airplane must be considered in designing the airplanes. The four forces (drag, thrust, lift, and gravity (mass)) must be balanced in order to have the plane fly any appreciable distance. Lift, opposing gravity, is accomplished when the pressure below the wings is greater than the pressure above them. Teachers can demonstrate this by blowing air across the top of a strip of paper. The paper rises from the difference in air pressure above it and below it. Wing design is an important factor in minimizing mass and maximizing lift. These opposing forces keep the plane in the air. The plane's mass must be kept to a minimum in order to maximize lift (NASA).

Another force, thrust, is responsible for pushing the plane through the air and enabling it to fly some distance. Modern airplanes use jet engines powered by expensive fuels. Paper airplanes are powered by the thrust from a child's arm launching it. How interesting to consider the effect of using a balloon to increase thrust! How can the power of rapidly-escaping air from a balloon (as demonstrated by allowing air to escape rapidly from the balloon) be harnessed? The wrestling with ideas generated by the design can bring students to a basic understanding of Newton's Laws of Motion. Learners will be required to research current information on flight. They will present written information from informational texts (Common Core English Language Arts), recognize the importance of space exploration (Social Studies Standards), and design an airplane with respect to aesthetics and function (Core Arts). They will use mathematics in collecting and analyzing data (Common Core Mathematics Standards). Principles in science will form the basis of instruction on the possibilities of flight (Next Generation Science Standards).

It is evident from classroom experience that such integration works. Knowledge should not be compartmentalized. All facets of an integrated curriculum need to be utilized in order to produce an individual with multiple varied investigation, collaboration, and scholarship skills.

CASE DESCRIPTION

Before planning simulated flights with paper and balloons, many questions could arise in students', and maybe even teachers', minds. How can a heavy airplane carrying hundreds of people and tons of baggage and freight actually fly? How can these massive rockets and spacecraft escape from the Earth? Teachers may elicit many questions like these from students while discussing the history of travel above and beyond the Earth. The interest and discussion that results from the research into flight's history can set the stage for research into its basic principles by simulating the flights of paper airplanes and balloon rockets. Students can build airplanes, consider their designs, and test their designs by holding competitions to see which is most successful in flight. Then they must consider the development of jet engines to increase the speed and accuracy of the paper aircraft which is a simulation of a spacecraft that could eventually escape the Earth's gravitational field. Students of all ages enjoy playing with paper airplanes that they make themselves. This is a motivating activity that teachers can capitalize on to generate ideas on variables, construction design, aesthetics and even practical results with their students. Teachers can use this intrinsically stimulating experience to practice skills of design, testing, collection and interpretation of data with students with investigations that support the Next Generation Science Standards and which provide opportunity for integration of mathematical and engineering concepts within the science curriculum. Artistic principles must also be incorporated in producing a paper aircraft with some degree of beauty. Students involved in designing and flying paper airplanes can perhaps replicate some of the excitement generated by the Wright brothers and all who followed in perfecting the flying machines that we have today.

The purpose of this experience is to use basic flight principles to design and test a paper airplane and to participate in a contest to determine which design and which materials build an airplane that flies the longest distance in the shortest time. This activity proved quite popular with preservice teacher candidates in a methods of teaching science course. It can be used from kindergarten through college depending upon the concepts in place, mathematics used, and the chosen design's complexity. It can be challenging at any grade level. This activity is cost effective since it only uses materials that are almost always available in schools. Students discuss variables that can affect the speed and distance a paper airplane travels in relation to forces and the materials available for construction. Discussion centers on what really matters. Students discuss the effects of mass, the sleekness of the design, the force behind the thrust, decorative effects, and the weather. They begin to demonstrate understanding of these variables' effects and come to realize the importance of varying one condition while keeping all others equal. The degree of understanding will be related to the ages and developmental levels of the students. Differentiated

Figure 2. Possible airplane designs from templates

Figure 2. Possible airplane designs from templates

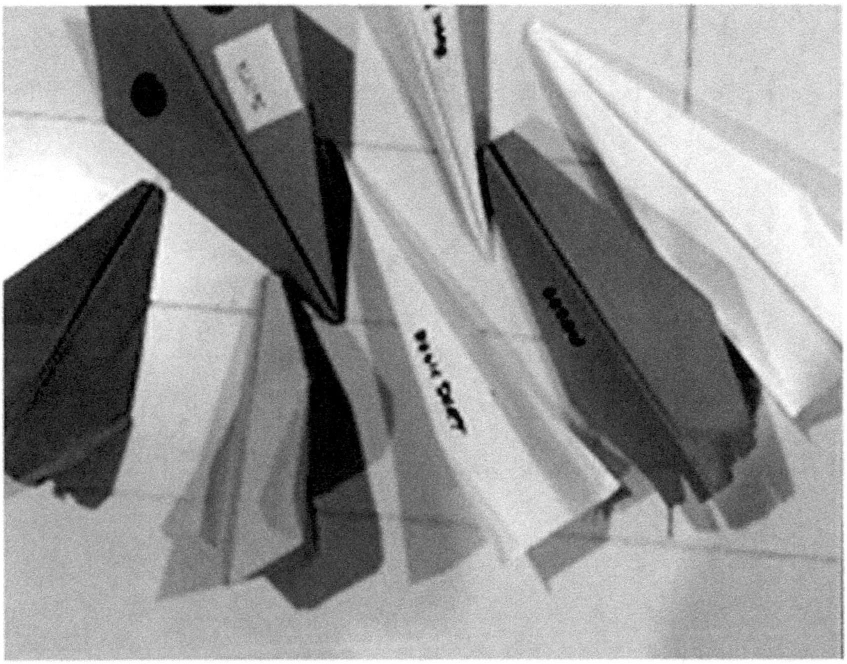

outcomes must be considered here. The idea of planning a fair race can be related to the concept of what really matters for most students or what is *fair*. Students at a higher developmental level will be able to demonstrate a deeper understanding of the importance of variables. A competition to determine which group builds the best airplane can be planned to include all students while taking into account students' developmental levels and specific talents. That will be the ultimate test of the various designs considered. All students can become excited about the race and be actively engaged in discussing conditions. They observe the designs of the airplanes; classify according to the variables discussed; make predictions based on size, mass, and streamlining to hypothesize; collect and analyze data; communicate results; and collaborate. Much discussion should take place concerning what conditions are important in ensuring a *fair* race. Some things that students have identified as necessary to control are:

Figure 3. Dart design constructed by grade 6 student

Figure 3. Dart design constructed by grade 6 student

1. Mass
2. Type of paper
3. Initial thrust
4. Friction
5. Effect of any identifying or decorative items (mass)
6. Streamlined design

These items are mostly controllable and can result in the design of a plane that will fly fast and relatively far. Thrust will probably be difficult to control since it will be accomplished by a student initially launching the plane with a push from the power of arm muscles. Students will see that the same student should be selected to give the initial thrust as the plane takes flight. Some students are stronger than others. This, of course, will vary from group to group and will be considered in the second part of this lesson plan. Students also volunteer additional items that they cannot control on the day of the race. These things include weather and wind speed. The race may be done in the gymnasium on days that the weather is not favorable. In fact, it might be better to use the gymnasium for the race if it is available. Discussing the importance of considering variables in an experiment, identifying what is important, and identifying controlled variables is vital to the scientific

Figure 4. Sample of planes built from templates

process. Younger students begin to show understanding of, all things being equal, which elements really produce an air airplane that will fly the farthest in the shortest time. Older students will recognize the relationship between the independent and dependent variables.

Lesson 1: Designing and Building the Airplane

Essential Questions to Be Answered:

1. What are the forces that enable an airplane to fly?
2. What design principles must be considered in designing an airplane that will fly?
3. How will flight time and distance be measured?

Enduring Understandings:

1. Science is a creative process that results in the development of new knowledge through experiment, data collection and analysis.
2. The importance of evidence in making generalizations about the natural world.
3. The value of aesthetics in design of a flying craft.

Students Must:

Ask questions that can be investigated within the scope of the classroom or outdoor environments.

1. Be encouraged to develop hypotheses based on observations
2. Plan and carryout experiments that involve multiple variables
3. Work collaboratively to identify independent and dependent variables as well as controls
4. Identify necessary controls

5. Identify variables, independent as well as dependent
6. Identify critical measurements
7. Tools to use
8. What data to record
9. Amount of data needed
10. How to communicate results
11. Identify important variables that need to be measured
12. Analyze and interpret data that is collected during an experiment

This experience ideally integrates the five disciplines: science, social studies, art, technology and mathematics. It can be used as an interdisciplinary activity in each of these content areas. Hopefully this activity will help participants realize that connecting evidence to explanations furthers scientific progress. The activity may also show the extent to which solutions to difficult problems require content from other curriculum areas. Students must engage in discussions based on evidence and data collected.

After discussing the research they have done on the history of flight, students should receive a packet of information about paper airplane design and construction. Groups of younger students are required to construct a basic airplane for each of the models selected based on the information in the packet. This is done before the time the race is run. Groups should be designed for maximum scaffolding and for differentiating instruction. The designs listed below were chosen to use with an elementary school class. They are simple to construct and yield good results. More complex designs can be used with older students. There is abundant information on the Internet about airplane design and construction. Students research these designs and build those that they are developmentally ready for.

1. Delta
2. Arrow
3. Classic Dart

For older designers, research on models must be done before the airplanes are actually constructed. Students present their ideas based on research to the class, which might span several classes. Construction can begin at home for all students, thus making the family part of the project. Different designs can be chosen according their complexity and the students' developmental levels. The discussion about the importance of considering variables in an experiment and identifying what is important and what can be controlled is vital to the scientific process. Students begin to understand, all things being equal, what is really important in constructing

Figure 5. Paper airplane shuttle and rocket constructed by Grade 6 student

Figure 5: Paper airplane shuttle and rocket constructed by Grade 6 student

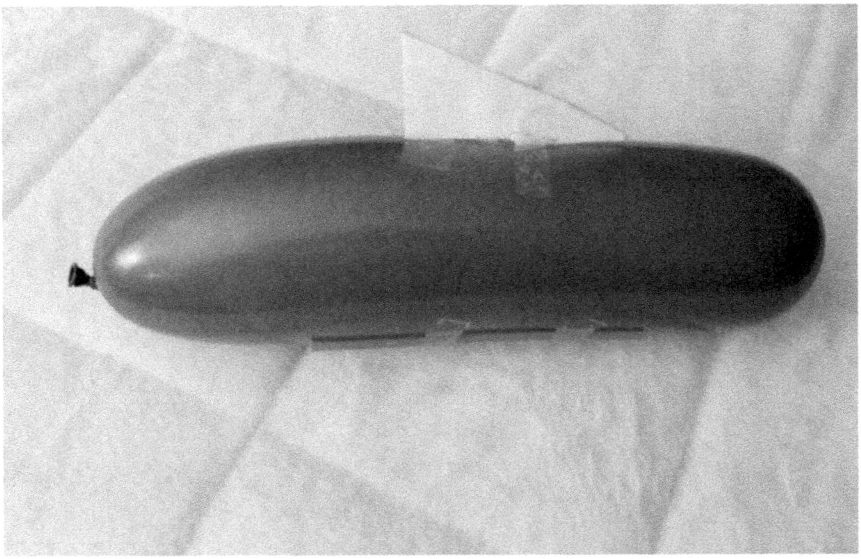

an airplane that will fly the farthest in the shortest time. Again, what will produce a fair race?

Safety Precautions

Students should be advised to be careful with scissors. Be sure to obtain permission of the principal to take the class outdoors to test the airplanes. Parents may also be enlisted in this activity to help keep track of the students as they spread around school grounds to set up their tests. Students should neither fly the airplanes in the classroom nor fly them towards other students. The same rules will apply if conducting the race indoors in the school gymnasium. Safety concepts must also be established for the construction phase.

Figure 6. Paper airplane shuttle and rocket constructed by Grade 6 student

Figure 6: Paper airplane shuttle and rocket constructed by Grade 6 student

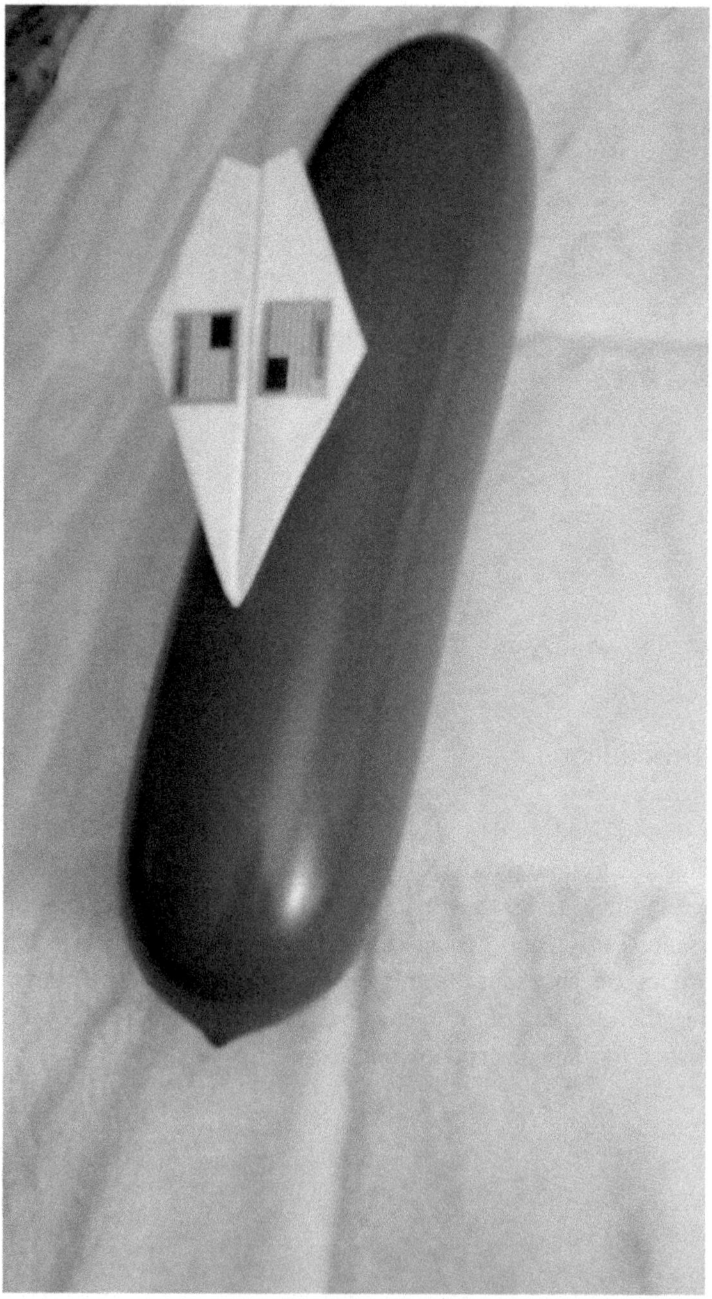

Table 1. Materials needed parts 1 and 3

Standard Copy Paper	Paper clips
Card stock paper	Crayons and stickers to decorate airplanes
Scotch Tape	Stop watches
Balance	Staple guns and staples
Clip boards	Measuring wheel or meter sticks
Video cameras (I pads)	Airplane templates
Balloon Pump	Computer

Procedure Lesson 1

Pre-Construction Phase: Research

1. Many students will have prior knowledge of paper airplanes but will need more detailed information on their construction. Begin with a search of the literature on paper airplanes. This can be done using search engines on a computer (Dewar). For younger students, the teacher can provide some appropriate reading material on paper (Collins).
2. The teacher can also provide templates for the children to use in construction.
3. Conduct a discussion on the results of the research.
4. Examine the work others have done.
5. How can previous work add to understanding of the process of flight?
6. What is an operational definition for fastest and furthest traveled by the planes?
7. After this discussion students can begin constructing their planes.

Construction Phase

1. Distribute the material and the templates before the actual race takes place. Give the students time to construct the airplanes according to the templates of the designs that they choose.
2. This is a collaborative effort. Teams of three to four work well. Scaffolding teams is recommended to ensure all students' individual strengths are considered.
3. Direct students to build a model of each airplane using the two weights of paper (copy machine paper and card stock) available.
4. Give students free rein to design their planes and to decorate them as they desire (always considering the variables and the purpose of the design).

Table 2. Sample note taking form for literature search

Names ______________________

What is the research question? ______________________

What have you found in your research that will help you construct an airplane that will fly?

1. ______________________
2. ______________________
3. ______________________
4. ______________________
5. ______________________

References

At least 5 resources should be listed.

1. ______________________
2. ______________________
3. ______________________
4. ______________________
5. ______________________

5. Allow students to decorate their airplanes as a method of personalization for each team.
6. Work with students to construct data tables that represent the variable data.
7. Consider the value of graphical analysis. Remember one picture is worth one thousand words.
8. On the day of the race direct each group to discuss the variables that should be considered and the construction of the airplanes.
9. Conduct the race under the conditions established by the class with consideration of all of the variables identifies previously in class discussion
10. Record the race on video so that students can evaluate results and look for ways to improve flight, which can be done in a post flight reflection.

Lesson Outcomes

Lesson outcomes can be differentiated to ensure that all students become involved and engaged in the endeavor and so that all students feel valuable to the project. At the conclusion of this activity most learners will be able to:

1. Research the literature on the construction of paper airplanes
2. Identify relevant information
3. Design and build an airplane using the three basic patterns available
4. Choose an alternate design based on Internet search
5. Demonstrate understanding of the forces that interact on the airplane that must be considered in its design through oral discussion of their designs

Figure 7. grade 6 child getting ready to launch rocket with spacecraft attached

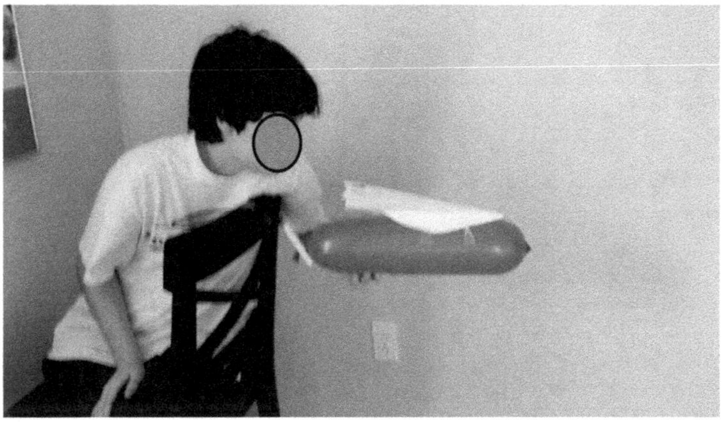

6. Use templates to construct a device that is attractive and that will possibly overcome the opposing forces
7. Consider all controllable and uncontrollable variables affecting the flight of the airplane
8. Extend the idea of the importance of variables in conducting an investigation
9. Conduct a perceived *fair* race of the airplanes constructed
10. Design a data table to present data
11. Demonstrate an understanding of graphical analysis's value
12. Demonstrate a basic understanding of Newton's Laws of Motion
13. Reflect on the procedures used in order to improve the designs in another race

Students must be involved in developing the data table necessary for comparing the speed and distance that the airplanes flew. This will represent averages of all the trials by the different collaborators involved.

This same table can then be used to summarize all data collected by the individual groups allowing for comparison of multiple trials of the airplanes constructed. The students will be able to identify the design of the airplanes according to the definition for fastest and shortest time developed before the race. A simple histogram can be used to present the data graphically. Students should also writ a reflective narrative description of the data in Table 2 and the histogram that presents the data visually.

Evaluation Rubric

Lesson 2: Increasing and Controlling Thrust Using a Balloon Rocket After conducting the race and seeing which design won out over the others, older students

Table 3. Student record keeping form; students keep procedural details using clipboards

Experimenters ______________ **Date and Time** ______________

Design Used ______________

Materials ______________

Weather ______________

Wind speed ______________

Distance flown

Trial 1 ______________

Trial 2 ______________

Trial 3 ______________

Average time in seconds' ______________

Time in the air

Trial 1 ______________

Trial 2 ______________

Trial 3 ______________

Average Time in seconds' ______________

Table 4. Data collection form for individual group results

Design	Material	Average Weight(g)	Average distance flown (m) (3 trials)	Average time in the air (Sec) (3 trials)
Design 1	Card Stock			
Design 2	Card Stock			
Design 3	Card Stock			
Design 1	Copy paper			
Design 2	Copy paper			
Design 3	Copy paper			

can consider the variables that were difficult to control, especially the amount of thrust by the individual students who were selected to launch the planes. Using the demonstration of the balloon that flew irregularly when the air was let out of it, the students begin to think about how to standardize thrust by controlling the air escaping from the balloon. The teacher can introduce the design of the space

Table 5. Evaluation rubric

Item	Not Met (0points)	Met(2 points)	Target (3points)
Preparation	The group did not prepare properly for the construction of the paper airplanes	The group did prepare but were not careful in following the templates provided	The group worked effectively and collaboratively to construct the planes and followed the templates as closely as possible
Construction	Construction was sloppy and did not meet the requirement of streamlining.	Construction was adequate but the air[planes produced were not attractive and streamlined enough	Airplane produced was streamlined well-constructed and shoed collaboration and good planning on team members.
Planning	Planning was not thorough and did not consider all of the variables mentioned and discussed prior to the race	Planning was adequate buy did not take into account all of the details.	Planning was thorough with consideration of most of the variables. Team members worked cooperatively in conducting the race.
Data	Did not complete all phases of the race. Data incomplete and not explained in a narrative or histogram	Completed all of the race but data table missing some measurements.	Completes all phases of the race and produced acceptable data that was capable of analysis
Explanation	Was unable to explain the results Data incomplete and not explained in a narrative or histogram	Produced an acceptable data table but was unable to write an explanatory narrative.	Produced an exemplar data table. Constructed a relevant histogram and explained the results in a narrative.
Reflection on results	Not able to produce or present a coherent reflection on the racing process.	Produced and presented an adequate analysis of what went well and what needed improvement in their procedures.	Produced and presented to the class a well thought out analysis of the race procedures and made valuable suggestions for improvement in procedures.

shuttles that were mounted on the Space Shuttle Solid Rocket Boosters, or SRB's, (NASA) that took the astronauts within the shuttle beyond the Earth's gravitational field and into orbit. These were the most powerful rockets ever built. Using the best airplane design identified in Lesson 1 as a result of the competition, the students can plan another race using balloons filled with air. Newton's Laws can be applied in planning the design of the paper airplane shuttles and balloon rockets. The Third Law, stating that for every action there is an equal and opposite reaction, provides the basis for this experiment. The SRB rockets burned fuel, and the hot gases that the burning fuel produced provided thrust because of the opposition created by the gases pushing against the rocket's walls of the rocket and the rapidly exiting gases pushing the other way. Pressure builds up within the walls of the rocket, and the only place that pressure can be released is at the relatively small opening at the base of the rocket. In our case, air will be used to exert pressure on the walls of the balloon and the small opening at the base of the balloon will provide rapid escape of air and consequent reduction in pressure within the balloon.

Procedure Lesson 2

This portion should possibly be used with older students. It could be handled by middle level and above. Examination of the photo of the Voyager flight should generate discussion about the thrust needed to have the Voyager spacecraft escape the Earth's gravitational field. Ask the students if it might be possible to simulate

Figure 8. Preservice teachers planning for the airplane race

such a flight by using paper airplanes attached to a balloon. Air would escape from the balloon and serve as the power to launch our simulated spacecraft. This then should begin construction phase 2. How is the airplane attached to the balloon?

Problems to Be Solved Through Class Discussion

1. What kind of balloon should be used?
2. How can we be sure that the balloons are inflated equally?
3. How does Newton's Third Law predict the movement of the balloon rocket?
4. How can we control the flight of the balloon rocket?

Discussion

1. Streamlined balloons should be used.
2. An air pump should be used to inflate the balloons as equally as possible.
3. Measurements of the lengths and circumferences of the largest part of the balloons should be taken to ensure that all start out containing the same amount of air.
4. Balloons should be mounted on a device that will control its flight and give it direction.

Table 6. Sizes of balloons

Balloon Trials	Circum-ference-widest part (cm)	Length (cm)	Number of pumps used with manual pump.	Weight of the assembled "rocket "balloon and paper air plane "shuttle".	Distance Flown	Problems Encountered
1						
2						
3						
4						

Safety

Be cautious in assembling and launching the balloons. Do not over inflate the balloons. Avoid loud noises from the popping of overinflated balloons. Do not point them at other students. Conduct this competition in a large room, perhaps the school gymnasium. If spectators are present, keep them out of the balloons' paths. Have at least one adult present to help ensure the safety of the participants and spectators.

Additional Materials

- Long balloons
- Clothespin (to keep air inside of the inflated balloons until they are released)
- Masking tape
- Fishing line (used for rocket guidance)
- Scissors
- Drinking straw (to attach balloon rocket to fishing line)

All balloons should be inflated to the same degree. Length and circumferences as well as number of pumps to fill the balloons should be recorded in the following sample table.

Preparing for the Competition

1. Choose a design for a paper airplane from Lesson 1
2. Use these design(s) for Lesson 2
3. Identify the variables in using balloons for thrust (recorded above).
    a. Size of balloon
    b. Degree of inflation of the balloons.
    c. Total weight of the assembled balloon and shuttle including ballast.

Conduct a similar competition as was done in Part 1 to determine the distance a plane attached to the "rocket" could fly when attached to a balloon. Students work in groups to construct the rockets and airplanes using different designs. The balloons should be fairly constant in size and pressure.

Further Questions During Construction

1. What shape of balloon should be used?
2. Does the color, texture or shape of the balloon matter?
3. How should the airplane be attached to the balloon?
4. How can balance be achieved in order for the model to fly straight with the airplane in the proper position?
5. How can damage to the airplane and the hypothetical occupants within the simulated shuttle be avoided?
6. How can the air be released from the balloon in a controlled manner?
7. How should the "rocket" be designed to maintain balance?

Once the questions above have been discussed as thoroughly as possible, students then construct the rocket launchers and test fly them. Data is collected by individual groups as in Table 2 above. Again, aggregate data tables should be constructed using individual group data. Graphical as well as narrative analysis must be used. A narrative reflection should be required explaining positive and negative aspects of the tests. Modifications in design are then discussed. These modifications may include:

- The use of a smooth string to guide the rocket balloons
- Method of holding the string straight and taut
- Length of the string needed
- Support of the string
    - Ring stand supports
    - Student support at launching point

ILLUSTRATION 1 APPARATUS THAT CAN BE USED IN BALLOON LAUNCH

Balloon on plastic straw with fine gauge string pulled through it. Tape is used to provide ballast and to keep the balloon attached to the straw. A clothespin is used to keep the air inside the balloon until the "rocket" is launched. The paper airplane is secured to the rocket by glue or scotch tape.

Figure 9. Example of 3D plane model

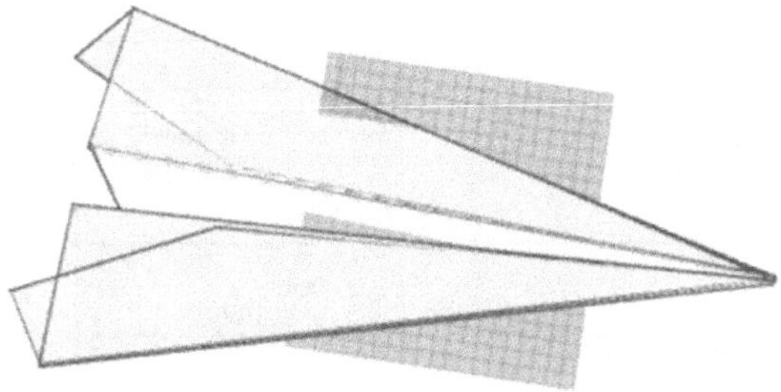

Outcomes

At the conclusion of this lesson, learners will be able to:

- List questions relevant to constructing a simulated rocket and shuttle launch
- Attempt, through researching and discussing design principles, to construct a simulated rocket/space craft launch
- Conduct a test fly with the designs that they produce
- Demonstrate understanding of the constraints of the test by outlining the variables involved
- Reach a conclusion based on this experience; which "space craft" and which "rocket launcher" was the best one based on the test fly results
- Suggest improvement in design that would provide a smoother and speedier flight

Lesson 3: Designing and Testing a 3D-Printed Airplane Engineers around the world use computer software to build 3-Dimensional (3D) models. The models allow builders to see their designs from all angles and make various adjustments prior to creating the design with tangible materials.

This is an extension activity to be completed after the initial race. The student teams will design their 3D model on the computer software and give the file to the instructor to print. The printing time can vary depending on the size and complexity of the airplane. After the 3D models are printed, the students will test models and compare the results to their paper airplanes in lesson 2 & 3.

Essential Questions to Be Answered

1. Does the material of the airplane change its ability to fly?
2. How does computer software assist engineers to design?

Table 7. Evaluation rubric

Item	Not Met	Met	Target
Discussion of Variables	Does not recognize the importance of control of variables.	Does account for some variables in the design	Accounts for as many variables in design as possible and recognizes these variables that are not possible to control.
Construction of Rocket	Poorly constructed	Adequately constructed but unable to assemble parts effectively.	Well-constructed, space craft attached securely. Ready for launching.
Selection of shuttle	Selects best design	Selects best design	Selects best design
Experimental set up	Does no demonstrate the importance of design in the launching of space craft and rocket launchers.	Adequately demonstrates the process of design but is unable to produce an acceptable design.	Completely demonstrates understanding of the necessities for construction. Actually constructs a rocket launcher that successfully launches.
Data collection	Data collection missing critical portion or no data collection design.	Has data collection plan but misses one or more critical measurements.	Complete data table and collection of critical data.
Graphic Analysis	No plans for graphic analysis	Plans incompletes	Identifies independent and dependent variables and where to place them on a histogram. Has enough data for an adequate analysis.
Explanation of results (reflection)	No explanation possible	Explains what s data has been collected but since plans were incomplete unable to explain results	Complete plans allow experimenters to initiate an explanation based on evidence.
Future improvements	None	limited	Recognizes problems with plans and able to initiate suggestions for improvement in design.

3. How will flight time and distance be measured and compare with the results in lesson 1 & 2?

Safety

Adult supervision is necessary when using the 3D printer, as the filaments that melt the plastic to create the print are extremely hot. Students should not touch the printer while a print job is being completed, and they should not be left alone with the 3D printer. Be cautious in assembling and launching the balloons. Do not over inflate the balloons. Avoid loud noises from the popping of overinflated balloons. Do not point them at other students. Conduct this competition in a large room, perhaps the school gymnasium. If spectators are present keep them out of the paths of the balloons. Have at least on adult present to help ensure the safety of the participants and spectators.

Table 8. Materials needed

Computer	Paper clips
Computer software	Long Balloons
3D printer and filaments	Stop watches
Balance	Masking Tape
Clip boards	Measuring wheel or meter sticks
Video cameras (I pads)	Fishing Line
Balloon Pump	Straw

Procedure Lesson 3

Choose a software platform to create or import the airplane to be 3D printed. thingiverse.com is a website with various imports of pre-designed airplanes that can be printed on any 3D printer on the market. Additionally, tinkercad.com is valuable resource. It is a free intro to AutoCAD software online. Students can import an airplane design and tinker with the construct or construct a model to be 3D printed from scratch.

Construction Phase: Using a Pre-Designed Import as Guide

1. Computers are needed at this time
2. Prompt students to create an account on tinkercad.com or grabcad.com. The software is free and open sourced.
3. Allow students time in their original teams to select an airplane to 3D print using their software's search bar. The design should be as similar as possible to the original paper airplane in Lessons 1 and 2.
4. Work with the students to modify the import to have the measurements to match their original airplane.
5. Upload the file to a 3D printer.
6. Print the model.

Test the 3D Printed Models

1. Work with students to construct data tables that represent the variable data.
2. Consider the value of graphical analysis. Remember one picture is worth one thousand words.

Figure 10. Illustration 1 apparatus that can be used in balloon launch

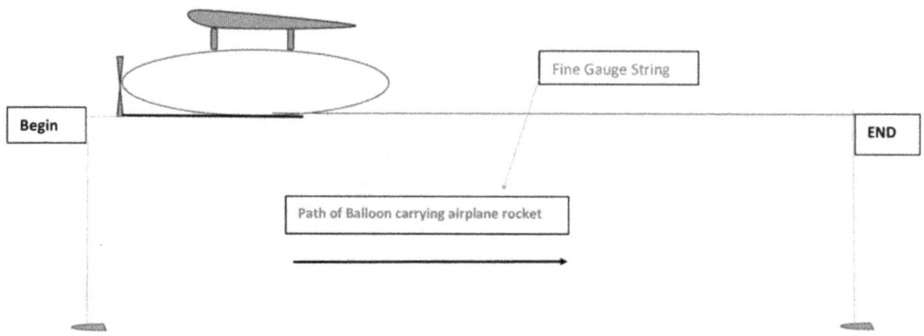

3. On the day of the race, have each group discuss the variables to be considered and the construction of the airplanes.
4. Conduct the race under the conditions established by the class with consideration of all of the variables identifies previously in class discussion
5. Video the race so that students can evaluate results and look for ways to improve flight. This can be done in a post flight reflection.

Lesson Outcomes

At the conclusion of this activity most learners will be able to:

1. Research the literature on the construction of paper airplanes vs 3D models
2. Identify why engineers use 3D modeling software
3. Design and build an airplane on computer software
4. Consider all controllable and uncontrollable variables affecting the flight of the paper airplane vs the 3D printed model
5. Extend the idea of the importance of variables in conducting an investigation to using computer software to build the design
6. Design a data table to present data of the paper vs the 3D model
7. Demonstrate an understanding of the value of graphical analysis
8. Demonstrate a basic understanding of Newton's Laws of Motion
9. Reflect on the procedures used in order to improve the designs in another race and/or the design of the plane on the computer software to print a better model
10. List questions relevant to constructing a simulated rocket and shuttle launch on computerized software

Table 9. Student record keeping form

Student Record keeping Form. Students keep procedural details using clip boards

Experimenters ______________ **Date and Time ________________**

Weather ___________________

Wind speed ________________

Table

Trial	Material	Average Weight(g)	Average distance flown (m) (3 trials)	Average time in the air (Sec) (3 trials)
Trial 1	Card Stock			
Trial 2	Card Stock			
Trial 3	Card Stock			
Trial 1	Copy paper			
Trial 2	Copy paper			
Trial 3	Copy paper			
Trail 1	3D printed Model			
Trail 2	3D printed Model			
Trail 3	3D printed Model			

Flying the Paper Airplane vs 3D Printed Airplane Without a Thruster

Flying the Paper Airplane vs 3D Printed Airplane With a Balloon Thruster

Compare the flight of a paper airplane and 3D printed airplane

1. Identify the following variables in using balloons for thrust:
    a. Size of balloon
    b. Degree of inflation of the balloons
    c. Total weight of the assembled balloon and shuttle including ballast

2. Weigh the different types of planes (copy paper, card stock and 3D model).

Students will conduct a similar experiment as was done in Part 2 and determine the distance each type of plane attached to the "rocket" could fly when attached to a balloon. Students work in groups to construct the rockets and airplanes using different designs. The balloons should be fairly constant in size and pressure.

Students must be involved in developing the data table necessary for comparing the speed and distance that each type of the constructed airplanes flew. This same table can then be used to summarize all data collected by the individual groups, allowing for comparison of multiple trials of the airplanes constructed.

Evaluation Rubric

Culminating Assessment

Students will be required to present the results of their investigations orally as well as in written form. A symposium on flight could be held in the school during which students could discuss all facets of their investigations to the other students. This might also be a means of attracting and motivating other students to STEAM careers.

Comments on the Lessons

Lessons 1 and 2 will help students recognize that designs need adjustments and improvements in order to increase chances for success. Lesson 3 takes the paper

Table 10.

Trial	Plane Type	Circum-ference-widest part (cm)	Length (cm)	Number of pumps used with manual pump.	Weight of the assembled "rocket "balloon and paper air plane "shuttle".	Distance Flown	Problems Encountered
Trail 1							
Trail 2							
Trail 3							
Trail 1							
Trail 2							
Trail 3							
Trail 1							
Trail 2							
Trail 3							

Table 11.

Team Members ________________________________

Item	Not Met (0points)	Met(2 points)	Target (3points)
Construction of 3D model	The group did not prepare properly for the construction of the 3D airplanes	The group did prepare but were not careful in following the templates provided	The group worked effectively and collaboratively to construct the planes and followed the imports as closely as possible
Data	Did not complete all phases of the race. Data incomplete and not explained in a narrative or histogram	Completed all of the race but data table missing some measurements.	Completes all phases of the race and produced acceptable data that was capable of analysis
Explanation	Was unable to explain the results Data incomplete and not explained in a narrative or histogram	Produced an acceptable data table but was unable to write an explanatory narrative.	Produced an exemplar data table. Constructed a relevant histogram and explained the results in a narrative.
Reflection on results	Not able to produce or present a coherent reflection on the racing process.	Produced and presented an adequate analysis of what went well and what needed improvement in their procedures.	Produced and presented to the class a well thought out analysis of the race procedures and made valuable suggestions for improvement in procedures.
Explanation of results (reflection)	No explanation possible	Explains what s data has been collected but since plans were incomplete unable to explain results	Complete plans allow experimenters to initiate an explanation based on evidence.
Discussion of Variables	Does not recognize the importance of control of variables.	Does account for some variables in the design	Accounts for as many variables in design as possible and recognizes these variables that are not possible to control.
Data collection	Data collection missing critical portion or no data collection design.	Has data collection plan but misses one or more critical measurements.	Complete data table and collection of critical data.
Graphic Analysis	No plans for graphic analysis	Plans incompletes	Identifies independent and dependent variables and where to place them on a histogram. Has enough data for an adequate analysis.
Future improvements	None	limited	Recognizes problems with plans and able to initiate suggestions for improvement in design.

airplane into the realm of the engineer by generating a computerized image and printing the design on a 3D printer. Experiences such as these can be used to engage students and serve as intrinsically motivational activities to develop engineering and design principles based on scientific thinking. They can be used to help students understand some of the principles of flight, especially space flight. The plans that the national Space and Aeronautics Agency for the future have are exciting and present inherent problems that must be solved before going further. Young citizens of the future should be introduced to the complications of design, problem solving skills, action and reaction of force, and testing procedures that scientists and engineers go through in planning such endeavors. Hopefully they may be encouraged to pursue careers in STEM that will be so crucial in ensuring a stable and successful world in the future.

CURRENT CHALLENGES FACING TEACHERS

The challenges facing teachers are based mostly on the lack of understanding of what science process skills really are, and how important these processes can be in developing students' critical thinking skills. This is something that takes interest, training, and commitment on the part of teachers and students alike. Some teachers may feel that physical science projects are not interesting to their students. Perhaps they have little interest in such projects themselves. Older teachers might deem these projects too difficult to take on and to manage. They may lack training in overcoming these attitudes. They may not see the value in the outcomes possible in such investigations. They also perhaps do not realize the need for scientists and engineers in the future. It is often said that we are living in a technological world. But what does this really mean? What are the implications for the future? There is a need for more scientists and engineers in the future to help to solve such problems as global warming, pollution, use of fossil fuels, alternative forms of energy, and ethics in biomedical advances to cite just a few. We are in the midst of technological change, and the implications of these changes are currently largely unforeseen. This era can be compared to the industrial revolution of the nineteenth century and its effects on daily life. Many of those living through the revolution of the nineteenth century did not realize the profound changes that were to occur as a result of the change from cottage-based manufacturing to machine age manufacturing. Those living now in the twenty-first century are in a similar predicament. They cannot really understand or predict the societal changes caused by advances in science and engineering. Just imagine the robot revolution and its effects in the work force.

Another challenge could also be found in the attitudes of school administrators who see little value in possibly increasing class time to work on physical science projects. They possibly cannot envision the ways in which these projects will develop their students into future scientists and technicians who will need these skills to tackle global problems with the environment, food supply, genetics, energy shortages, population growth, climate change, land and water supply, and more. Perhaps more importantly they do not realize the ethical concerns society will face in the future. School administrators must accept the fact that a firm foundation in STEM will help future citizens manage problems that will most certainly arise. Curriculum designers must consider integrating science topics within the entire curriculum. Scientific knowledge and processes depend upon skills pertinent to reading, mathematics, and social studies. In fact, scientific knowledge impacts all facets of the curriculum. In addition, a properly functioning democracy requires a scientifically literate populace of voting citizens to understand the impact that scientific development can have on lifestyle and living standards. America needs scientifically literate citizens as well as scientifically literate politicians to work together to improve conditions on the Earth that we all share.

The current emphasis on standardized testing in mathematics and literacy means that schools assign large blocks of time for these subjects while minimizing the amount of time during the school day for science content and activity. Schools, administrators, curriculum writers, and legislators do not realize, but must be made to realize, the importance of cross curriculum projects in a real-world learning environment. It is understandable that the emphasis on test scores in literacy and mathematics are important in developing an educated populace capable of following a variety of career paths. Nevertheless, scientific literacy is just as important, and it must be integrated within the entire school curriculum. This is not difficult to do as it might seem.

Another challenge could be the perception that projects are too expensive, thus taking up too much of the schools' supply budgets. There are many projects that can be done using ordinary materials commonly found in schools. Of course, there are projects that can be expensive, but a good start on developing concepts of design and consequent problems can be achieved at all instructional levels using ordinary materials as suggested in the lessons described.

Challenging also is the current realization that all children, despite disabilities, are entitled to an education in inclusive classrooms. Science is for *all* children. All children are equally valuable in classes but not all children are equally competent. This is especially apparent in science classes where some ideas can be too abstract for disabled students to master. In fact, they can often be too abstract for average students to master. We need to work together with special education teachers to develop differentiating methods so that all children become scientifically literate to some degree.

SOLUTIONS AND RECOMMENDATIONS

Perhaps all of the challenges and faulty perceptions in the previous section could be overcome by teacher training programs that emphasize the importance of incorporating such topics within interdisciplinary curricula. Techniques of incorporating science concepts throughout the curriculum must be further explored. Science can be integrated seamlessly within the mathematics, language arts, social studies, and even art curricula. Preservice teacher candidates and experienced teachers must have access to programs that demonstrate methods of incorporating science concepts into all curricula. Candidates in Schools of Education must be trained in science methods that consider the processes as well as the factual science content.

Preservice teachers in science methods classes should be practicing interdisciplinary techniques that do not isolate scientific concepts and processes from other curricula. This must be done in science methods classrooms at the college and university level. In supervision of student teachers from Monmouth University, integration of science concepts within mathematics, reading selections, and social studies is encouraged. Perhaps having preservice teachers study interdisciplinary subjects will lessen some of the challenges facing current teachers and future members of the profession.

Experiences in college level methods courses have been successful in generating ideas and interest in preservice teachers. The lessons cited in this paper were carried out towards the end of the course after the students conducted a series of experiments designed to instruct them in a science process approach to elementary and middle level science instruction. Hopefully they will be inspired to take this approach in their future science classrooms.

Teachers currently in service do a valiant job of educating students under increasing pressures of testing and accountability. It is understandable that they, as well as administrators, feel that the value of their institutions is being tied to standardized test scores and are therefore reluctant to change curricula and time slots designed to maximize scores. In-service programs conducted in universities by science trained professionals could be made available to in service teachers and administrators. With the adoption of the Next Generation Science Standards, this becomes imperative. If these standards are to have an impact on science curricula throughout the nation, it is imperative that educators on all levels join together as a profession to assist teachers and administrators to comprehend the implications of these standards and to develop methods of incorporating them within current curricula. This will take individual as well as state-wide effort. Those of us currently involved in science education need to take initiative.

The challenge presented by the idea that science is for all children needs to be faced by science educators and special educators. Special needs children will be just as impactful on the future as their peers in general education classrooms. Methods of making science concepts comprehensible for all children must be explored by partnerships between special educators, general educators, and science educators. Incorporating the strategies, techniques and art of co-teaching needs to expand. Co-teaching is being implemented widely in New Jersey, and its benefits must be recognized in all venues. Teachers have to be afforded the opportunity to collaborate using tried and true strategies already employed with special needs children. The concept of Universal Design suggests that educators use multiple methods of presenting knowledge, multiple methods of engaging students, and multiple methods of assessing students to have effective inclusive classrooms for all despite learning complications that might exist.

The solutions to these challenges are not easy to implement. There can be resistance on the part of preservice as well as in-service teachers to an emphasis on science in the curriculum. Questioning students in my science methods teacher-preparation classes if they enjoyed science in previous schooling resulted in a not-so-surprising negative response. They spoke of science courses with reliance on mountains of facts, and that memorization of these facts as well as a whole new vocabulary was daunting and boring. Students mostly said that loss of interest in science began in middle and high school courses. Science in the secondary schools is often lacking in inquiry methods. This is understandable since teachers at the secondary level often have classes that are college preparatory and are concerned with covering information that might be asked on achievement tests. Perhaps a whole new curriculum needs to be developed that stresses scientific literacy for the citizenry of the future. Hopefully the emphasis on problem solving and engineering design in elementary schools will continue to motivate those planning on careers in science and engineering to take the more challenging courses that still should remain in the secondary curriculum.

In summary some solutions to the challenges include:

1. In-service training for in-service teachers and administrators
2. More cooperation among special educators and science educators
3. Emphasis on interdisciplinary curricula
4. Methods of integrating the Next Generation Science Standards within current science curricula
5. Exploration of the secondary curricula to develop more science-society courses for the students who might have other talents besides science

These solutions suggest even further challenges. Teachers must be equal to these challenges.

REFERENCES

Christa McAuliffe Quotes. (n.d.). Retrieved from https://www.brainyquote.com/quotes/christa_mcauliffe_134582

Collins, J. M. (2013). *The new world champion paper Airplane book*. New York: Crown Publishing.

Dewar, A. (2014). *One minute paper airplanes kit*. Rutland, VT: Tuttle Publishing Company.

Farthest Flight by a Paper Aircraft. (n.d.). Retrieved from http://www.guinnessworldrecords.com/world-records/farthest-flight-by-a-paper-aircraft

Grant, R. G. (2007). *The complete history.* London, UK: Dorling Kindersley New Jersey Department of Education Standards for all curriculum area. Retrieved May 4, 2016 from http://www.state.nj.us/education/educators/standards/

Ross-Nazzal, J. (2004). *NASA Johnson space center oral history project.* Johnson Space Center History Collection. Retrieved from http://www.jsc.nasa.gov/history/oral_histories/BlackburnGC/BlackburnGC_5-4-09.htm

Shapell, B., & Willen, S. (2012). *The Wright brothers: Inventing the first successful airplane.* Retrieved from http://www.shapell.org/collectionfilters/the-wright-brothers

Voyager. (n.d.). Retrieved from http://voyager.jpl.nasa.gov/mission/fastfacts.html

KEY TERMS AND DEFINITIONS

Aerodynamics: A science that studies the movement of air and the way that objects (such as airplanes or cars) move through their: the qualities of an object that affect how easily it is able to move through the air.

Drag: The force of wind or air resistance pushing in the opposite direction to the motion of the object.

Gravity: The force that attracts a body toward the center of the earth, or toward any other physical body have mass. For most purposes, Newton's laws of gravity apply with minor modifications to take the general theory of relativity into account.

Lift: The force that directly opposes the weight of an airplane and holds the airplane in the air. Lift Is generated by every part of the airplane, but most of the lift on a normal airliner is generated by the wigs. Lift is a mechanical aerodynamic force produced by the motion of the airplane through air.

Mass: The property of matter that measures its resistance to acceleration. Roughly, the mass of an object is a measure of the number of atoms in it. The basic unit of measurement for mass is the kilogram.

Template: A form, mold, or pattern used as a guide to make something. Examples: A ruler is a template when used to draw a straight line. A document in which the standard opening and closing parts are already filled in is a template hat you can copy and then fill in.

Thrust: The force used to overcome the drag of an airplane and to overcome the weight of a rocket.

Chapter 9
Bee Pollination

Kerry Carley Rizzuto
Monmouth University, USA

John Henning
Monmouth University, USA

Catherine Duckett
Monmouth University, USA

EXECUTIVE SUMMARY

The purpose of the chapter is to provide an exemplar of an inquiry-based unit on pollination for designing and implementing constructivist instructional practices while simultaneously providing outstanding teacher preparation. The unit on pollination was developed by preservice teachers through a partnership between the Monmouth Conservation Foundation and the Monmouth University School of Education. Through collective action, these institutions were able to enhance student learning on a vital part of the science curriculum, provide a rich clinical experience for pre-service teachers, and to familiarize teachers with a more constructivist approach to pre-school science instruction.

INTRODUCTION

Science education for preschool children has recently received considerable attention. Recent research has shown that children who engage in scientific exploration in early childhood have a better understanding of science concepts later in life (Patrick, 2009). Children are natural scientists. They are inquisitive and enthusiastic with a predisposition for investigation. Children live in a world of things that are constantly beckoning to them: some living, some hard and cold, some large, some metallic,

DOI: 10.4018/978-1-5225-9631-8.ch009

and some dirty (Suransky,1982). Some refer to it as an innate curiosity to discover everything within their reach (Bredekamp & Copple, 2009, Piaget, 1955). Preschool children engage in the study of science long before they have entered any school building (Grennon-Brooks, 2011). As soon as children realize that they can discover for themselves, their first encounter with science has occurred (Bredekamp, 2014; Tu, 2006).

Current research indicates that young children have the capacity for constructing conceptual learning and the ability to use the practices of reasoning and inquiry (NRC 2007, 2012). Many educators tend to underestimate children's capacities to learn scientific concepts and practices in the early years; therefore, young children are not provided the opportunities for them to explore science-based understanding (NRC 2007). Additionally, when children are directly engaged in authentic science explorations, they are able remain focused for longer periods of time than educators typically estimate. Effective science investigations can deeply engage young children for extended periods of time beyond a single activity or session.

Despite its importance, teachers typically do not provide high-quality science experiences in early childhood classrooms (Gerde, Schacter, & Wasik, 2013; Fitzgerald & Schneider, 2013; Tu, 2006). Many early educators are hesitant about introducing science in their own classrooms, citing that they themselves did not enjoy studying science as students, and they feel ill-prepared to teach it (Conezio & French, 2002). As a result, many teachers teach science through a series of isolated experiments without any connection to a meaningful and broader science curriculum. In one study, preschool teachers reported that the science curriculum consisted of children watering plants and displaying a random assortment of stones and shells in the back of their classrooms (Fitzgerald & Schneider, 2013).

Children have the capacity to engage in scientific practices and develop understanding at a conceptual level. Current research shows that young children have the capacity for conceptual learning and the ability to use the skills of reasoning and inquiry as they investigate how the world works (NRC 2007, NRC 2012). For example, their play with blocks, water, and sand shares some science-relevant characteristics. Additionally, young children can learn to organize and communicate their learning, and differentiate between concrete and abstract ideas (Carey 1985). Adults who engage children in science inquiry through the process of asking questions, investigating, and constructing explanations can provide developmentally appropriate environments that take advantage of what children do as part of their everyday life prior to entering formal school settings (NAEYC 2013, p. 17; NRC 2007). These skills and abilities can provide helpful starting points for developing scientific reasoning (NRC 2007, p. 82).

Therefore, it is not surprising that several studies have discovered that preschool students are not ready to learn science when they enter kindergarten (Conezio &

French, 2012). According to some studies, children in kindergarten are already forming negative views about science that might negatively affect their dispositions towards science throughout their school years. When researchers interviewed kindergartners from typical classrooms, scarcely a third of the children showed any knowledge of science. Further, student interest in science fades through the elementary years, which has been attributed to "less investigative science practices" employed by their classroom teachers (Watters and Diezmann 2007, p. 351). This is especially true for girls, who even before entering kindergarten are less likely to say they like science (Sharapan, 2012).

The purpose of the chapter is to provide an exemplar of an inquiry-based unit on pollination for designing and implementing constructivist instructional practices while simultaneously providing outstanding teacher preparation. The unit on pollination was developed by preservice teachers through a partnership between the Monmouth Conservation Foundation and the Monmouth University School of Education. Through collective action, these institutions were able to enhance student learning on a vital part of the science curriculum, provide a rich clinical experience for preservice teachers, and familiarize teachers with a more constructivist approach to preschool science instruction.

LITERATURE REVIEW

In this section, the authors provide a framework for understanding the purpose, goals, and outcomes of the case study project. There will first be a discussion about the importance of pollination to the environment's future and hence to the science curriculum. To facilitate learning, instruction on pollination can begin as early as preschool while the brain is undergoing rapid development. This instruction is best accomplished through a constructivist approach to teaching and learning. Critical features of this approach include a caring, play-centered approach that fosters a sense of wonder in the child. This section concludes with an explanation of the importance of providing preservice teachers with an experience in creating inquiry-based instruction.

Importance of Pollination to the Curriculum

Pollination is vital part of the early childhood and elementary science curriculum primarily because of its significance to food production. Although approximately 75% of crop species rely on animal pollination to assure or maximize seed and fruit set (Klein et al. 2007), global crop production by weight from animal pollinated crops is 35% (Klein et al. 2007). These data support the widely quoted figure that bees are responsible for one in three bites of every food that we eat (MacGregor, 1976, ars.usda.gov/SP2UserFiles/Place/20220500/OnlinePollinationHandbook.pdf, Klein et al 2007). However, this figure has been disputed as exaggerated, explicitly (geneticliteracyproject.org/2015/08/12/pollinator-myth-bees-responsible-one-third-global-food-heightening-crisis-like-7/#link) and implicitly (Aizen et al. 2009). The values that Aizen et al (2009) produced from studying actual production trends indicate that animal pollination yields a real crop value closer to 10%, but this does not recognize nutritional value of crops for a balanced diet beyond calories and proteins. However, biologists and agriculturalists agree that animal pollinators are responsible for significant services, without which some crops would not be available (Klein et al. 2007), others would have significantly reduced yields (Aizen et al 2009, Klein et al. 2007), and more land would be required to feed the world (Aizen et al 2009).

Due to its role in food production, the economic role of animal pollinators is also highly important. In the USA the principle agricultural pollinator at this time is the honeybee, although other systems are being actively researched. Honeybee pollination services in the USA were valued at approximately $240 million in 2006 (FAO) and at $15 billion in 2014 (White house). Native wild pollinators services were valued at approximately $9 billion (White house 2014). Moreover, some crops such as almonds and squash are almost exclusively dependent on bee pollination for fruit set (USDA). Other crops will give higher yields in the presence of animal pollinators due to the greater quantities of pollen transferred (Klein et al. 2007).

However, honeybee colonies have declined precipitously and steadily in America during the last four decades (Aizen et al 2009 and Whitehouse 2014). Unfortunately, there is no clear cause of this decline although many forces (mites, viruses, pesticide residues, loss of non-agricultural habitats) clearly have negative impacts on honeybees and other pollinator populations. Whatever the precise causes (and here scientific uncertainty needs to be discussed with preservice teachers so that they have a clear understanding of the recursive nature of science), pollinator populations are declining sharply. This problem is world-wide and will require societal attention if food security for most is to be achieved (Long & Krupke 2016, Whitehouse 2014). There has been extensive discussion in both popular press and scientific literature on the effects of land use on both honeybees and locally native pollinators (Koh et al 2016, Long & Krupke 2016, Palma 2016).

Apart from its vital role in the production of food, pollination is also a part of the science curriculum. An understanding of pollination often makes related topics seem more relevant and less boring. The authors' lessons include topics like botany and plant anatomy, life history, ecological interconnectedness, and biodiversity. Knowing the parts of the flower necessary for successful pollination is essential for advanced botany study and the most basic identification of plants. Understanding the parts of the plant that are eaten will also aid the student in plant identification. Furthermore, knowing the edible parts of plants affords students' understanding of ecological relationships between animals and their plant 'prey' beyond humans and our food. Understanding the collection of pollen as a food source for the bees' larvae is important a preparation for learning life history, an essential concept for understanding biological diversity, and basic conservation measures. Life history is generally a poorly taught concept although it is recognized as essential for understanding biodiversity (Fabian and Flatt 2012).

In summary, pollination is critically important to the science curriculum. This discussion has implications for citizen action, and interested conservation organizations are creating potential tools for teachers at all levels (See: the Pollinator Partnership, Xerxes Society, Monmouth Conservation Foundation, etc). Indeed, the establishment of pollinator gardens as a citizen science project and teaching 'lab' has been promoted by many organizations as part of the Pollinator Partnership (http://www.pollinator.org/). The authors created this lesson plan with the final objective that it be paired with the creation of a pollinator garden. Encouraging students and teachers of all ages to make a positive difference in the world, no matter how small, is empowering and engaging to them.

Constructivist, Inquiry-Based Learning

To interest preschool children in learning science, teachers must shape their instruction in order to follow the scientific interests of the children while providing activities to engage and challenge children. Inquiry-based, constructivist instruction has been the most effective approach (American Association for the Advancement of Science, 1993; Bianchini & Colburn 2000). Constructivist theorists argue that both teaching and learning science is a result from observing the natural world. They also argue that teachers should scaffold information with prior knowledge and allow children to interact with more capable peers to construct new understandings (Bredekamp, 2014 Llewellyn 2002). Inquiry approaches are especially important in preschool classrooms, where children are developing foundational knowledge on which all future science understanding will be based (Conezio & French 2012). In addition, science education that begins during the early years of schooling can capitalize on the sensitive time for brain development (Grennon-Brooks, 2011). In early childhood classrooms, teachers have the opportunities to create brain-stimulating experiences by designing curricula that will enhance the wiring of the brain. The brain is a pattern-seeking machine, and science is the pursuit to recognize and classify naturally occurring patterns (Gerde, Schacter, & Wasik, 2013).

Features

To successfully implement a constructivist approach to science education, teachers must understand its features (Llewellyn 2002). The National Research Council (2000) provided five essential features which form the core of inquiry-based approaches to science teaching and learning. The learner: (1) engages in scientific questioning, (2) searches for evidence to support ideas, (3) hypothesizes possible explanations based on evidence, (4) connects those explanations to science understanding, and (5) shares findings and explanations with larger classroom community. These features support students in becoming science learners who are comfortable in formulating answers to scientific questions.

Wondering

Constructive teaching principles are innately connected to modeling and engaging young children in wondering (NAEYC, 2012). Therefore, the first step in creating scientifically-thinking preschoolers is to create environments in which young children feel inspired and motivated to wonder like scientists. An environment that fosters scientific thinking is one that gives young children the time, space, and materials to exercise their wonder and curiosity (Grennon-Brooks, 2012). It also gives them the freedom to engage in child-centered explorations, experimentations, and time to engage in social play situations. To become engaged in scientific thinking, children need access to materials that they can take apart and the tools to assist them in doing so. They need places where they can play in mud and examine flowers and bugs. They will also require the tools of the trade: ample time to explore outside, measuring tools, magnifying glasses, pails, and shovels.

Play-Centered

Moreover, teachers need to be aware of the many ways science can be naturally integrated into a play-centered curriculum. Science should not be viewed as a separate part of early childhood programs. Chaille and Britain (2003) coined the termed *sciencing,* to refer to what children already do and how they think about what they do. For example, as a child experiments with a mixture of water and sand, she is making observations and predictions. The child is also building theories and testing those theories, perhaps even sharing the theory with a classmate. When preschool children are engaged and using all three of the early childhood domains (social and emotional, cognitive, and physical), they are engaged in a well-rounded scientific experience.

Social Learning

According to developmental practice, children need direct social experience because they are active leaners who are active in their own learning activities. Without the environment to engage in a group, the children would not have the opportunity to develop different social skills. All of life is a social endeavor, and science is a social endeavor as well (Grennon-Brooks, p. 45). A child's active construction of knowledge is an interactive process that is promoted by social contact, with the teacher, or more knowledgeable other playing a key role (Vygotsky, 1978). Children's construction of knowledge can increase by sharing their observations and ideas with each other. Children should be encouraged to work together "in building theories, testing those theories, and then evaluating what worked, what didn't, and why" (Conezio & French, 2002, p. 13). Shared inquiry, where children work together can be especially beneficial in fostering curiosity and stimulating new ideas (Chaille & Britain, 2003). One way to involve children in shared inquiry is having them solve problems that focus on a specific situation: How can we move this heavy box? Shared inquiry generates a number of diverse ideas about the problem and challenges children to communicate the reasoning behind their ideas. There's no doubt that "problem solving and reflective thinking play an important role in children's science learning" (Lind, 1999, p. 80).

Caring

According to Noddings (1984), the teacher is the caregiver of a classroom. It follows that the first task a teacher needs to accomplish in the classroom is making the students feel cared for and happy. This requires the fulfillment of objective needs and subjective wants through the cultivation of students' character imbued with intellectual and moral virtues such as curiosity, openness, integrity, thoroughness and care for others. Teaching these prosocial skills helps children achieve positive interactions and reduce tensions that exist in schools. If the students are sure that somebody cares, they feel they can rely on someone to help them avoid building aggression that can result from unhappiness and struggles. This is done when the teacher attends to expressed needs of the students and assumed needs of the class.

In this way, teachers must learn to be flexible and unafraid to take advantage of the teachable moment; this is also a key element of the constructivist paradigm. Teachable moments are abundant during science experiences. Teachers need to become careful and caring listeners, capturing those moments that children wonder aloud. Teachers should have a large, rich repertoire of ideas depending on what catches their students' hearts and minds. This idea also suggests that students should be given many choices and different options. In some cases, Noddings (1986) posits that it is a good idea to have the teacher and students collaborate on the lesson plans and units to be covered in class.

For example, an excellent way to create a caring environment within the STEAM curriculum is to begin the science lesson or experience with a poem or a discussion of special moment that the class shared in order to connect prior knowledge to the new lesson. In addition, art activities such as sharing observations and outcomes of science experiences through dance, painting, and yoga can be creative methods of teaching children that all learning is interconnected. By allowing children to explore science and art together, (along with the other STEAM fields), we are able to teach students that the world of ideas is capable of response (Noddings & Shore, 1998).

Fostering Preservice Teacher Learning

To ensure that preschool children are exposed to constructive teaching strategies, appropriate professional development opportunities and rich clinical experiences should be offered for preservice teachers. For this project, a partnership between the Monmouth Conservation Foundation and Monmouth University provided an opportunity for both in-service and preservice teachers.

Declaring a common goal of increasing P-12 student learning as the primary motivating force for clinical experiences brings schools and universities together in common cause. If executed well, devising a plan to improve P-12 student learning will result in greater buy-in from schools, increased collaboration, and the most effective teacher preparation. Ultimately, it will lead to a shared responsibility between schools and teacher preparation programs for P-12 student learning and teacher preparation.

It also ensures the best quality teacher preparation because the best contexts to learn to teach in are those with high engagement and learning levels. In such environments, there is a synergy between increased student and teacher candidate learning. Three strategies for enhancing this strategy are co-teaching, sustained clinical experiences, and assessing P-12 student learning. Co-teaching can lead to richer contexts for teacher candidate learning through more high engagement learning strategies. Sustained clinical experiences can lead to more recursive learning by spending more time in the field. A third approach is to gather student learning data to gather assessment data to ascertain the impact of the teacher candidate on student learning. Incorporating student learning data can be done during either a sustained or diverse experience. Relevant data could include large scale assessments, benchmark data, or student growth objectives, among others.

A significant challenge for designing effective clinical experiences is the variability among teaching contexts. Thus, the development of teacher candidates in clinical settings can be highly individualized and the opportunities for learning very uneven. An important part of program design is ensuring consistency across clinical experiences. To put such a curriculum in place requires identifying the target practices that teacher candidates are to learn, charting a clear map for accomplishing those practices, and identifying specific instructional strategies for fostering teacher candidates' development in clinical settings. Professional preparation programs typically have elements of such a curriculum in place, although not well organized and articulated across the program. This was illustrated in an interdisciplinary study of three professional preparation programs. Findings indicated that professional preparation programs use representations, decompositions, and approximations as key elements of teaching a profession. Representations refer to the way we conceptualize teaching practice. Decompositions involve breaking down teaching practice into smaller, more manageable components that can be individually mastered and gradually integrated into more complex teaching practices. Approximations are instructional practices that provide teacher candidates exposure to teaching practices in safe environments.

The Partnership

In November of 2015, Monmouth University School of Education was contacted by board members from the Monmouth Conservation Foundation (MCF). The MCF is Monmouth County's only county-wide land trust. The MCF is a not-for-profit organization that has been preserving land and protecting the natural habitat of Monmouth County for over thirty-eight years. In an effort to partner with the youth, the MCF founded the Kids for Conservation project in 2013. It is primarily an academic enrichment program whose purpose is to teach young children about the environment. The curriculum is based upon age-appropriate themes that teach children about land conservation, wildlife protection, and the preservation of farmland and open space.

The goal of the partnership was to develop a new curriculum that would focus on teaching preschoolers about the importance of pollination. In order to determine which students would write this original curriculum, the authors held several meetings with ten junior and senior preservice teachers who were pursuing their early childhood teaching endorsements. Several meetings were held during December of 2015 to form teams of preservice teachers. Once the teams were established, a meeting was held with MCF members to form common goals and ensure mutual expectations. A meeting for writing lessons plans and forming the curriculum was scheduled for winter break.

Preschool Goals, Activities, and Assessments

In early February of 2016, the MCF contacted Monmouth University with a list of five schools that they wanted preservice teacher-led teams to visit and teach their pollination lessons. Additionally, the preservice teachers, in conjunction with their cooperating teachers, decided to teach the pollination curriculum in their clinical placements. A tentative schedule was made for early January to meet again and begin writing curriculum.

Goals for the Preschool Students

The overarching goal of the curriculum for the preschoolers was to have them understand the importance of the pollination process. Detailed goals are as follows:

- Students will demonstrate knowledge of healthy pollinators by speaking, writing, or drawing (with an emphasis on honeybees).
- Students will be able to demonstrate knowledge of the pollination process by speaking, writing, or drawing.
- Students will demonstrate that we eat different parts of flowers and plants by speaking, writing or drawing.

Preschool Student Activities

In order to assess each class, the unit would begin with what is commonly known as a KWL chart. These types of charts are large three-column charts; the first column is labeled **K** for What I **K**now, the second **W** for What I **W**onder, and the third **L** for What I **Learned**. KWL charts are useful before beginning an activity to activate students' background knowledge, learn their questions, and get familiar with their interests. KWL charts are also useful at the end of an activity to help children remember what they learned. KWL charts are one of the best ways to assess what preschool children have learned (Bredekamp, 2014). Additionally, according to NAEYC (2010), KWL charts help children learn to use nonfiction texts to find answers to their questions, learn that nonfiction texts are read differently than works of fictions, and learn about text features such as table of contents, captions et al.

A typical lesson would be conducted with a whole-class meeting in the rug area. The lead preservice teacher reads a book such as *Bee and Me* (McGuiness, 2008). The teacher would pause every few pages and check for understanding, asking questions such as, why is the boy afraid of the bee? Have you ever been afraid of a bee? What does the bee tell the boy that makes him unafraid? What important information did this book teach us? After the reading, talking, listening, and speaking about the book, the preschoolers have a chance to visit each of the four learning centers. One preservice teacher attends to each center, so that all children are able to understand the center's learning goals and participate to their best abilities. For example, in this lesson, one of the learning centers instructs students to create the act of pollination, using moist cotton balls and various colors of Kool-Aid powder to represent pollen. As the cotton balls drag through the different colored balls, they represent bees pollinating.

Students will participate in the HoneyBee Waggle Dance. The Honeybee Waggle Dance helps students learn how bees communicate with each other. It also teaches students important facts about honeybees e.g. bees create 43% of the United States' food supply. The teacher marks an area in the middle of the room to represent the hive. Then, the students practice waggling (shaking their abdomens!) while moving around. This is an extremely important part of the waggle dance. Have students practice doing a Round Dance. Remind them that this dance is used when a food source is near to the hive. Then the students practice dancing towards entrance, or dancing upwards in a hive, which bees use to communicate as "fly towards the sun." Dancing away from entrance mean or dancing downwards in hive is bee-speak for "I will fly away from the sun."

Each child will pick their own seeds (either Nasturtium, Sweet Pea or an Aquilegia) and plant their individual flower. Students will create replicas of the different parts of flowers: roots, stems, leaves, seeds, stamens, and stigma. Each class will list the plants that we eat. The instructor will say that all the foods listed below depend on bee pollinators, and engage in a dialogue that goes something like the following. Are these foods part of your favorite meal or snack? Think back on what you ate for breakfast or lunch. Could you have eaten all those foods if bees were no longer around to pollinate our crops? Children will discover that we eat all parts of a plant: roots, vines, fruits from trees, seeds et al. Children will sort real food by the parts of the plant it represents. For example, flax seeds, celery, spinach leaves, apples, and carrots would all be sorted into different categories.

In the Kool-Aid pollinator activity, the children recreate the pollination process by dipping moist cotton balls in different colors of Kool-Aid, which represent different flowers. This activity illustrates the act of pollination through tangible experiences in real-time and becomes a physical model for students to refer to later in their learning.

Students will also take part in the Pollination Relay Race. Each class is divided into two teams. Each team holds a honeybee and races to buckets filled with pollen coins. Each student (holding the bee) must visit the three flower buckets and deposit pollen coins into each flower, then return to the line. Another useful source is Pollination Song (Missouri Botanical Gardens, 2008), which teachers can play in their classrooms. Students may even learn the lyrics and sing along.

Teachers found that of all three activities, the Honey Bee Waggle Dance, the Seed Sort, the Kool-Aid Pollination, and the Pollination Relay Race, the children were most engaged in the Kool-Aid activity. At times, the dance and the race provided classroom management issues, which were directly related to teachers' years of classroom experience. For example, the novice teachers had more difficulty in maintaining the children's focus in tasks, which required moving about the classroom.

Assessing Preschool Students

Although there is an increased interest in early childhood science, no formal or informal assessment tools exist. The National Research Council (NRC) report on assessment in early childhood (Snow & Van Hemel, 2008) established that science assessments could not be included in their discussion because there "simply was not a basis in theory, research, or practice to include…science, despite its obvious importance" (p. 59). Instead, researchers indicate that the current evaluation process involves identifying evidence of children's scientific learning during everyday classroom activities by collecting data over time from multiple sources. These sources include actions, talk, and any form of literacy that children are able to do by themselves or in groups.

In addition to using the KWL charts, the preservice teachers had the students use science journals. Journals provide a motivation for children to observe carefully. There is a purpose to observing details because these will be incorporated into a drawing and description. Although the variance in artistic abilities may limit the details some children can symbolize clearly, teachers can engage in follow-up discussions so that children can describe their observations. When the children record, they choose among many possible features to represent. The actual features included in the drawing provide a window onto children's conceptual understandings. For example, even when children are observing and documenting the outside of a flower, a number of them will include the stem or stamen after being engaged in lessons regarding the pollination process.

Therefore, the preservice teachers used the data from the Learned column from their KWL charts as well as the information from the science journals to collect assessment data. They were also encouraged to be careful observers and take anecdotal notes while in the classrooms with the preschool children. They brought those qualitative findings back to the group once all the teaching sessions had been completed.

Two years following the preschool implementation, the cooperating school districts reported on the success and suggested modifications for the Pollinator Curriculum. Further, three of the original preservice teachers are now the teachers of record in their own preschool classrooms. All participants reported some similar feedback for the program implementation: provide opportunities each day for students to engage in science inquiry and learning by intentionally designing a project approach for thematic learning. All teachers found that when the pollination activities were interwoven through thematic learning, they were able to guide and focus children's natural interests and abilities better. They did this through carefully planned open-ended, inquiry-based explorations in all content areas, including literacy, social studies, and mathematics. A few of the newly minted teachers who designed the original unit invited guest-speakers into their classrooms. Local bee-keepers and farmers spoke about the importance of pollination for the good of their communities. Those teachers who did not have access to guest-speakers buoyed their curriculum using information gathered from interactive Internet sources such as YouTube and Skype.

In one instance, a teacher looped/moved up with her preschool class. This teacher chose to continue working with the theme of pollination. However, she posed new challenges to her students. Capitalizing on her students' burgeoning engineering skills, which the children had displayed during their initial exposure to the pollination lessons, this teacher challenged had students create a bee-hive. This required that the art teacher work in the classroom rather than having the children attend art as a 'special.' It was noted that during the creation of the bee-hive, children with special needs as well as the children who were English Language Learners (ELLs), became more actively involved. This heightened level of engagement was attributed to the deeper-levels of peer assistance during the creation of the hive. Both native English speakers as well as the ELLs developed richer interpersonal relationships during the creation of the hive.

Goals and Assessment of the Preservice Teachers

When the partnership between the MCF and the Monmouth University School of Education began, one goal was to form learning objectives and assessments for preservice teachers so that they might benefit along with the preschool students. Current research on practicing and preservice teachers demonstrates that teachers' attitudes toward science teaching have been understood as an important component for effective science education in early childhood classrooms (Kim & Cho, 2008). Further, preservice teachers' self-efficacy regarding teaching science in early childhood classrooms has a direct impact on how much science content they teach and how well they teach it (Sharapan, 2012).

Therefore, the authors thought it important for the preservice students to develop a high degree of self-efficacy and positive attitudes about teaching early childhood science content. This was measured by administering semi-structured interviews to each participant ($N = 8$) before the curriculum's writing and at the conclusion of the teaching sessions. In addition, the early childhood faculty member took anecdotal notes during the preschool teaching sessions after the preservice teachers held short debriefing sessions.

Preservice Teachers' Activities

The MCF held a raffle for schools in Monmouth County to have the preservice teachers come out and teach their pollination lessons. The preservice teachers decided that one preservice teacher would lead the lesson e.g. introduce the learning goals in a whole group context and then describe the goals for each learning center. The lesson leader would also be responsible for the overall assessment of each lesson. Four other preservice teachers would also accompany the lesson leader in order to facilitate each learning center and assess each center's learning goals. All of the materials for each lesson and learning center were purchased and paid for by a grant from the MCF. Materials included art supplies and manipulatives such as: seeds, soils, bucket, tangible and realistic food play items, age appropriate nonfiction books about bees, paper plates, Popsicle sticks, Kool-Aid, crayons, markers, and glue.

First, the group needed to educate themselves on the steps of pollination as well as the various sources of pollinators. Next, they had to decide exactly what information was developmentally appropriate for their three- and four-year-old audience. Finally, a consensus was reached, and the following goals were developed: Why are bee important to helping our plants grow? Can we help bees pollinate? What are the different animals that pollinate plants? Can we explain pollination process of flowers and bees? What can I do to help [the bees]?

Results

At the unit's end in each school ($N = 8$) the preservice teachers and the early childhood faculty member held a debriefing meeting to discuss their experiences and share their results. The three team leaders all reported that their "what we learned" columns on the KWL charts they completed in their clinical placements had much more information on them than the charts in the three schools designated by the MCF. They also reported that the students' science journals in their clinical placements showed greater details and depth of knowledge of the pollination process in the same schools versus the MCF designated schools. The faculty member examined samples of the science journals and concluded with the preservice teachers opinions. All of the children grasped the concept of pollination, but the students in the clinical placement schools used greater details in 60 – 70% of their total drawings (out of a total of three schools with a total of fifty-eight students). See example #1 and example #2.

Additionally, to measure whether or not the project's goals were met, the KWL charts and the science journals were analyzed. The goals set at the beginning of the unit were examined, and the group looked for evidence in the KWL charts and journals: Why are bee important to helping our plants grow? Can we help bees pollinate? What are the different animals that pollinate plants? Can we explain pollination process of flowers and bees? What can I do to help [the bees]? Students were able to answer all of the questions with an 80% degree of accuracy. However, once again, when looking at the overall spectrum of schools, the sites in which the preservice teachers had clinical experiences drew the highest numbers of details in science journals and the most numbers of answers under the "what we learned" column in the KWL charts. For example, each learning goal in the three clinical sites drew at least six responses in the L column. In the MCF affiliated schools, three out of the five schools had four as their highest responses in the L column. All responses were rated by the early childhood faculty along with another faculty member from the Monmouth University School of Education. None of the charts had the schools' names listed on them for the sake of anonymity.

Evidence of Preservice Teacher Learning

During the first round of interviews, the preservice teachers responded to the interview prompts such as: Did you enjoy learning science when you were very young? What do you remember of the teaching and learning experiences that you had? How do you feel about teaching science now? Are you aware of the scientific process? Typical responses were: "I don't really remember a lot of learning science when I was young, learning science when I got older (like middle and high school) was okay, as long as we did experiments," "I am afraid to teach science," "I don't feel like I will be good at teaching science," "I am not sure what the scientific process is, I would have to look it up," "I think that a big part of me volunteered to be a part of this project in order to help me feel more confident about teaching science."

Since the responses demonstrated poor past experiences, which had led to low levels of self-efficacy, the early childhood faculty member decided that the first aspect of the project needed to be education. Specifically, the education of the preservice students regarding the scientific method and the necessity of the STEAM curriculum in early childhood classrooms needed to be a priority. During the first meeting over winter break, the group conducted a very simple science experiment, making home-made bubbles. At the lesson's end, the group outlined their scientific method together and then compared it to the traditional method. The net result was that all of the participants ($N = 8$) reported feeling more confident working with the scientific method.

At the next meeting, we reviewed the scientific meeting and reviewed relevant content for the pollination curriculum. This is the time when the faculty member first noticed a difference in the group's self-efficacy levels. Preservice students who had seemed less interested in taking writing and teaching roles in the past few meetings were now more comfortable volunteering for more responsibilities.

During the meetings, the faculty member would also review key terms with the group such as constructivism, developmentally appropriate practice, and inquiry-based learning. These terms were looked at carefully in the context of teaching early childhood science. The group discussed subtle but powerful shifts in their own experiences and learning of science. It was important to develop and teach the pollination curriculum from a place of active exploration, with the teacher as facilitator, connecting the content to the children's experiences. This helps the content become flexible and integrated with other curricular areas. Using methods in science similar to those that the preservice teachers already knew how to work with seemed to create even a greater level of self-efficacy amongst the group.

While the preservice teachers followed the schedule for implementing the pollination curriculum, the early childhood faculty member met with the group twice to unpack their experiences. When asked individually if they were feeling stronger about teaching science in early childhood, each preservice teacher responded affirmatively. However, three of the group ($N = 3$) reported that they thought that their confidence might be connected in some way to being familiar with the curriculum.

At the conclusion of the pollination unit, the semi-structured interviews were administered again. This time, the responses to the questions were more positive. Some examples to some of the prompts were: "I actually enjoy the thought of planning for science – I look at it so differently now," "I can see how the science was there all along, the children are the ones doing all of the work," "I just have to help them figure some of the stuff out and they can do the rest. I am so impressed with the level of their thinking, they literally blew me away," "I really do believe that some of these kids are going to save the world," "I think that even we, as early childhood student teachers even gave them enough credit," "I have a whole different feeling about teaching science," and "Now I can see how it connects to everything and how it was all already there and how important it is."

DISCUSSION

Preschool children want and need to learn about their worlds, and as developmental research (Bredekamp & Copple, 2009, Piaget, 1955, Vygotsky, 1978) puts forth, young children become engaged when they are given chances to discover the environments and the natural world around them. The goal of science education in early childhood should be to encourage and refine children's instinctive passion for discovering their natural worlds. Another goal should be to assist them in channeling their joyful need to explore the world around them and assist them in shaping their energies and help them grow enthusiastic true scientific literacy (Editors, Scientific American, 2010). Children are biologically prepared to learn about the world around them, just as they are biologically prepared to learn to walk and talk and interact with other people. Some particularly notable points follow.

1. This case showed how kids can be engaged in STEM learning in early child.
2. It also showed how teacher preparation can be enhanced through experiential learning.
3. Most importantly is showed how accomplishing those goals together can create synergy for learning at both levels.

Developing an appropriate constructivist practice that combines the social, cognitive, and content related aspects that meets the needs of all students is an exceedingly difficult task (Bredekamp, 2014). More research is needed in how to facilitate teachers as they struggle to enact the inquiry-based classroom approaches called for in reforms (Sharapan, 2012).

REFERENCES

Bianchini, J., & Colburn, A. (2000). Teaching the nature of science through inquiry to prospective elementary teachers: A tale of two researchers. *Journal of Research in Science Teaching*, *37*(2), 177–209. doi:10.1002/(SICI)1098-2736(200002)37:2<177::AID-TEA6>3.0.CO;2-Y

Bredekamp, K., & Copple, S. (2009). *Effective practices in early childhood education: Building a foundation* (3rd ed.). Upper Saddle River, NJ: Pearson.

Chaille, C., & Britain, L. (2002). The young child as scientist (3rd ed.). Boston: Allyn & Bacon.

Conezio, K., & French, L. (2002). Science in the preschool classroom. *Young Children*, *42*(3), 1–5.

Fitzgerald, A., & Schneider, K. (2013). What teachers want: Supporting primary school teachers in teaching science. *Teaching Science*, *59*(2), 7–14.

Gerde, H., Schachter, R., & Wasik, B. (2013). Using the scientific method to guide learning: An integrated approach to early childhood education. *Early Childhood Education Journal, 41*(5), 315–323. doi:10.100710643-013-0579-4

Grennon-Brooks, J. (2011). The courage to be constructivist. *Educational Leadership, 57*(3), 18–24.

Guteral, F., & ... . (2010). How we are not learning science. *Scientific American*, 23–30.

Johnson, J. (1999). The forum on early childhood science, mathematics, and technology education. In Dialogue on Early Childhood Science, Mathematics, and Technology Education. Washington, DC: AAAS.

Llewellyn, D. (2002). *Inquire within* (2nd ed.). Thousand Oaks, CA: Corwin Press.

National Research Council. (2000). Retrieved from http://www.nationalacademies.org/nrc/

National Research Council (NRC). (2012). *A framework for K–12 science education: Practices, crosscutting concepts, and core ideas*. Washington, DC: National Academies Press.

Patrick, H. (2014). *Engaging young children with informational books*. Thousand Oaks, CA: Corwin. doi:10.4135/9781452284903

Piaget, J. (1955). *The construction of the reality in the child*. Basic Books.

Sharapan, H. (2012). From STEM to STEAM: How early childhood educators can apply Fred Rogers' approach. *Young Children, 67*(2), 36–40.

Snow, K., & Van Hamel, C. (2008). *Early childhood assessment: Why, what, and how*. The National Research Council.

The White House. (2014). *Fact Sheet: The economic challenge posed by declining pollinator Populations*. Retrieved from: https://obamawhitehouse.archives.gov/the-press-office/2014/06/20/fact-sheet-economic-challenge-posed-declining-pollinator-populations

Tu, T. (2006). Preschool science environments: What is available in an early childhood classroom. *Early Childhood Research Journal, 33*(4), 245 – 251.

KEY TERMS AND DEFINITIONS

Assessment: The process of determining the level of student dispositions, skills, and understandings.

Discovery Learning: A constructivist-based approach to learning that engages children in an inquiry of their environment.

Environmental Education: Engages young children with the natural world to promote their understanding of conservationism.

Inquiry-Based Learning: Challenges children to learn through investigating and reasoning about a problem.

Project-Based Learning: An approach to teaching that engages children in exploring their environment.

Science Education: Engages students in learning opportunities that capitalize on their natural curiosity for observing, exploring, and discovering the world around them.

Scientific Method: Involves teaching young children how to identify a problem, collect data, and test their hypotheses.

Chapter 10
Cases on STEAM Education in Practice Catapults and History of Catapults

Warren James DiBiase
University of North Carolina at Charlotte, USA

Judith R. McDonald
Belmont Abby College, USA

Kellan Strong
University of North Carolina at Charlotte, USA

EXECUTIVE SUMMARY

The catapult was one of the most effective and deadly weapons in medieval siege warfare. They have been critical to warfare since ancient times and have been used by Greek, Roman, and Chinese warriors to vanquish their enemies. This case will present a problem-based scenario where students will assume the identity of an astronaut stranded on Mars. Like the character in the Disney film The Martian, the astronaut only has a small collection of "spare parts" at their disposal. In an effort to fend off predators, they must design and construct a catapult.

LITERATURE REVIEW

Engineering design activities can be a powerful entry point into science learning. Design-build projects are a great way to encourage students to use creativity within projects to tackle a problem (Wicklein, 2006). As students develop an understanding of design-build projects, teachers can challenge them by presenting

DOI: 10.4018/978-1-5225-9631-8.ch010

Figure 1. Engineering design triangle: next generation science standards, grades 9-12

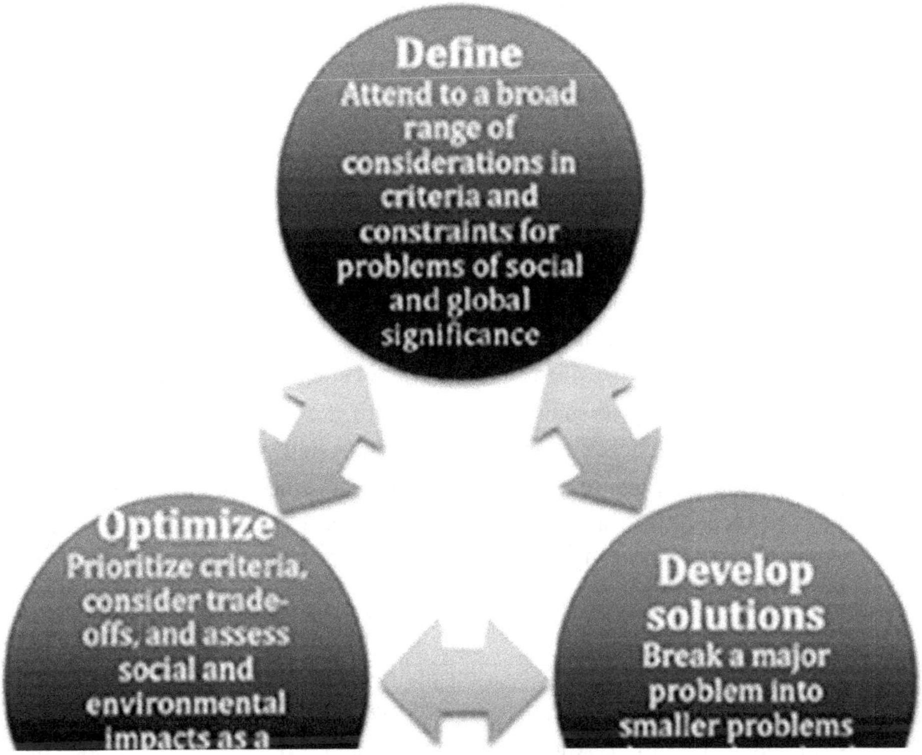

them with a local community-based engineering issue to solve. Selecting a local issue means that it becomes relevant to students. Engineering design-build projects, "hands-on," or "learning by doing" is grounded in constructivist theory (Fortus, Krajcikb, Dershimerb, Marx, & Mamlok-Naamand, 2005) that is shown to improve student achievement in higher level cognitive tasks, such as scientific processes and mathematic problem solving (Satchwell & Loepp, 2002). Current research indicates that when students are given a project-based task, their interest in science, technology, engineering, and mathematics (STEM) can be increased because it requires them to solve genuine, real-life problems (Fortus, Krajcikb, Dershimerb, Marx, & Mamlok-Naamand, 2005). Teachers today are challenged in many ways ranging from administrative tasks to having students reach state mandated test scores. This form of instruction (STEM) provides an engaging methodology that allows the curriculum to be relevant while enabling teachers to provide students with important skills. Students will use critical thinking and problem-solving skills, collaboration, communication, and creativity all while learning required content. The integration of art and engineering embodies creativity. Applying art and design to physics as a means to efficiently solve a problem epitomizes innovation because the two disciplines may seem incompatible on the surface. Students will experience or be introduced to STEM careers while being challenged to solve a problem.

CONTENT INFORMATION

This engineering design-build project integrates many content areas. Students are first asked to research catapults throughout history and identify effective designs. Students will use high quality, reliable Internet resources while utilizing quality search terms to seek the information necessary to design an effective catapult. Students will also identify the time period in history when catapults were used and their purposes. Some students may ponder how this device could be used today. As the students continue designing their catapults, they will need to collaborate and communicate with other group members, sharing their thoughts and ideas.

Mathematics is an important component in this activity. Students will investigate many variables, and perhaps identify many more during this investigation. The first variable that students will most likely want to investigate is the angle of launch. If the angle is too great, the projectile will go too high which will result in a shorter distance launched. On the other hand, an angle that is too small may overshoot the target. It is important to allow students time to test various angles. Figure 2 explains the angle of launch. Another variable that could be tested is rubber band resistance. If possible, provide various tensions of rubber bands which will affect the projectile's speed. The greater the tension, the greater the projectile's speed. An additional variable that students can experiment with is the distance between the catapult and the target. One final variable available for student exploration is the length of the catapult's launching arm. How will the launching arm's length, in conjunction with varying launch angles, change the catapult's accuracy? Encourage students to explore all these variables and create a data table to collect the changes in their variables.

A general summary of the content covered with this activity is:

English/LA

- Using accurate and reliable online sources
- Identifying helpful search terms and keywords
- Communicating findings

History

- Identifying time period and uses of catapult

Figure 2. Angle of launch using the laws of trigonometry which relate the initial velocity to the maximum height and distance of the projectile. V= velocity and X is the launch angle

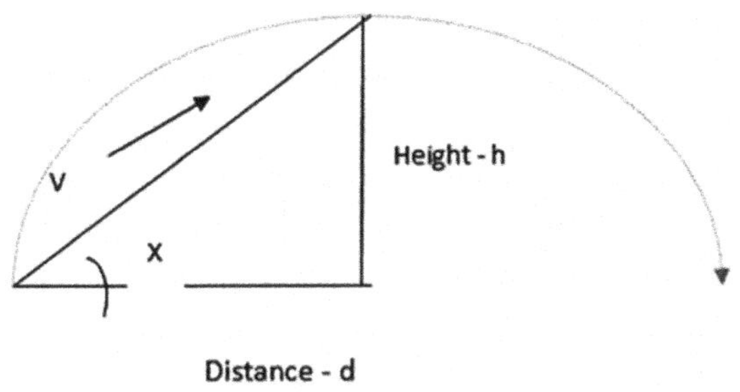

Mathematics

- Identify/change variables
- Angle height vs. distance
- Object's mass

Art

- Design effective catapult
- Explore materials in the design

Physical Science

- Force and Motion – movement of catapult

INTEGRATING STANDARDS

Using Common Core State Standards Connections

For the Middle Grades

Art is incorporated into the lesson with the catapult's design and creation of an object that will launch another object as far as possible or to a target. Leonardo da Vinci saw no boundaries between the arts and science. During this activity many art skills will be utilized and acquired. One is exploring different ways to design and create a working catapult. Another skill is researching designs and materials for maximum effective, efficient use (i.e., determining what makes most sense). Finally, the skill of creating a catapult that is as economic as possible (using the suggestions in the extension part of the lesson plan) will be acquired (Sousa & Pilecki, 2013).

Engineering Standards Met by This Activity

Students who demonstrate understanding can:

- MS-ETS1-1. Define the criteria and constraints of a design problem with sufficient precision to ensure a successful solution, taking into account relevant scientific principles.
- MS-ETS1-2. Evaluate competing design solutions using a systematic process to determine how well they meet the criteria and constraints of the problem.
- MS-ETS1-3. Analyze data from tests to determine similarities and differences among several design solutions to identify the best characteristics of each that can be combined into a new solution to better meet the criteria for success.
- MS-ETS1-4. Develop a model to generate data for iterative testing and modification of a proposed object, tool, or process such that an optimal design can be achieved.

English/Language Arts Standards Met by This Activity

Students who demonstrate understanding can:

- ELA/Literacy - RST.6-8.1 Cite specific textual evidence to support analysis of science and technical texts. (MSETS1-1),(MS-ETS1-2),(MS-ETS1-3)
- RST.6-8.7 Integrate quantitative or technical information expressed in words in a text with a version of that information expressed visually (e.g., in a flowchart, diagram, model, graph, or table). (MS-ETS13)
- RST.6-8.9 Compare and contrast the information gained from experiments, simulations, video, or multimedia sources with that gained from reading a text on the same topic. (MS-ETS1-2),(MS-ETS1-3)
- WHST.6-8.7 Conduct short research projects to answer a question (including a self-generated question), drawing on several sources and generating additional related, focused questions that allow for multiple avenues of exploration. (MS-ETS1-2)
- WHST.6-8.8 Gather relevant information from multiple print and digital sources, using search terms effectively; assess the credibility and accuracy of each source; and quote or paraphrase the data and conclusions of others while avoiding plagiarism and following a standard format for citation. (MS-ETS11)
- WHST.6-8.9 Draw evidence from informational texts to support analysis, reflection, and research. (MS-ETS1-2) SL.8.5 Integrate multimedia and visual displays into presentations to clarify information, strengthen claims and evidence, and add interest. (MS-ETS1-4)

Mathematics Standards Met by This Activity

Students who demonstrate understanding can:

- MP.2 Reason abstractly and quantitatively. (MS-ETS1-1), (MS-ETS1-2), (MS-ETS1-3), (MS-ETS1-4)
- 7.EE.3 Solve multi-step, real-life, and mathematical problems posed with positive and negative rational numbers in any form (whole numbers, fractions, and decimals), using tools strategically. Apply properties of operations to calculate with numbers in any form; convert between forms as appropriate; and assess the reasonableness of answers using mental computation and estimation strategies. (MSETS1-1),(MS-ETS1-2),(MS-ETS1-3)
- 7.SP Develop a probability model and use it to find probabilities of events. Compare probabilities from a model to observed frequencies; if the agreement is not good, explain possible sources of the discrepancy. (MS-ETS1-4)

Social Studies Standards Met by This Activity

Students who demonstrate understanding can:

- D2.His.13.6-8. Evaluate the relevancy and utility of a historical source based on information such as maker, date, place of origin, intended audience, and purpose.

High School

Engineering Standards Met by This Activity

Students who demonstrate understanding can:

- HS-ETS1-2. Design a solution to a complex real-world problem by breaking it down into smaller, more manageable problems that can be solved through engineering.
- HS-ETS1-3. Evaluate a solution to a complex real-world problem based on prioritized criteria and trade-offs

English/Language Arts Standards Met by This Activity

Students who demonstrate understanding can:

- ELA/Literacy - RST.11-12.7 Integrate and evaluate multiple sources of information presented in diverse formats and media (e.g., quantitative data, video, multimedia) in order to address a question or solve a problem. (HSETS1-1),(HS-ETS1-3)

- RST.11-12.8 Evaluate the hypotheses, data, analysis, and conclusions in a science or technical text, verifying the data when possible and corroborating or challenging conclusions with other sources of information. (HS-ETS1-1),(HS-ETS1-3)
- RST.11-12.9 Synthesize information from a range of sources (e.g., texts, experiments, simulations) into a coherent understanding of a process, phenomenon, or concept, resolving conflicting information when possible. (HS-ETS1-1),(HS-ETS1-3)

Mathematics Standards Met by This Activity

Students who demonstrate understanding can:

- MP.2 Reason abstractly and quantitatively. (HS-ETS1-1),(HS-ETS1-3),(HS-ETS1-4)
- MP.4 Model with mathematics. (HS-ETS1-1),(HS-ETS1-2),(HS-ETS1-3),(HS-ETS1-4)

Social Studies Standards Met by This Activity

Students who demonstrate understanding can:

- D2.His.5.9-12. Analyze how historical contexts shaped and continue to shape people's perspectives.

CASE DESCRIPTION

This case will present a project-based scenario in which students will take the place of an astronaut stranded on Mars. Like the characters in the Disney film *The Martian*, the astronauts only have a small collection of "spare parts" at their disposal to ensure survival. In this scenario the astronauts meet predators. In an effort to fend them off, they must design and construct a catapult. The students will be engaged in a deep dive. This deep dive exploration should take a maximum of eleven days, but it can be adapted to suit students' learning needs.

This science, technology, engineering, arts, and mathematics (or STEAM) inquiry activity is a deep dive exploration. It is a deliberate "dive" into a content area of focus. In this case the area of focus is catapults. During this process students will research the history, use, and structure of catapults. As stakeholders in the problem, students will assess their situations, take an inventory of "spare parts," and brainstorm with the group on how to creatively design a catapult using the best combinations of materials. The deep dive continues as the students test, revise, and retest their designs. Students must not only assess the situation with critical thinking; they must be able to communicate their ideas to their group and think creatively. This activity will actively engage students and integrate many content areas.

Materials (or "spare parts") per group:

- 1 cardboard shoebox
- 4 rubber bands
- 2 popsicle sticks
- 1 6-inch piece of masking tape
- 1 plastic spoon
- 1 ruler
- 1 pair of scissors
- 1 marshmallow

Engage: Day 1

Begin the lesson by sharing the scenario:

Your team has been selected to explore Mars for the possibility of colonization. You and your team have been on Mars for approximately 6 months. Regular and routine communication has been occurring between you and NASA. Things have been mostly proceeding according to plan. At 3:00 in the morning your team has been awoken to an alarm, which means that something very troubling has been discovered by NASA. Unfortunately, a group of aliens are approaching at a slow pace, but they are dangerous and must be stopped. Because NASA didn't foresee this danger, you and your fellow astronauts must create a defense. After taking inventory of the materials you have and consulting with the experiences at NASA, it seems that the only spare parts you have are materials to create a catapult. It is impossible for NASA to send supplies in the time that the aliens could approach you because you are forty-nine million miles from Earth, and the aliens could approach within the next month. You and your team will need to research catapult design, figure out the best use of your materials, test, redesign, and prepare for an alien assault.

Next, using some of the resources at the end of this chapter, show pictures and review the history of the catapult contained in this chapter. There are several YouTube videos available to use to introduce catapults. This would also be a good time to have the students share facts they are aware of, their hunches, and questions they have about this engineering task. Also try to elicit the information they might need in order to design, construct, and test a catapult.

Explore: Days 1 and 2

Place the students into groups of three or four. Provide groups with a list of the spare parts available to them. Have the students spend the remainder of day one researching catapult designs and how to best use their materials. Teachers may want to spend a few minutes at the beginning of day two to discuss some of the mathematics associated with this activity such as the angle of catapult, the force needed to hit a target, and the length of the catapult arm. Allow more research time if necessary.

During a whole-class discussion, have the students share what they learned from their research, perhaps in a jig-saw or a gallery walk. Working in groups of three or four, have the students brainstorm and sketch a design for their catapults. During this step it is important to focus on the details of how materials will be used. Equally important is making predictions about how material use will impact the catapult's length and force. Allow plenty of time for this step, and demand details on the design sketches. Once students' designs are completed and have been approved by the head of NASA design team (the teacher), the students will share their ideas. Remind them that scientists often collaborate to gather the best information.

At this point have the students post their groups design and do a graffiti walk. This is a strategy similar to a gallery walk; students will post their work, and others walk around with Post-It notes asking questions about their work or giving compliments. Instructors may need to model appropriate comments for the Post-It notes so that students are providing constructive criticism during the activity. Allot students time to read the comments and perhaps make changes to their design.

More Engage: Days 3 – 5

Students will begin building their catapults. Tell students that when engineers, or in this case astronauts, are given a certain amount of materials, they must be used to their best ability. Allow enough time for groups to design and create a working catapult. If any group is struggling allow them to consult with NASA (the teacher) to assist them with their construction. Take the students into the hall, gymnasium, or outside to test their catapults and determine the effectiveness of their designs. At this point they are looking for the catapult to launch the marshmallow the farthest, keeping the aliens away. This is a good time to allow some of the variables, such as angle of launch or length from the target be explored. Record all data. Examine the sample data table for students to copy into their lab notebooks.

Have the teams once again brainstorm and sketch a redesign of their catapult on poster paper. These redesigns are informed by their experience testing their catapults. Refer to the data table in the appendix. Review some of the variables prior to the redesigning activity. After the students have completed their redesigns, have them post these designs around the classroom. Have the students do another graffiti walk, placing their Post-It note comments on each redesign. Review the students' comments with the class.

Explain: Days 6 – 8

The teacher will assign a number, 1- 4 to each member of each group to create teams. These teams will be determined from the prior week's observation of students' strengths. See the table below to get ideas for some characteristics for each type of team member. Create a design team, composed of the #1 member from each group, and charge them with sketching a design of a catapult incorporating students' feedback and the best components of the entire classes' designs. Create an engineering materials team, composed of the #2 member from each group, and charge them with gathering all the materials needed to build the catapult. Group #2 will also collect the data during the testing of the catapult. Create a building team composed of the #3 member from each group, and charge them with building the catapult designed by the design team. Create a testing team composed of the #4 member from each group, and charge them with testing the catapult to see how far the marshmallow is projected.

Extension: Days 9 and 10

Gather the class, and challenge all groups to build upon the class design to craft a catapult that will launch the marshmallow even farther. They are to use the current design and add to it, with the goal of launching the marshmallow farther in mind. Allow each group one class period to design and one to build the object. Follow the protocol used before in testing the projectile.

Evaluation: Day 11

Compare the new catapult's data to the recorded data of previous designs. Debrief the activity with students. Have the students compare and contrast their designs with the pictures shared at the beginning of this learning experience. Ask them to write about a scenario in which a catapult could be used today.

STEAM TEACHING

What is STEAM education, and why is it important? Switching from STEM to STEAM is not a large leap at all. After all, well-executed STEM education already

Table 1.

Design Team Number	Skills Necessary	Assigned Student
1 -Sketching Design	Drawing, collaborative, flexible, creative	
2- Materials Manager and Data Manager	Organized, detailed, good with data tables	
3 – Building Team	Follows plans, asks good questions, creative	
4 – Testing team	Flexible, gets team excited	

involves writing, communication, design, and often creativity that engages the arts. STEAM includes the arts which is not limited to the visual arts but inclusive of theatre and music. The National Art Education Association (NAEA) defines arts education as music, drama, dance, media arts or visual arts, including folk arts. Ruth Catchen (2013), an artist turned STEAM curriculum researcher, is asking the question, what is the learning impact on students? She is currently investigating this question using a new STEAM program located in Colorado. She believes that introducing the arts into STEM can benefit the underrepresented students as an on-ramp to the STEM curriculum. Traditionally, STEM learning is focused on qualitative work and memorization, and STEAM allows the arts to integrate with STEM which opens more paths for diverse learners without fear (Jolly, 2014). In order for democracy to function, twenty-first century students need students to become global citizens who are problem solvers and critical thinkers, even if they choose a career that does not directly involve STEM. STEAM education address this for all students.

Art and science have a common theme, discovery! In each of the disciplines we are discovering. The power of the arts aligns with science objective outcomes, discovery and exploration. Both art and science promote curiosity, require observation to ensure accuracy, construct meaning through observation, and allow collaboration. As educators, integrating content areas with the arts makes teaching more interesting, develops cognitive growth, and improves long term memory (Sousa & Pilecki, 2013).

A study conducted at John Hopkins University concluded that arts integration significantly improved retention of learning. They cited seven effects:

1. practicing the learned information or skills
2. expanding meaning to learning
3. seeking out more information
4. acting out their learning
5. talking more about the content

6. connecting emotionally to the learned material
7. representing the material through drawing

All these effects help students learn through multiple instructional strategies using different parts of their brain (Rinne, Gregory, Yarmolinskaya, and Hardiman, 2011).

Examining this activity, *Constructing a Marshmallow Catapult*, it becomes clear that many skills from the arts align with the practices of science and engineering. When students are asking questions and defining the problems associated with the designing and creating the catapult, they are fully integrating STEAM into this engineering activity. This unit asks the students to design and create a model, which blends art and STEM skills together. Today's Next Generation Science Standards (NGSS) and math Common Core standards require the quantitative skills necessary to developing a model while using mathematics and computational thinking. These STEM practices dovetail nicely with the art skill of researching the material used for the design and what will work the best for the purpose of the catapult.

The bottom line is that educators need to create individuals who are FULLY literate in the twenty-first century using the four C's – collaboration, communication, critical thinking and creativity. STEAM is the perfect way to achieve this!

CURRENT CHALLENGES FACING TEACHERS

Along with many obvious challenges including additional preparation time for teachers, the need for additional materials and resources may seem insurmountable. This is a problem for all science classrooms, yet students benefit from design-build activities. Some of the resources may include computer lab time and perhaps library time, which is often a scheduling nightmare. With a supportive administration and collaborative team approach, these challenges are manageable. This collaborative approach would include not just the science department but perhaps the English, mathematics, technology, and social studies departments as well. Collaboration allows each teacher to share and communicate their content expertise along with the standards that could be highlighted with this activity.

Addressing teacher attitudes towards shifting teaching practices may not seem obvious, but learning a new pedagogy and how to implement that change can cause a lot of controversy among faculty. Additionally, in integrated programs, teachers from different departments or areas of expertise find themselves forced to learn new content, which is often challenging material. Research indicates that implementing these collaborative types of programs had a direct impact on student learning. Again, all the aforementioned challenges can be minimized through a supportive administration, continuous staff development, and perhaps even a contracted specialist or university professor who focuses on the specific needs of teachers transitioning to a new way of teaching (Satchwell & Loepp, 2002). It is very important for the teachers who are implementing the new program to be given the opportunity to practice this instructional strategy prior to teaching it to a group of students through comprehensive professional development. Leading teachers through the student materials as a teacher development strategy goes a long way toward preparing and motivating teachers to embrace the pedagogical approach (Diaz & King, 2007).

SOLUTIONS AND RECOMMENDATIONS

Research indicates that teachers tend to teach as they were taught; change is difficult to manage; and once a new innovation is adopted, increased student learning is all that is required for the teacher to sustain the change. This project-based-learning (PBL) deep dive is an inquiry learning experience. The students, in the role of a team stranded astronauts, work to find a solution to a problem. Furthermore, it is a STEAM activity. While exploring this activity, the students will experience a deep dive in engineering and the science, technology, art, and mathematics of designing and redesigning a catapult. As such, students will not only be exposed to the process of engineering, but also how all the components of STEAM are interrelated. As for teachers, if they decide to take a risk and do an inquiry activity, they will be pleased to see that the students are not only engaged but learning the content and concepts.

THE HISTORY OF CATAPULTS

An informal definition of a projectile is any object that is cast, fired, flung, heaved, hurled, pitched, tossed, or thrown. The path of a projectile is called its trajectory (Elert, 2018). Some examples of projectiles include:

- a baseball that has been pitched, batted, or thrown
- a bullet the instant it exits the barrel of a gun or rifle
- a bus driven off an uncompleted bridge
- a moving airplane in the air with its engines and wings disabled
- the space shuttle or any other spacecraft after main engine cut off (MECO)
- an object hurled by a catapult.

Gravity is the primary force acting on a projectile. In Aristotle's theory of motion, projectiles were pushed along by an external force which was transmitted through the air. He called this force impetus. This impetus caused the object to move in a straight line until it was expended, at which point the object fell straight to the ground.

Under closer inspection and when viewing projectiles traveling greater distances, it becomes clear that projectiles do not behave in this manner. With the spread of the cannon in warfare (beginning AD 1248), the study of projectile motion had taken on greater importance. With more careful observation and more accurate representation came the realization that projectiles did not move the way Aristotle stated. The path of a projectile did not consist of two consecutive straight-line components; instead, it was a smooth curve. Galileo (born 1564) realized that falling bodies picked up speed at a constant rate due to gravitational force and called that force acceleration. Using an inclined plane, Galileo had performed experiments on uniformly accelerated motion, and he now used the same apparatus to study projectile motion. He reasoned that a projectile shot from a cannon is not influenced by only one motion, but by two. The motion that acts vertically is the force of gravity, and this pulls the projectile down by the times-squared law (acceleration due to gravity). While gravity is pulling the object down, the projectile is also moving forward horizontally at a constant velocity (Wilson, 2016). Galileo placed an inclined plane on a table and provided it with a curved piece at the bottom which deflected an inked bronze ball into a horizontal direction. The ball thus accelerated, rolled over the table-top with uniform motion, then fell off the edge of the table where it hit the floor, and left a small mark. The mark allowed the horizontal and vertical distances traveled by the ball to be measured (Drake, 1974). By varying the ball's horizontal velocity and vertical drop, Galileo was able to determine that the path of a projectile is parabolic. According to Elert (2018),

An essential characteristic of a projectile is that its future has already been preordained. Batters may apply "body English" after hitting a long ball, but they do so strictly for psychological reasons. No amount of leaning to one side will make a foul ball turn fair. Of course, the pilot of a disabled airplane may regain control before crashing and avert disaster, but then the airplane wouldn't be a projectile anymore. An object ceases to be a projectile once any real effect is made to change its trajectory. The trajectory of a projectile is thus entirely determined the moment it satisfies the definition of a projectile. (Para. 2)

The catapult was one of the most effective and deadly weapons in medieval siege warfare. They have been critical to warfare since ancient times and have been used by Greek, Roman, and Chinese warriors to vanquish their enemies. War was common throughout Europe and the Middle Ages. Thick, protective stone walls were often built around cities and castles to prevent attacks from outsiders. This made catapults ideal for launching missiles. Although many different objects were utilized as missiles,

Figure 3. Projectile motion

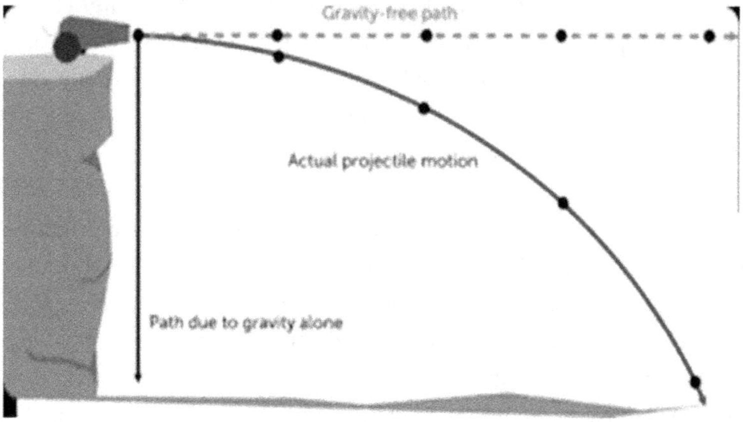

the function of the catapult was to launch the object either directly at the wall to cause maximum damage or over the wall to strike the population within the walls' protective barriers. A website documenting the history of catapults explains that:

Catapults have been integral to siege warfare since antiquity. Though ancient Catapults were one of the most effective weapons in siege warfare. Various types of Catapults have been used by the Greeks, Romans, and Chinese. The first catapults were early attempts to increase the range and power of a crossbow. Diodorus Siculus, a Greek historian, was the first to document the use of a mechanical arrow firing catapult (early ballista) in 399 BC. Catapults as we think of them were introduced to Europe during the Middle Ages. Catapults made their exhibition in England in 1216 during the Siege of Dover, the French crossed the Channel and were the first to use Catapults on English soil. With war prevalent throughout Europe during the Middle Ages the popularity of fortified castle and city wall rose significantly. This made Catapults an essential. Catapults were used to launch missiles (many different objects were utilized). These missiles were either launched directly at the wall to cause maximum damage to the fortifications or were launched over the wall to lay siege on the population within the protective walls. Catapults were also used to throw missiles at soldiers. The first accounted acts of Biological Warfare involved catapults. Bodies of the diseased were hurled over the city walls to infect the residing citizens. Catapults were used extensively throughout Europe (predominantly by the French) until AD 885-886 when new defense systems rendered catapults ineffective. (sites.google.com/site/physicsofcatapults/home/history-of-catapults).

There are three main types of catapults that were used in medieval warfare. The first is the ballista, which was an attempt to increase the range and power of another weapon, the crossbow. The word *ballista* comes from the Greek word *ballistes,* which means "to throw." Invented by the Greeks and modified by the Romans, the ballista was created to amplify the range and power of the crossbow. The early ballista in ancient Greece was developed from two weapons called oxybeles and gastraphetes (Wilson, 2016). The gastraphetes ('belly-bow') was a handheld crossbow. It had a composite prod and was spanned by bracing the front end of the weapon against the ground while placing the end of a slider mechanism against the stomach. The operator would then walk forward to arm the weapon while a ratchet prevented it from shooting during loading. This produced a weapon which, it was claimed, could be operated by a person of average strength but which had a power that allowed it to be successfully used against armored troops. The oxybeles was a bigger and heavier construction employing a winch and was mounted on a tripod. It had a lower rate of fire and was used as a siege engine. With the invention of torsion spring bundle technology, the first ballista was built. The advantage of this new technology was the fast relaxation time of this system. Thus, it was possible to shoot lighter projectiles with higher velocities over a longer distance. For an oxybeles, the rules of a torsion weapon demanded that the more energy could be stored, the thicker the prod and the heavier the projectile had to be to increase the amount of stored energy delivered to the projectile. The earliest form of the ballista is thought to have been developed for Dionysius of Syracuse, circa 400 BC (Wikepedia https://en.wikipedia.org/wiki/ballista) The ballista was an invaluable Medieval siege attack weapon. The ballista design was similar to a giant crossbow and worked by using tension. The ballista was designed to aim huge wooden, iron-clad darts or arrows which were powered by twisted skeins of rope, hair, or sinew. The ballista design was based on a huge dart-throwing machine. The ballista loosed heavy bolts, darts and spears along a flat trajectory. The force of the missiles launched from the ballista was designed to have great penetration and was capable of skewering several of the enemy at one time

The second type of catapult is the *mangonel* (from the Latin word "manganon," meaning "engine of war"), which is what many modern people think of when they think of a catapult. A mangonel, also called the traction trebuchet, was a type of catapult or siege engine used in China starting from the Warring States period, and later across Eurasia in the 6th century AD. Unlike the earlier torsion engines and later counterweight trebuchet, the mangonel operated on manpower pulling cords attached to a lever and sling to launch projectiles. Although the mangonel required more men to function, it was also less complex and faster to reload than the torsion powered ballista and onager which it replaced in early Medieval Europe (https://en.wikipedia.org/wiki/Mangonel).

Figure 4. Ballista weapon

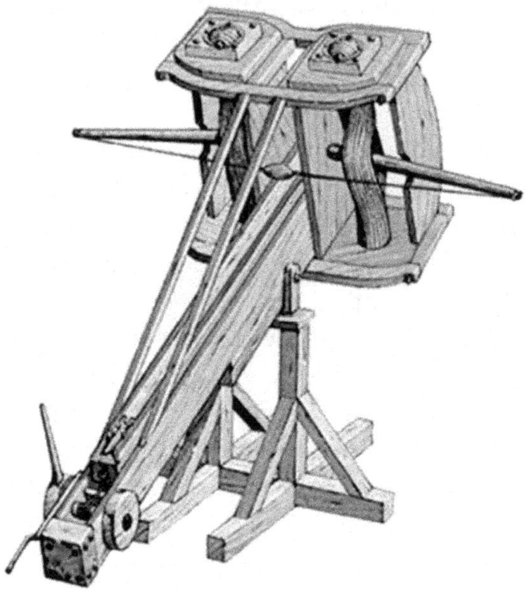

The mangonel was invented by the Romans in 400 BC. The missiles fired with mangonels included anything from stones, human excrement, or rotting animal carcasses. It consisted of a long wooden arm with a bucket (though earlier models used a sling) with a rope attached to the end. The arm is then pulled back from a 90-degree angle, the bucket loaded, and the energy stored in the tension of the rope and arm would be released. Upon release, the mangonel's arm would return to its position. When it came into contact with the beam, the arm would stop, but the missiles stored in the bucket would continue on its path toward the enemy. The mangonel was capable of firing 1,300 feet and was fairly easy to construct. Wheels were sometimes added to the design to increase mobility.

The third type of catapult is the trebuchet, which was designed for maximum force. Created by the Chinese in 300 BC, the trebuchet was the most powerful of the catapults. It arrived in Europe around AD 500, and the stones it hurled had the destructive force that could demolish city and castle walls. The most commonly used ammunition was stones, but darts and sharp wooden poles could be used if necessary. This device consisted of a long arm up to sixty feet long, balanced on a fulcrum that was off center. The short arm was attached to a counterbalance, a heavy lead weight, or a pivoting ballast box filled with dirt, sand, or stones. A sling was attached to the end of the long arm, and a rope was attached to the long arm and pulled down until the counterbalance was high in the air. The sling was then loaded

Figure 5. Mangonel

with projectiles, then the rope released and the energy in the long arm brought to an abrupt stop. The projectile, however, continued its path to its target. The trebuchet was a feared siege weapon and the target of reconnaissance missions sent to burn them to the ground.

Catapults were used extensively throughout Europe (predominantly by the French) until AD 885-886 when new defense systems rendered catapults ineffective.

Figure 6. Trebuchet

REFERENCES

Catchen, R. (2013). Reflections: How STEM becomes STEAM. *The STEAM Journal*, *1*(1), 22. doi:10.5642team.201301.22

Diaz, D., & King, P. (2007). *Adapting a post-secondary STEM instructional model to K-5 mathematics instruction*. Clemson, SC: Clemson University.

Drake, S. (1973). Galileo Gleanings XXII: Galileo's experimental confirmation of horizontal inertia: Unpublished manuscripts. *Isis*, *64*, 291–305. doi:10.1086/351124

Elert, G. (2018). *The physics hypertextbook*. Retrieved from https://physics.info/projectiles/

Fortus, D., Krajcikb, J., Dershimerb, R., Marxc, R., & Mamlok-Naamand, R. (2005). Design-based science and real-world problem-solving. *International Journal of Science Education*, *27*(3), 855–879. doi:10.1080/09500690500038165

Fortus, D., Krajcikb, J., Dershimerb, R. C., Marx, R. W., & Mamlok-Naamand, R. (2005). Design-based science and real-world problem solving. *International Journal of Science Education*, *27*(7), 855–879. doi:10.1080/09500690500038165

History of Catapults - Physics of Catapults. (n.d.). Retrieved January 17, 2016 from https://sites.google.com/site/physicsofcatapults/home/history-of-catapults

Jolly, A. (2014). STEM vs. STEAM: Do the arts belong? *Education Week Teacher*. Retrieved February 21, 2016, from http://www.edweek.org/tm/articles/2014/11/18/ctq-jolly-stem-vssteam.html

Rinne, L., Gregory, E., Yarmolinskaya, J., & Hardiman, M. (2011). Why arts integration improves long-term retention of content. *Mind, Brain and Education: the Official Journal of the International Mind, Brain, and Education Society*, *5*(2), 89–96. doi:10.1111/j.1751-228X.2011.01114.x

Satchwell, R., & Loepp, F. L. (2002). Designing and Implementing an Integrated Mathematics, Science, and Technology Curriculum for the Middle School. *Journal of Industrial Teacher Education*. Retrieved February 14, 2014 from http://scholar.lib.vt.edu/ejournals/JITE/v39n3/satchwell.html

Sousa, D. A., & Pilecki, T. (2013). *From STEM to STEAM: Using brain-compatible strategies to integrate the arts*. Thousand Oaks, CA: Corwin Publishing Company.

Wicklein, R. C. (2006). Five good reasons for engineering as the focus for technology education. *Technology Teacher*, *65*(7), 25–29.

Wilson, M. (2016). *The ancient history of projectiles*. Retrieved from resources21.org/cl/files//project492.../Ancient%20History%20of%20Projectiles.pptx

KEY TERMS AND DEFINITIONS

Catapult: A device designed to launch a projectile.

Deep Dive: An in-depth examination into a content topic or problem.

Project-Based Learning: An inquiry instructional strategy in which students work in teams as collaborators on an ill-structured problem.

Science: Both a body of knowledge and a process of studying the natural world.

Technology: Using science to solve problems.

Chapter 11
A Mathematical Approach to Designing Insulators

Kathryn Electa Pedings-Behling
College of Charleston, USA

EXECUTIVE SUMMARY

How do we keep hot drinks hot and cold drinks cold? Companies such as Tervis, YETI, and Thermos spend their time researching and designing products around that very question. In this lesson, students will discover, through mathematical modeling, which materials provide the best insulation and be tasked with designing their own insulator. This lesson has been designed at two different levels for students from grade three through high school with an optional extension activity for more advanced students. Students will use technology to explore the rate of change of the temperature of hot water over two minutes using different insulation materials. After this exploration, students will use the data they have collected to determine the best materials for designing their own insulator. This insulator will then be judged based on the ability to keep a hot drink hot and on the aesthetic value.

LITERATURE REVIEW

The history of insulated drinkware begins around 1896 when Sir James Dewar invented what would become the first Thermos. When Thermos began making bottles in the US in 1907, it quickly gained popularity with famous users such as President William Howard Taft and the Wright Brothers (Insulated Bottle, 2006). It seemed as though everyone was looking for a way to keep their hot drinks hot and cold drinks cold!

DOI: 10.4018/978-1-5225-9631-8.ch011

The process for manufacturing commercial insulated drinkware is still very similar to what it was in the early 1900s. Most cups made for hot drinks are a vacuum: sealed between two walls of the same material. Most bottles for cold drinks are lined with a liquid foam that is allowed to harden (Insulated Bottle, 2006). Although this process began in the early twentieth century, it hasn't kept people and companies from trying to perfect the insulated bottle. If one were to look up "insulated drinkware patents" on Google Scholar around 906 results would be returned. The search is never-ending to make consumers happy with their drink temperature.

When one thinks of insulated drinkware, the disposable coffee cup often comes to mind. These have become ubiquitous with the morning commute of many Americans. The history of the disposable coffee cup is riddled with controversy over the battle between material that is good at insulation (Styrofoam) versus material that is better for the environment (paper). This argument was ultimately won by Howard Schultz, owner of Starbucks, in 1987 when he chose the paper cup. In order to insulate more like a foam cup, manufacturers have worked to develop paper cups with walls two or three times as thick as normal paper cups (Park, 2014).

However, the quality of the insulation is not the only consideration for disposable coffee cups anymore. A popular food blog, Eater.com, suggests that starting in about 2014, the visual design of the cups became incredibly important for consumers to feel like they are getting their money's worth for expensive coffee (Baird, 2015). In fact, unique designs of coffee cups have become immortalized on the internet by way of specialized Instagram accounts (Hargreaves, 2016), design blogs (Durso, 2014), and food culture websites (Hohenadel, 2014).

In this activity, students will explore the quality of insulation of different types of materials in order to make educated decisions about how to design an insulator that can compete with the best being manufactured. This is not a solved problem as over 367 of the patents for insulated drinkware on Google Scholar were filed as of 2015; people are still searching for the perfect solution! Not only will teams of students have to think about the quality of their insulation, but they will have to consider the aesthetic design qualities as well since aesthetically pleasing cups sell better.

CONTENT INFORMATION

This activity aligns with many national standards across several different subject areas. Most notably, this activity puts to use several standards from the Common Core State Standards for Mathematics (CCSSM). Certain questions in this activity are written at two different levels with an optional extension for students beyond the Algebra II level. The questions specifically written for grades 3-5 align with CCSSM standards related to differences and numerical comparisons such as 3.NBT.A.2,

4.NBT.A.2 and 4.NBT.B.4. The questions specifically written for grades 6-12 draw from standards about ratios and rate of change including 6.RP.A.3, 7.RP.A.3, 8.F.B.4, and HSA.CED.A.1. Finally, the optional extension activity relates to standards involving modeling exponential functions found in HSA.CED.A.2 and HSF.LE.A.1 (Common Core Standards Initiative, 2010).

Although this activity takes a mathematically-forward approach to designing insulators, standards from the other STEAM (science, technology, engineering, art, and mathematics) fields are integral to the well-roundedness of the learning experience. From the Next Generation Science Standards, the activity specifically aligns with standards relating to thermal energy transfer and engineering including 3-5-ETS1, MS-PS3-3, MS-ETS1, HS-PS3-3, and HS-ETS1 (NGSS Lead States, 2013). This activity also heavily relies on technology and draws from the International Society for Technology in Education (ISTE) standards 1.c, 4.b, and 4.c (ISTE, 2007). Finally, the arts component of this activity aligns with the creating anchor standards 1, 2 and 3 from the National Core Arts Standards (2014).

This mathematical approach to designing insulators integrates the components of STEAM in many ways. Mathematical modeling is the core of the activity, using data collection with a technology component. Science concepts involving energy and heat loss provide the background behind why the system is functioning in a specific way and how it can then be improved. Engineering design is incorporated with the arts to design and create an insulator that is effective in retaining heat and aesthetically pleasing. The activity culminates in students being required to justify their final design through scientific explanation using data they collected.

CASE DESCRIPTION

This lesson is adapted from a PASE (Portable Advanced STEM Exploration) Math lab developed by the South Carolina Governor's School for Science and Mathematics titled *How do we keep it hot?* (Pedings-Behling, 2015).

Lesson Plan Information

Technology Overview

Throughout this activity, students will perform data collection a minimum of five times, finding the temperature of water at least twelve times during each collection. This data collection can be done one of two ways:

1. Data Collection Technology: Companies such as Vernier, Pasco, and Texas Instruments have created hardware and software specifically for data collection with a temperature probe. If your school has access to this technology, this activity is the perfect opportunity to make use of these items. The experiment would need to be set up to collect data for two minutes, and then an analysis of the data can be done using this technology.
2. Thermometer: If your school does not have access to any of this data collection technology, each team of students can use a thermometer and a stopwatch to fill out the table in the lesson. This data would then need to be graphed and analyzed by hand.
    - *Materials Per Group*2 disposable plastic cups (e.g., Solo)
    - Temperature collection technology (see above)
    - Piece of approximately 6" x 6" mesh fabric
    - Piece of approximately 6" x 6" cotton fabric
    - Piece of approximately 6" x 6" wool fabric

Materials Per Classroom

- Electric Kettle with constant temperature control
- Water
- Markers
- Glue/tape
- Scissors

Collection of mesh, cotton, and wool fabric

Safety Concerns

This activity requires students to work with hot water. Depending on their age, teachers may instruct students to wear goggles or protective gloves. The teacher may also want to designate an adult to supervise the hot water station. Only about one inch of hot water is needed in the bottom of the cup for the temperature probe to collect correct data. None of this hot water should be used for consumption after using the temperature probe.

Lesson Plan

The following procedures will be carried out five times using the specifications in the following section.

Figure 1. Graph of temperature versus time for both predicated data and actual data (Pedings-Behling, 2015)

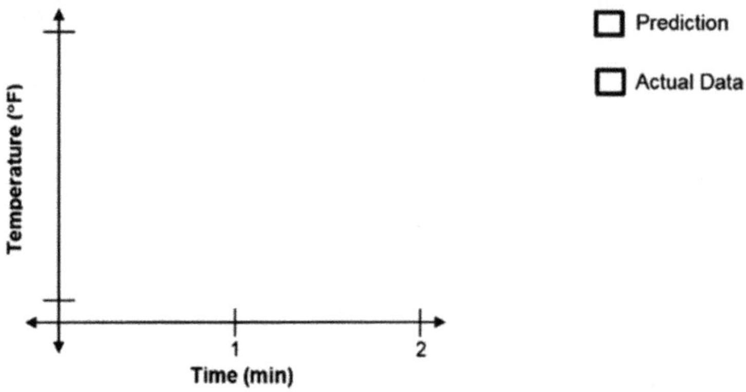

Procedures

Think about the following scenario: You make hot chocolate in the cup you are testing and accidentally forget it on the kitchen counter for two minutes. If you were monitoring the temperature of the hot chocolate the entire time, what would a graph of those temperatures look like? Sketch the prediction of your graph below. Make sure to color in the box next to "Prediction" with the marker you are using.

You will now simulate this scenario using hot water with the following procedures.

1. Gather about 1 inch of hot water in the cup.
2. Insert the thermometer and wait until the temperature finishes rising.
3. Start your timer and collect data on the temperature every 10 seconds for 2 minutes using Table 1.
4. Using a different color marker on the graph in Figure 1, graph the actual data on top of the predicted graph. Make sure to color in the box next to "Actual Data" with the marker you are using. Also, make sure to label the markings on the y-axis with the maximum and minimum values from your data collection.

Data Analysis

1. How well does the graph of your actual data match your predicted graph? Explain any differences between the two graphs.
2. (Grades 3-5): Find the difference between the highest temperature and the lowest temperature. Record this information in Table 2.

Table 1. Two-minute temperature data collection using a thermometer

Time	Temperature (°F)
0:10	
0:20	
0:30	
0:40	
0:50	
1:00	
1:10	
1:20	
1:30	
1:40	
1:50	
2:00	

3. (Grades 6-12): Find the average rate of change in the temperature of the water over the 2-minute period. Record this information in Table 2.
4. What are the units of measurement for your answer? What are some other units that could have been used?
5. What does the value from question #2 mean in the context of the problem?
6. What could you do to retain more heat in the water?
7. What would happen if you collected data for longer than two minutes?
8. Approximately how long would it take for the water to get back to room temperature? Justify your answer using mathematics.

Advanced Activity

Use Newton's Law of Cooling, either as an exponential function

$$T(t) = T_s + (T_0 - T_s)e^{-kt}.$$

or as a differential equation to create a more accurate model of the cooling of the water.

$$\frac{dT}{dt} = -k(T - T_s)$$

$T(t)$ = temperature at time t

Figure 2. Visual process of nesting fabric between two plastic cups

Figure 2. Visual process of nesting fabric between two plastic cups

T_s= temperature of surroundings
T_0= initial temperature
k= constant

Use this model to predict how long it would take for the water to reach exactly 0.1°F above the current room temperature.

Part A: Control Data

Follow the procedures using two nested plastic cups.

Part B: Testing Different Materials

Repeat the procedures three times, once for each type of material (mesh, cotton, wool). Take two plastic cups and place the material between the two cups to create an insulator. See Figure 2.

Additional Questions:

1. Before collecting data in Part B, which of the materials do you think will be the best at insulating the hot water? Why?
2. How is each graph different from the graph of the control data?
3. (Grades 6-12): Look at the four graphs of the data you have collected so far. Is this data linear? What is the difference between this data and linear data?

Table 2. Fabric type and cost for insulator design

Fabric Type	Cost Per Square Inch	Number of Square Inches	Total Cost
Mesh	$0.25		
Cotton	$0.50		
Wool	$1.00		
TOTAL			

4. Which fabric retained the most heat? Justify this mathematically and scientifically. Did that surprise you?
5. (Advanced): How can you tell from the exponential models which fabric retained the most heat?

Part C: Insulator Design

Each group will need to design their final insulator. Insulators must be freestanding, meaning that they do not need to be held together with your hands, and must hold one of the plastic cups which will contain the hot water. These insulators may include a top, but it must be removable to allow for hot water to be added to the cup.

This activity can be done with each group having unlimited access to the materials, or each group can have a price limit, like $20, to buy supplies where the costs are as follows:

- Wool: $1 per square inch
- Cotton: $0.50 per square inch
- Mesh: $0.25 per square inch

Groups not only need to focus on retaining heat, but also the aesthetics of their insulator which will also be judged. Groups may design and test as many insulators as their time and money allow. Use Table 3 to keep track of spending. All insulator designs must begin with a sketch.

Part D: Final Insulator Testing

Follow the same procedures from above to test the heat retention of your insulator.

Table 3. Aesthetic judging for insulators using different design qualities

Design Quality	Score (1 - 5)
Texture: Is the insulator pleasant to hold?	
Superficial Quality: Does the design look like a quality insulator?	
Functionality: Can the insulator be easily held with one hand?	
Transportability: Can the insulator be easily transported in a car cup holder?	
Marketability: Could this insulator be designed with logos or different colors?	
Durability: Does it appear as though the insulator would last for a long time?	
Total Score	

Aesthetic Judging

Option 1: Have the students vote on their favorite insulator design that is not their own. Tally the votes to determine the winner.

Option 2: Either the teacher, the students, or both rate each insulator based on the design qualities in Table 4 with 1 being the lowest score and 5 being the highest score.

Final Analysis

1. Justify the design of your insulator citing information gained from your data collection.
2. Who made the best insulator in terms of heat retention? Justify this mathematically. What does it look like and how is it different than yours (if it isn't yours)?

Table 4. Aesthetic judging for insulators using different design qualities

Design Quality	Score (1 - 5)
Texture: Is the **insulator** comfortable to hold?	
Superficial Quality: Does the design look like a quality **insulator**?	
Functionality: Can the **insulator** be easily held with one hand?	
Transportability: Can the **insulator** be easily transported in a car cup holder?	
Marketability: Could this **insulator** be designed with logos or different colors?	
Durability: Does it appear as though the **insulator** would last for a long time?	
Total Score	

3. Who made the best insulator in terms of aesthetics? What does it look like and how is it different than yours (if it isn't yours)?
4. If you could redesign your insulator, what would you change?

Optional Extension Activities

The following could be used to further extend the idea of cup insulator design.

- *Explore Other Materials:* Have students bring in materials other than mesh, cotton, and wool to test as an insulator. Some examples are packing peanuts, sponges, paper, aluminum foil, cotton balls, etc.
- *How do we keep it cool?* Flip the problem on its head by having students discuss how they would design a cup to keep the drink inside as cool as possible. Would the insulator look the same? Test it with a pitcher of ice water!
- *Building Bigger Objects:* Have students expand on the knowledge gained from this activity to build bigger insulated objects like a coffee carafe or an ice box. How would they test the effectiveness of their insulator for these objects?
- *Insulating Other Objects:* How does this activity apply to insulation for something as large as a house? Have the students brainstorm all of the different places they see (or know about) insulation around their house to keep it cooler in the summer and warmer in the winter. Some ideas they may come up with are insulation (obviously), curtains, window shades, reflective material in the roof, etc.
- *Shape of the Graphs:* Have students collect data for a longer period of time (like five minutes). What do they notice about the shape of the graph? Have the students put their hot water in the refrigerator and collect data in longer intervals (like every thirty seconds instead of every ten seconds). How does that change the shape of the graph compared to the original experiment?

STEAM Content

The overall goal of this activity is to design an insulator in light of the mathematical information gained about the hot water's change in the temperature using different materials for insulation. Achieving this goal incorporates all aspects of the STEAM model. For the science aspect, students have to question why some materials are better insulators than others. Technology is utilized in this activity through the use of the temperature probes for data collection. The students are asked to design their own insulator which incorporates both the engineering and art components since they have to consider both goals of it retaining as much heat as possible and being as aesthetically pleasing as possible.

The heart of this activity is the mathematical integration. Students in grades 3-5 will explore differences in the temperatures of the hot water over the two-minute collection period. Students in grades 6-12 will explore the average rates of change in the temperatures of the hot water over the two-minute collection period. If the students have studied exponential models, they can then complete the extension activity about Newton's Law of Cooling. These explorations will lead them to make decisions about what material they will use in making their insulator. The activity culminates in having the students provide a scientific explanation of why they chose the materials they did using their mathematical information as justification.

CURRENT CHALLENGES FACING TEACHERS WITH SOLUTIONS AND RECOMMENDATIONS

My School Does Not Have Any of This Technology

There are many solutions to this challenge. Multiple teachers at a school or in a district can share this technology and create a checkout system for use in their classrooms. The teacher can also choose to collect data as an entire class demonstration or at one station that each group rotates through so that only one set of technology is needed. One final option would be for the students to bring in basic thermometers that they may have at home to use for data collection.

I Don't Have Enough Class Time to Complete This Activity

Depending on the length of your class, this project can span over multiple days or class periods. When considering the large number of standards this activity addresses, this time in the classroom is justified, but can still be difficult to complete in light of other material needing to be covered in a specific unit. The activity is easily broken into many small portions that can each be completed using much less class time. For instance, data for each type of material can be collected on different days. This will allow the activity to be spread over several classes or weeks. The design portion of the activity can be assigned as out of class work to lessen the amount of class hours spent on the project.

I'm Concerned About the Safety of My Students During This Activity

This is particularly a concern for teachers of younger students especially in handling the hot water. The teacher could choose to collect data as an entire class which would address this challenge since the teacher would be the only person handling the hot water. For very young classes, it may be best to bring in additional adults (teachers, administrators, or parent volunteers) to assist in both pouring hot water and handling the hot cups for each of the groups. Alternatively, a teacher could choose to switch the activity to keeping ice water cold as opposed to retaining the heat of boiling water.

I Like Math Because Everything Has One Solution. I'm Scared to do Data Collection Because Everyone's Data Will Look Different and I Might Not Know the "Right" Answer

The beauty of data collection is that each group of students will have their own unique data with which to work. As one may suspect, the wool *should* be a better insulator than the cotton or the mesh. What if a group's data does not reflect this? This provides a wonderful opportunity to have conversations in your classroom with questions such as: *What could have caused the data to reflect a result that was different from your hypothesized outcome? Was the data collected the same way each time? Do you suspect an issue with your data collection?* Once a teacher embraces the fear that there is no one exact answer with this activity, it may be more freeing to give students the ability to create truly incredible products.

REFERENCES

Baird, S. (2015). *How Disposable Coffee Cup Design Became High Art*. Retrieved September 30, 2016, from http://www.eater.com/drinks/2015/8/14/9153849/how-disposable-coffee-cup-design-became-high-art

Blinder, S. M. (2007). *The Coffee Cooling Problem*. Retrieved June 07, 2016, from http://demonstrations.wolfram.com/TheCoffeeCoolingProblem/

Bryant, J. (2008). *Newton's Law of Cooling Demonstration*. Retrieved June 07, 2016, from http://demonstrations.wolfram.com/NewtonsLawOfCooling/

Dubson, M., & Loeblein, T. (2010). *Curve Fitting*. Retrieved June 07, 2016, from http://phet.colorado.edu/en/simulation/legacy/curve-fitting

Durso, C. (2014). Coffee Cups of The World, A Photo Series of Beautiful Takeaway Cups. Retrieved September 30, 2016, from http://www.foodiggity.com/coffee-cups-of-the-world-a-photo-series-of-beautiful-takeaway-cups/

Hargreaves, H. (2016). *Coffee Cups of the World (@coffeecupsoftheworld) • Instagram photos and videos.* Retrieved September 30, 2016, from https://www.instagram.com/coffeecupsoftheworld/

Hohenadel, K. (2014). *Traveling the World in Search of the Most Unique Disposable Coffee Cup Designs (Photos).* Retrieved September 30, 2016, from http://www.slate.com/blogs/the_eye/2014/07/03/photographer_henry_hargreaves_coffee_cups_of_the_world_highlights_the_best.html

Howcast. (2013). *How to Demonstrate Insulation Principles.* Retrieved June 07, 2016, from https://www.youtube.com/watch?v=8I96BzSNNqE

Insulated Bottle. (2006). Retrieved June 09, 2016, from http://www.madehow.com/Volume-4/Insulated-Bottle.html

International Society for Technology in Education. (2007). *National educational technology standards for students.* Retrieved June 7, 2016, from http://www.iste.org/standards/iste-standards/standards-for-students

Khan, S. (2016). *Linear vs. exponential growth.* Retrieved June 07, 2016, from https://www.khanacademy.org/math/algebra/introduction-to-exponential-functions/comparing-exponential-and-polynomial-functions/v/linear-exponential-models

National Coalition for Core Arts. (2014). *National core arts standards.* Retrieved June 7, 2016, from http://www.nationalartsstandards.org/

National Governors Association Center for Best Practices & Council of Chief State School Officers. (2010). *Common Core State Standards for Mathematics.* Washington, DC: Author.

Newton's Law of Cooling. (2015). Retrieved June 07, 2016, from https://www.khanacademy.org/math/differential-equations/first-order-differential-equations/modeling-with-differential-equations/v/newtons-law-of-cooling

NGSS Lead States. (2013). *Next Generation Science Standards: For States, By States.* Washington, DC: The National Academies Press.

Park, M. Y. (2014). *A Brief History of the Disposable Coffee Cup.* Retrieved June 09, 2016, from http://www.bonappetit.com/entertaining-style/trends-news/article/disposable-coffee-cup-history

Pedings-Behling, K. E. (2015). *How Do We Keep it Hot?* Hartsville, SC: South Carolina Governor's School for Science and Mathematics.

The Science of Insulation Explained. (2019). Retrieved January 05, 2019, from https://www.knaufinsulation.com.au/homeowners/energy-efficiency/science-of-insulation-explained

KEY TERMS AND DEFINITIONS

Average Rate of Change: The change in the dependent variable divided by the change in the independent variable. For this activity, students will find the rate of change of the temperature of the water with respect to the time of the data collection.

Difference: In this activity, students will find the difference between the highest and lowest temperatures using subtraction.

Exponential Function: A function where the variable is located in the exponent. The rate of change is not constant and is either increasing or decreasing across a graph.

Insulator: A device that does not easily allow for the transfer of heat either from a hot beverage or to a cold beverage.

Linear: A descriptor for a function that indicates that it has a constant rate of change.

Newton's Law of Cooling: A mathematical model that exhibits the physical process that the rate of heat loss of an object is proportional to the difference between the temperature of the object and the temperature of the surroundings.

Temperature: The measure of heat in an object. In this activity, it is specifically the measure of heat in the water.

Thermometer: A tool used to measure temperature.

Chapter 12
Engineering and Art:
Putting the EA in STEAM

Sara B. Smith
Goose Creek High School, USA

EXECUTIVE SUMMARY

This chapter discusses challenges faced by first year pre-engineering students. Also discussed are several topics taught within the curriculum including: the engineering design process, sketching, measurement, the elements and principles of design, and three-dimensional modeling. The chapter proposes a design project for engineering students that would tie all of these concepts together to provide an additional learning opportunity for students and more relevant practice of skills like isometric sketching, creating three-dimensional computer-aided design models, and measurement. Samples of student work from the project are included.

LITERATURE REVIEW

I became a teacher after working in industry as an engineer. I chose to enter a teacher education program to become certified through a traditional route. Once I entered the classroom, aspects about my prior life became more known, and I was given the opportunity to teach classes such as engineering and energy in high school. To me, engineering is the application of math and science concepts to solve problems. This definition parallels that of the Next Generation Science Standards (NGSS, 2013). While contemplating the differences between teaching engineering and chemistry, I discovered a few notable distinctions. One of those distinctions is that engineering solutions to problems could be an invention, an innovation, or a more efficient process. Engineers design, build, conduct tests, market, and communicate

DOI: 10.4018/978-1-5225-9631-8.ch012

with others in the pursuit of solutions. Engineering requires creativity and ingenuity. Art is an important and integral part of the creativity involved in engineering. It lends a hand to the engineer's process of developing solutions. John Maeda does a great job of demonstrating this. Maeda is former president of the Rhode Island School of Design where he was very involved in adding art to the STEM (science, technology, engineering, and mathematics) education movement. He holds degrees in Electrical Engineering and Computer Science from the Massachusetts Institute of Technology as well as a Ph.D. in design from Tsukuba University in Japan. In his piece for Scientific American in 2013, Maeda said that "STEAM and arts integration are crucial in K-12 education, engaging students in the STEM subjects and ensuring that creativity doesn't fall by the wayside as we chase innovation..." (Para. 5).

Integration is also important at the college level, so much so that multiple colleges and universities have designed courses that blend art and engineering into one. "Enduring Design: The Art of Engineering" is one such course being offered at the Rose-Hulman Institute of Technology in Terre Haute, Indiana. The course was described by Mirth and Findley (2015) as being designed by both humanities and engineering faculty. The faculty members saw that the school's engineering students did not take many liberal arts courses and felt that this caused a "potential handicap in their ability to see opportunities from a broader perspective" (p. 1). While the course is open to all students, a majority of those enrolled in the first offering were engineering students. The course is an art course that incorporates examples of engineering. This course is still new, and data is still being collected, but initial data show that the art and engineering connections taught throughout the course helped students to become more creative in their ideas for improving both form and function of objects.

At the University of South Florida in Tampa, an engineering professor created a course that "merges his research world with the world of fine art" (National Science Foundation, 2006, Para. 1). The course itself teaches concepts like electromagnetism and optics; lab activities incorporate lessons including everything from examining works of fine art to using pinhole cameras and learning how to spot a forgery.

As these schools created courses to strengthen the minds and creativity of engineering students by incorporating art, the project described in this chapter is an attempt to do the same for high school students enrolled in pre-engineering programs, most specifically the Project Lead the Way (PLTW) students that I teach.

Project Lead the Way

PLTW began at Shenendehowa High School in upstate New York in 1987. There was a decrease in engineering program enrollment, creating a predicted shortfall of over 600,000 engineers in the United States by the year 2000. Many engineering undergraduates were changing their goals and dropping out of engineering (Kelley & Wyanski, 1987). Richard Blais, the Director of Occupational Education in the school district at the time, set out to address these problems by instituting a pre-engineering curriculum in high school. The original program consisted of four courses that were designed with input from higher education institutions and local industry personnel. The introductory course included technical drawing both on paper and using computer-aided design (CAD) technology. Students also built scale models from those drawings. Another course consisted of "Digital Electronics," which involves logic and controls, and electrical circuits. Additionally, a "Principles of Engineering" course was created in which students built mini manufacturing plants that simulated the automated manufacturing processes used in industry. The courses also taught students about pneumatics, hydraulics, electric motors, automation, and robotics. All concepts were delivered through project-based learning. A young man in Kelley and Wyanski's video (1987) said, "It's half teaching ourselves and half of the teacher teaching."

The program's development continued with input and grant funding from post-secondary institutions like Rensselaer Polytechnic Institute, and from local business and industry partners. Even in 1987 the high schoolers were able to earn college credit for their pre-engineering courses. To help spark interest in STEM at a younger age, high school students went to local elementary schools and conducted small projects with students.

In 1997, PLTW launched its Pathways to Engineering program in twelve New York State high schools. A partnership with High Schools That Work and the Southern Region Educational Board helped to spread the program to an additional thirty states (Chautauqua Lake School District, n.d.). Today, PLTW reaches over two million students in all fifty states, the District of Columbia, and some U.S. territories. It has trained over 35,000 teachers through its two-week summer training institutes and on-going professional development (PLTW, 2016). There are courses in aerospace engineering, civil and architectural engineering, environmental sustainability, and digital electronics among others for the high school level students. "Gateway to Technology" is one of the courses available to middle school students. In the past few years, PLTW has extended its mission into elementary schools. The curriculum provides research-based, problem-based learning experiences for students. After successfully completing my training, I began teaching the "Introduction to Engineering Design" course and have been doing so since 2013. This course introduces students to many concepts and skills that they will need throughout their pre-engineering and college engineering programs. Course concepts and skills include various types of sketches, the engineering design process, 3D modeling with Autodesk Inventor, and reverse engineering. Of these, many of my students have the most difficulty with sketching. Students are required to generate isometric sketches of objects. Isometric sketches show three faces of a 3D object. The sides extend at 30-degree angles to give the object its 3D appearance. These sketches are completed on isometric graph paper as illustrated in Figure 1.

Some students can look at that paper and see where the lines go and how to draw the shapes. Others, however, cannot visualize making a 3D shape on this 2D surface. My pondering about this issue raises one question: Why is seeing three dimensions in a 2Dspace so difficult for high school students? In the Project Lead the Way - Where It Started video (Kelley & Wyanski, 1987), Richard Blais stresses the importance of this skill by saying "Drawing is a language of industry, and before they can really take other courses in technology, they need to understand that language."

The Mathematics Link

I taught a homebound student geometry recently and was excited to see that he had to generate isometric sketches from the 2D multi-view sketches as well as the reverse process. This was included in chapter one of the textbook, Prentice Hall's Geometry: Common Core 12th edition (2012). The rest of the curriculum we worked through was traditional geometry: similarity, congruence, triangles, parallel lines and transversals, polygons, etc. While there was overlap between some of this

Figure 1.

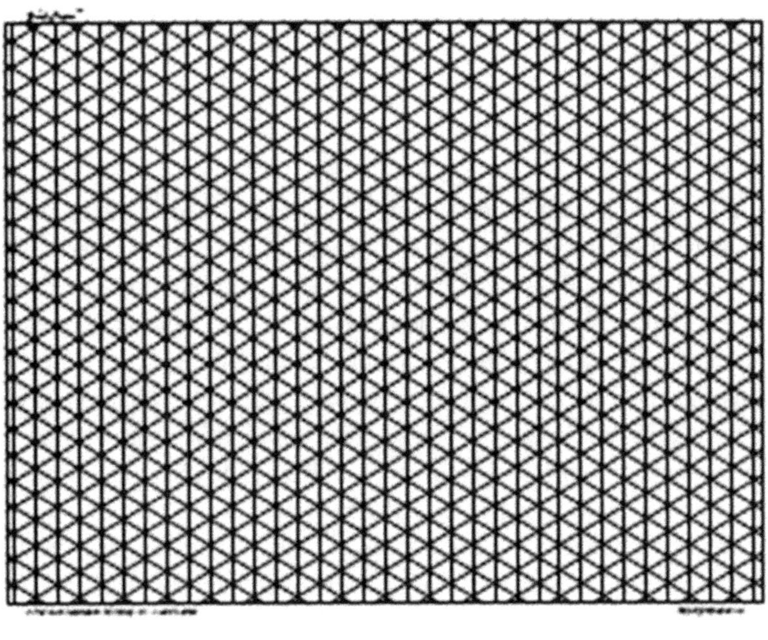

content and the content of the "Introduction to Engineering" course, none was as helpful to the goal of being able to construct isometric sketches of 3D objects. The high school at which I teach reversed the order of geometry and algebra II at the beginning of the 2016-2017 school year. What does this mean for "Introduction to Engineering Design"? The course will have both ninth and tenth graders who are enrolled in algebra I, algebra II, or geometry. The combination of this information indicates that mathematics courses will not provide ample support of the skills being taught in this course.

The Fine Arts Link

My high school is the site for the district's School of the Arts. The school has students from throughout the district that have an aptitude for music, art, dance, creative writing, drama, or a combination of those. Art students tend to excel at the sketching skills required for the engineering course. They can create 3D sketches on isometric paper, graph paper, and, really, any paper they're provided. It seems as though they come equipped with an unteachable skill: spatial reasoning. They already have the capacity to see the form on paper. They know how to shade an object to provide a more realistic appearance. These students can take 3D forms and draw them in 2D renderings, and cross-sectional views without dissonance. Students that come to the course without this skill often struggle and describe sketching as "stupid," "difficult," or "useless." I find that humor and flexible expectations aid me in teaching students how to sketch. I modify my lessons each time I teach the course to meet my students' various needs in the class, as they are not all equally prepared when they arrive.

The Engineering Design Process

Why are the abilities to sketch three-dimensionally and to resolve a 3D object into 2D views so important? The engineering design process requires them. There are a variety of acceptable layouts of the design process, and the combination of geometry and 3D drawing are key components. The process flow taught in the introductory course includes identification of a problem, the generation of concepts, the development of solutions, the construction and testing of a prototype, the evaluation of solutions, and the presentation of solutions. These steps can and will be iterated as many times as necessary for successful completion of a project. Sketching is incorporated in at least four of the six steps listed. Sketches need to be detailed and accurate for communication between client and designer, designer and manufacturer, and between the multiple designers working on a project team. Art and geometry help with the visualization process and the physical creation of the sketches. Sketches are turned into CAD parts and assemblies in Autodesk Inventor. From those parts and assemblies, annotated drawings are generated. These drawings include overall size dimensions, dimensions for the size and location of part features, and can also show how individual parts interact with each other to form the final product.

Reverse Engineering and Elements and Principles of Design

In reverse engineering, students are required to fully document an item like a kitchen appliance, a toy, or an Automoblox vehicle. Automoblox is a line of wooden toy vehicles that have removable parts and sections. The removable parts allow students to deconstruct them and learn how the parts interact with each other. The vehicles come in a large variety of models from classic cars to modern day SUVs to emergency response vehicles. I used an Automoblox pickup truck in my summer training institute to learn reverse engineering concepts, and the program at my school has several for students to use in the course. Reverse engineering requires students to perform structural, functional, and visual analyses on an item. Students document the materials that make up the parts, how the parts interact with each other, parts' dimensions, and the product's individual components. The students hand-sketch the parts in their notebooks, build and assemble the parts in Inventor, and generate annotated drawings of each part. They also create an explosion drawing with a parts list (see Figure 2) and a video showing how all of the parts go together.

One of the introductory lessons in reverse engineering is the elements and principles of design. These are used in product design, art, architecture, and even landscape design to attract the eye and convey feelings through the work. The elements and principles are used in product design to capture consumer attention and increase the sale of the product. This knowledge is used by the Introduction to Engineering Design students in their visual analysis of an object. Students can conduct an Internet search to learn what these elements and principles are and what each invokes in a design. The PLTW curriculum details eight visual design elements: point, line, color, value, shape, form, space, and texture. The curriculum also details seven design principles: balance, emphasis, contrast, rhythm, proportion, unity, and economy. Any combination of these can be found in monuments, buildings, homes, advertisements, gardens, toys, cars, and even tennis shoes!

Students are asked to find a variety of objects and analyze them for these elements and principles. They ultimately present the information in a poster. I require students to complete an analysis of six different objects, including information on at least three elements and three principles used per object and how those affect their selected designs. In their reverse engineering project, this would comprise the visual analysis portion. One of the main points that most students struggle with is describing the emotions or feelings these elements and principles invoke. They can

Figure 2.

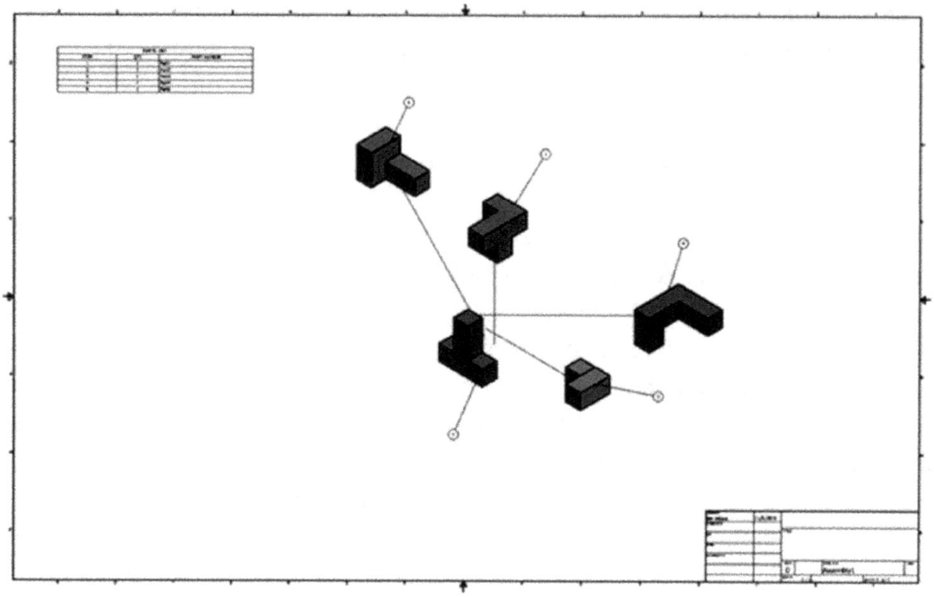

amply identify which are present but not why. The visual arts students seem to excel at these visual analyses; they are able to verbalize the analysis with more depth and detail. For example, given a photograph of ice cream cones, one student listed the required quantity of elements and principles, but failed to describe the aesthetic impact of the elements. An equally motivated student with artistic experience not only gave the same elements and principles, but also described how she was engaged by the photograph because the color and symmetry enticed her to want to eat an ice cream cone.

Computer Aided Design: Autodesk Inventor

Throughout the course, students learn how construct parts in Autodesk Inventor, a CAD program. The parts start out simple and then increase in difficulty. The students create the parts using additive modeling (adding features to a base shape) and subtractive modeling (starting with a large generic shape like a rectangular prism and cutting away to achieve the part required) or a combination of both. For many parts, students are given isometric sketches that they use to guide them in generation of the parts. As they progress through the course, parts are generated by following dimensioned drawings. In one of reverse engineering's final projects, students generate parts in Inventor based on their own drawings and measurements.

As this is a group project, all students are expected to use their measurement and Inventor skills to contribute parts to the final assembly that represents their item. If the parts are not measured and generated correctly, they will not fit together well in the CAD assembly. This inconsistent measurement results in parts of unequal scale causing gaps between parts or one part being wider than its adjacent parts.

Measurement

The final concept with which students struggle is measurement. As a teacher of American high school students, I assumed that students would have had a lot of exposure to measurement in the U.S. Customary System of Units (inches and feet), and that is why they struggle with the metric system taught in my Physical Science and Chemistry courses. Engineers must be able to work and communicate in both systems of units, and, as such, both are taught in the Introduction to Engineering Design course. Students must be able to measure dimensions with the necessary level of precision.

On a ruler with sixteenths of an inch increments, students have to be able to measure to the thirty-second of an inch. This takes a bit of practice, but students seem to quickly gain an understanding of more precise increments of measurement. In addition to a ruler, students learn how to measure with dial calipers. This is a very new instrument for high school students, and it introduces an odd concept: inches with decimal places rather than fractions. This is the tool that students use to measure the parts for their reverse engineering project. Precision on a dial caliper extends to the ten-thousandths place. If not calibrated appropriately prior to making a measurement or not read correctly, errors in scale find their way into the project. While measurement is crucial in the fields of science and engineering, it is also essential in art. Artists use measurement in the form of scales, ratios, and proportions. While I consider these three forms of measurement very mathematical, they add so much beauty to a work of art. Whether it be a canvas painting, a sculpture, a monument to celebrate impactful historical figures, or even a beautiful garden, the use of perspective, scale, rhythm, and color provides a pleasing aesthetic. When aesthetics and engineering principles align, it becomes clear that a student is fully seeing a product's necessary aspects.

The engineering design process, sketching, and measurement are all taught in the first half of the course. Elements and principles of design are taught near the end of the course with reverse engineering. Would students be better served and better practiced in all of the skills described if the latter topic of design were introduced earlier in the curriculum? Teachers don't need to rewrite curricula when there are excellent materials available. These excellent materials help students master skills and concepts. Why not provide students with their own design opportunities that will afford them a chance to use the engineering design process, sketching, measurement, and design elements and principles to design their own house or building or monument or product?

Spatial Intelligence

Howard Gardner's theory of multiple intelligences includes spatial intelligence. People who possess this intelligence can easily visualize in three dimensions. They can visualize rotations of objects in space and are able to notice very fine details. Inspiring Breakthrough (2005) lists three skills associated with this intelligence: spatial awareness, working with objects, and artistic design. Navigators, sculptors, and architects are known to have high spatial intelligence. People with high spatial intelligence like puzzles, drawing, and daydreaming. People begin to develop visual-spatial skills as infants and toddlers. Along with verbal and mathematical abilities, visual-spatial intelligence is a strong predictor of future success. Park, Lubinski, and Benbow (2010) describe work done with Jonathan Wai. Together they reviewed data from two longitudinal studies: Project Talent and the Study of Mathematically Precocious Youth (SMPY). Through Project Talent data, Wai and his colleagues found that teens with strong visual-spatial intelligence are very drawn to working with their hands. Those same students went on to pursue education and careers in science and technical fields. They also noted that there are not many programs in high school through which these students can pursue their interests. Similar findings stemmed from the SMPY data.

Engineers employ visual-spatial intelligence when they design parts. The task requires them to understand the part features from multiple perspectives and how the part works in conjunction with other parts (Tseng & Yang, 2011). The sketching and CAD modeling of objects in the Introduction to Engineering Design course as described before employ visual-spatial skills. Students must be able to look at an isometric sketch of an object from one angle and understand the features of that object in all three dimensions separately. This understanding is then used to construct multi-view sketches that can be used to generate a 3D model in Inventor. The reverse process is also sometimes necessary. Students will receive a multi-view sketch from which they will draw the isometric view. This is particularly difficult as features appear differently in each view. For example, a curved surface on the top of an object may appear as a rectangle in the top view of the object. The only way to know that it is curved is to use the front view of the object at the same time. Visual-spatial intelligence is also strong for artists, especially sculptors. They are able to visualize their piece in the space occupied by the raw material. They can see in their mind's eye from what areas to remove material. The artist can visualize the form and make mental measurements regarding scale and proportion.

It is in the possession and use of strong visual-spatial intelligence that artists and engineers are similar. While the products of their work serve different purposes, the approach to their design and creation is the same. Maeda writes that "Artists and scientists tend to approach problems with a similar open-mindedness and inquisitiveness - they both do not fear the unknown, preferring leaps to incremental steps" (2013, para. 7).

The Standards

There are many sources for engineering education standards. The Next Generation Science Standards include engineering standards as they relate to science education. The International Technology Education Association and its Technology for All Americans Project first published the Standards for Technological Literacy in 2000. The standards were written to "address the mismatch between dependency and understanding" (ITEA, 2007, p. v) of the technology that surrounds us. PLTW developed standards that guide their curriculum and are only accessible to school systems under contract with PLTW. Individual states may even have their own standards for engineering courses. Rather than list the many relevant standards here, I have listed several key words and phrases that relate to standards addressed in the included project. These key words include attributes of design, design process, engineering design, design principles, 3D modeling, isometric sketching, and CAD.

The Project

The first three units in PLTW's Introduction to Engineering Design Curriculum are the design process, technical sketching and drawing, and measurement and statistics. One key concept to include within these units is the concept of scale. It is important for students to realize that they will be designing a house, building, or monument of some kind, but it will have to be sketched and generated in the CAD software on a much smaller scale. By this point in the curriculum, my students have also started using Autodesk Inventor. It breaks up the monotony, and students really want to learn how to use the program. I would incorporate the elements and principles of design, which is included in reverse engineering, immediately after students learn how to measure in the metric and U.S. Customary systems of units.

After giving clear and pertinent instructions, the educator should assign the project to students. It can take as many class periods as the teacher would like to allow. I asked my students to conduct a five-minute brainstorming session to generate concepts. Then we talked about the criteria and constraints of the project. They had to generate an original design. They had to be able to identify the scale they used in their sketches and CAD renderings. The students also had to be able to explain the incorporation of the elements and principles of design that they used in their design. If they put columns on their monument, for example, they had to be able to explain that the vertical lines of the columns added a sense of formality and distinction to their monument. The students then employed a decision matrix to choose the most viable option. The criteria for their decision matrix (see Figure 3) included elegance, aesthetics, time required to complete design requirements, skill level required to complete the design project, and ease of including a wide variety of elements and principles of design. As a class, students determined which of these should be weighted more heavily. They then scored each concept and used the decision matrix results to select a concept.

Once the concept is selected, students should generate an annotated isometric sketch on paper. They should generate sketches for multiple views if they would best show the design details. These sketches need to be drawn to scale using a scale that they have selected (see Figures 4 and 5).

The students will then be able to generate a 3D model in Autodesk Inventor. If Autodesk Inventor is unavailable, students can use Google Sketch Up which is a free program with multitudes of tutorials available online.

If the time and supplies are available, students could also build a scale model of their design. This would incorporate more art and make the project more kinesthetic as well as increase student engagement.

A very important piece of the engineering design process is communication with peers, management, and clients. While requiring students to create a PowerPoint presentation does not provide students much of a challenge, requiring them to present the PowerPoint's information before the class certainly does. The students should include at minimum the results of their brainstorming session, the decision matrix, a picture of their hand-drawn sketches, and a picture of their computer-generated design. I like to have my students include a brief biography slide as well. So many

Figure 3.

	Skyscraper	Total	Statue	Total	House	Total	Fort	Total
Elegance-1	2	2	1	1	2	3	1	1
Aesthetics-2	3	6	3	6	4	8	1	2
Time-4	3	12	3	12	2	8	2	8
Skill-4	3	12	4	12	3	8	2	8
Elements and principles-5	4	20	3	15	3	15	3	15
Total		52		46		41		34

Figure 4.

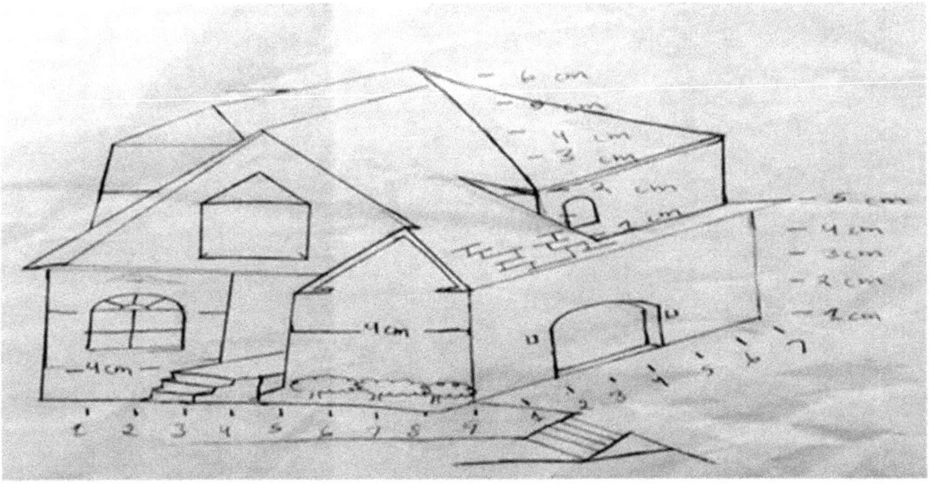

Figure 5.

students make it through entire semesters and do not even know their classmates' names! Students can use this opportunity to learn how to make a five to ten-minute presentation based on slides from which they cannot read.

Figure 6.

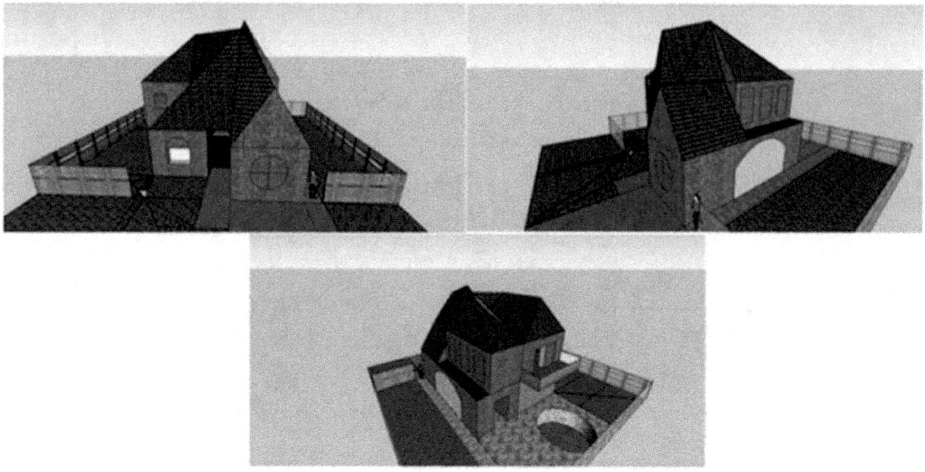

Figure 7.

Inviting guests to participate in this project would be great too. Guests could consist of other teachers, especially art teachers, administrators, or even representatives from community organizations and businesses. The guests can be included for as much of the project as the educator would like. They could be the client who requires a design and provides criteria and constraints of their own. The students could then be required to communicate professionally with the client throughout the process to get design approval or feedback on design concepts. The guests could then attend the presentations. This will provide a more formal atmosphere for students and a more genuine experience.

Lessons Learned

Every time I teach this lesson, I discover new things about the content, its delivery, and students' abilities to learn it all. Some aspects into which I've gained insight include logistics, aligning the engineering and aesthetics skills, and the importance of questioning. One way I've improved upon logistics is by incorporating the use of a document camera to better teach sketching and shading. I walk students through each step of construction in Inventor, asking questions to guide them through my thought processes in hopes of inspiring them. I have also reached out to my school's art teachers and obtained digital copies of student work. To better align engineering and aesthetic skills, I use the student work provided by the art teacher as examples for analyzing the elements and principles of design. Using student art for examples has improved my ability to question students about how those elements and principles make them feel upon consideration. I ask deep, open-ended questions to activate their critical-thinking about the engineering and aesthetics at the same time.

In the future, I plan to improve on the basic elements and principles lesson even further by assigning the six objects so that all of students will be sharing their perspectives about the same things. Those objects will include all forms of art and architecture so as to expose the similarities between art and engineering design worlds. After they have had this experience to practice the engineering and aesthetic skills together on a common set of objects, students will be able to choose and analyze their own objects.

REFERENCES

Chautauqua Lake Central School District. (n.d.). *History of project lead the way*. Retrieved from http://www.clake.org/District/Class/28-Engineering-and-Technology-Department/1447-HistoryOfPLTW.html

Inspiring Breakthrough. (2005). *Visual spatial intelligence*. Retrieved from http://www.inspiring-breakthrough.co.uk/learning-styles/visual-spatial-learning.html

International Technology Education Association. (2007). *Standards for technological literacy: Content for the study of technology* (3rd ed.). Retrieved from https://www.iteea.org/File.aspx?id=67767&v=b26b7852

Kelley, J. A. (Producer & Director), & Wyanski, J. (Producer & Director). (1987). *Project lead the way - Where it started 1987 Dick Blais* [video]. Retrieved from http://youtu.be/iIdomsEhmYM

Maeda, J. (2013). Artists and scientists: More alike than different [Web log comment]. Retrieved from http://blogs.scientificamerican.com/guest-blog/artists-and-scientists-more-alike-than-different/

Mirth, J., & Findley, A. (2015). Enduring design: Examining the relationship between art, engineering, and creativity. *Proceedings from ASEE '15: 2015 ASEE Zone III Conference.* Retrieved from https://www.asee.org/documents/zones/zone3/2015/Enduring-Design-Examining-the-Relationship-Between-Art-Engineering-and-Creativity.pdf

National Science Foundation. (2006). *The art of engineering* [press release]. Retrievedfromhttp://www.nsf.gov/news/news_summ.jsp?cntn_id=107990

NGSS Lead States. (2013). *Next Generation Science Standards: For States, By States*. Washington, DC: The National Academies Press.

Park, G., Lubinski, D., & Benbow, C. (2010). Recognizing spatial intelligence. *Scientific American.* Retrieved from http://www.scientificamerican. com/article/recognizing-spatial-Intel/

Post Illuminance. (n.d.). *Custom graph: Isometric paper*. Retrieved from http://customgraph.com/piart.php?art=590

Project Lead the Way. (2016). *Engaging students from New York to Honolulu.* Retrieved from https://www.pltw.org/about-us

Tseng, T., & Yang, M. (2011). The role of spatial-visual skills in a project-based engineering design course. *Proceedings from ASEE 2011 Annual Conference & Exposition*. Retrieved from http://web.mit.edu/~mcyang/www/papers/2011-tsengYang-ASEE.pdf

KEY TERMS AND DEFINITIONS

Computer-Aided Design (CAD): Using a software program to generate a 3D representation of a design that can then be used for documentation or 3D printing a physical model.

Design Principles: How an artist or designer combines the elements of design in their work.

Design Process: Process that guides the development of solutions and designs aimed at solving the world's problems.

Elements of Design: The fundamental aspects of paintings, drawings, and designs.

Engineering Design: The design process followed by engineers in their pursuit of answers and innovation.

Isometric Sketching: Creating a picture of a design on paper while presenting three faces to show the three dimensions.

Three-Dimensional Modeling: Converting a 2D representation of a design into a 3D representation.

Visual-Spatial Intelligence: The ability to visualize and manipulate an object or design in 3D space.

Chapter 13
Developing and Applying the Law of Cosines:
Using Star Maps as a Context

Vecihi S. Zambak
Monmouth University, USA

Budi Mulyono
Sriwijaya University, Indonesia

EXECUTIVE SUMMARY

In history, geometry was founded more as a practical endeavor than a theoretical one. Early developments of the branch portray philosophers' attempts to make sense of their surroundings, including the measurement of distances on earth and in space. Such a link between earth and space sciences and geometry motivated us to develop and implement a multidisciplinary lesson focusing on the conceptual understanding of the law of cosines in the context of astronomy. In our content specific STEAM lesson, the authors aimed to facilitate an understanding of the law of cosines in ninth grade students, and then apply the law in a star map task to find approximate distances between stars. The second part of the lesson also included the use of an instructional technology to support students' work with the star map task. In the conclusion, the authors discuss possible ways to improve the quality of their STEAM education efforts for the given context.

DOI: 10.4018/978-1-5225-9631-8.ch013

INTRODUCTION

As research indicates, students experience significant gains and achievement from interdisciplinary instructional approaches in developing mathematics and science content knowledge; as a result, STE(A)M Education has globally become the ultimate goal of educational programs (Bybee, 2010; Marginson, Tytler, Freeman, & Roberts, 2013; Morrison & McDuffie, 2009; Venville, Rennie, & Wallace, 2004). With an emphasis on creativity and artistic processes, STEAM Education is now defined as the intentional integration of concepts, practices, and design features of science, technology, engineering, art, and mathematics subjects (Bequette & Bequette, 2012; Gess, 2017). Yakman (2010) discusses how the application of STEAM Education, in the context of a classroom, occurs in five different levels: 1) Universal level, 2) Integrated level, 3) Multidisciplinary level, 4) Discipline Specific level, and 5) Content Specific level. In this chapter we define *Content Specific STEAM Education* as the detailed study of a content area with the integration of two or more STEAM disciplines, one of which is the main focus. In the first part of our lesson, ninth grade students were guided to collaboratively discover *the law of cosines*. Then they solved a mathematics problem dealing with *Earth and space sciences*. In the second part of our lesson, students used an applet built in a dynamic geometry environment. In other words, we aimed to integrate the disciplines of mathematics, science, and technology using *discovery learning* as the main instructional strategy.

The law of cosines can be called the extension of the Pythagorean Theorem for acute and obtuse triangles. Even though the concept of cosine was not developed during the Euclid's time, the Elements includes the same equation with different notations and representations. The law makes it possible to find an unknown angle or side when two sides and the angle between them are known or when three sides are known.

Since the law of cosines includes several notations and concepts, many learners find it difficult to retrieve it from their memory and use it when it is needed to solve a problem. There are strategies to reduce the learner's memory load for the distance formula and the law of cosines. For example, McMullin (2003) used computer algebra systems to reduce students' memory load while they were solving a mathematical problem using the law of cosines. Whereas instructional technology usage in his article might have enabled students to find the correct answer for the problem quite promptly, McMullin's approach does not guarantee students' conceptual understanding of the law. To be able to conceptually understand the law of cosines, we believe that students need to synthesize several connected terms and notions in trigonometry and geometry.

Another way to reduce memory load is to use discovery learning approaches that enable learners to analyze and synthesize laws from triangles through deductive reasoning. Discovery learning has varying meanings, from the use of open-ended student projects to inquiry oriented activities. In general terms, it can be defined as a curriculum structure through which students are guided by questions to discover targeted concepts, terms and notions (Hammer, 1997). This approach was found to be more beneficial for learners who had a familiarity with the concepts and facts covered during the instruction (Tuovinen & Sweller, 1999).

CONTENT INFORMATION

The lesson we present in this chapter aims to address many content and practice standards active in the U.S. education system. The following Common Core State Standards for Mathematical Practice have been targeted within the lesson: MP1. Make sense of problems and persevere in solving them; MP2. Reason abstractly and quantitatively; MP4. Model with mathematics; MP5. Use appropriate tools strategically; and MP7. Look for and make use of structure (Common Core State Standards Initiative [CCSSI], 2010). In addition, two Common Core Content Standards shaped the mathematical tasks for our lesson:

High School: Geometry » Similarity, Right Triangles, & Trigonometry » Apply trigonometry to general triangles:

10. (+) Prove the Laws of Sines and Cosines and use them to solve problems.

11. (+) Understand and apply the Law of Sines and the Law of Cosines to find unknown measurements in right and non-right triangles (e.g., surveying problems, resultant forces). (CCSSI, 2010)

Our lesson is also aligned with the *International Society for Technology in Education (ISTE) Standards for Students*. More specifically, the second part of our lesson and the use of the applet aimed to enable students to become Empowered Learner(s), Knowledge Constructor(s), and Computational Thinker(s) (ISTE, 2016). The following specific ISTE Standards for Students were central to our lesson:

1c. Students use technology to seek feedback that informs and improves their practice and to demonstrate their learning in a variety of ways.

3d. Students build knowledge by actively exploring real-world issues and problems, developing ideas and theories and pursuing answers and solutions.

5b. Students collect data or identify relevant data sets, use digital tools to analyze them, and represent data in various ways to facilitate problem-solving and decision-making. (ISTE, 2016)

Finally, the "Universe and its Stars" was tangentially used as the Disciplinary Core Idea(s) of Earth and Space Sciences adapted from the *Next Generation Science Standards* (Next Generation Science Standards Lead States, 2013). That being said, it is worth stating that mathematical content and practice standards were the primary focus of our lesson; technology and science standards were secondarily integrated into the instruction. (To make sense of our rationale for the intentional centralization of mathematics in our lesson planning and implementation, please see the definition of Content Specific STEAM Education in the Introduction section of this chapter.)

CASE DESCRIPTION

The Goals and Settings

We established two main goals for the development and implementation of our interdisciplinary lesson: 1) effectively integrate the content of Earth and space sciences and Information Communication Technologies (ICT) in a middle school setting, and 2) implement a discovery learning approach as the leading instructional method. To do that, we developed our lesson plans around student-centered tasks. The authors collaboratively designed the lesson plans, interdisciplinary tasks, and a conducive applet.

We applied our lesson plans in an international school located in Hilversum, the Netherlands. The cooperating teacher behaved as an external advisor for the development and revisions of the lesson plans providing support for possible technical issues that may have emerged during the implementation of lesson plans. The school environment and technological infrastructure also enabled us to use ICT during the instruction. Every student had their own personal laptop.

After negotiating with the cooperating teacher, we selected the law of cosines as the major content focus of our lesson. We considered the logical construction and conceptual understanding of the law of cosines as precursors to student application of the law. For this reason, we developed the first part of our lesson, *My Theory* task. As we mentioned before, Earth and space sciences were chosen as the theme to apply and integrate the law of cosines to the science content. This theme allowed us to develop the second part of our lesson, *Star Map* task. These two tasks were implemented during two class meetings (i.e., a total of 90 minutes) in a ninth-grade class, including seventeen students.

Lesson Plan

Learning Objectives

The following three learning objectives and goals were set for our lesson:
The students will be able to…

1. Construct the law of cosines using trigonometric ratios and the Pythagorean Theorem within various triangles (i.e., obtuse or acute triangles).
2. Identify the characteristics of triangles requiring the use of the law of cosines to find unknowns in problem situations.
3. Solve realistic and contextual problems by applying the law of cosines and using ICT to collect and make sense of data.

Materials

My Theory Task:

- PowerPoint Slides (*My Theory*)
- Teacher PC Station
- Projector
- Student Worksheet Part 1
- *My Theory* Solution Sheet[1]

Star Map Task:

- PowerPoint Slides (Star Map)
- Teacher PC Station
- Projector
- Student Worksheet Part 2
- Star Map Extension Task
- PC/Laptops
- Calculator

Tasks

For the first part of our lesson, *My Theory* task, we aimed to give students an opportunity to construct the law of cosines by themselves instead of directing them to memorize without conceptually understanding the content). The *My Theory* task included two parallel problems (see Figure 1 and 2). While *My Theory* Problem 1

Figure 1. My Theory Problem 1

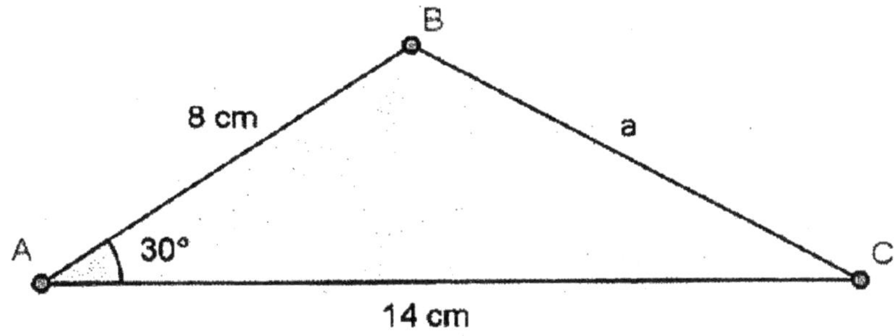

involved the lengths of two line segments and the angle between them, *My Theory* Problem 2 was the abstraction of the first problem which required students to use algebraic reasoning and manipulations. To be able to solve the second problem, the students needed to recall the trigonometric ratios and the geometrical theorems they used while solving the first problem, and then convert them into related letters.

The Student Worksheet Part 1 also includes hints and guiding questions to support and scaffold students' problem solving processes. For example, in *My Theory* Problem 1, students were first asked to find the height of the triangle drawn perpendicular to the $|AC|$ line segment, which would allow them to apply their previous knowledge about trigonometric ratios and the Pythagorean Theorem. This would also make it possible for the students to find the unknown length a from the givens. *My Theory* Problem 2 was specifically designed for students to write down the law of cosines

Figure 2. My Theory Problem 2

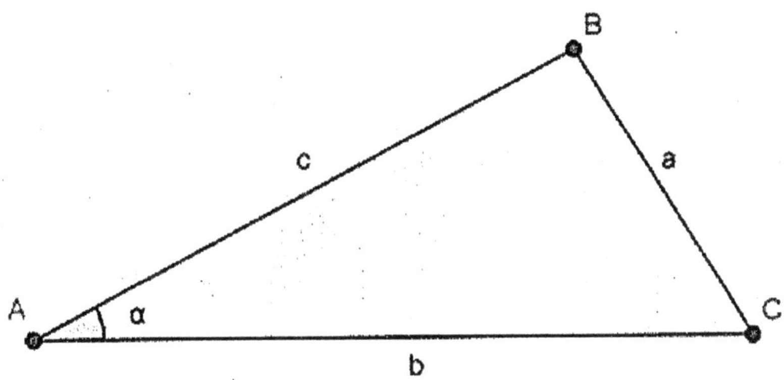

according to their conceptual understanding, based on their solutions for the first problem given.

My Theory task was built as a small-group activity, and three different versions of the task were therefore created. Versions differed according to the shape of the triangle in *My Theory* Problem 1. While the triangle was acute in two versions, it was obtuse in the third version (see Figure 3).

For the second part of our lesson, the *Star Map* task, we aimed to present a real-world context connected with the history of the subject. The Student Worksheet Part 2 represented a 2D star map including some stars and the Earth at the center (see Figure 4). A table informing the actual distances of the stars in the star map to the Earth was additionally provided (see Table 1). To be able to solve the main problem in the Star Map task, students were asked to use the given information on the star map, the table, and an applet.

Students were expected to find the distance between two stars (e.g., Alpha Centauri and Barnard) by applying the law of cosines. To do that, students needed to know distances of each star from the Earth, which were given in the table, and also the angle between the line segments from each star to the Earth. The angle was not given in the table; students were asked to use the applet to find the missing information in the question.

Technology Integration

The applet was designed to find the angle between the line segments from each star to the Earth by playing with three points that should be placed on two stars and the Earth (see Figure 5 for screenshot). The applet was designed with GeoGebra, a dynamic geometry software. The applet was delivered to each student's laptop through a USB stick. The Student Worksheet Part 2 explicitly guided students to play with the applet on their laptops before beginning to work on the problem. We did not provide students with any additional instruction explaining how to use the applet.

Figure 3. Three different versions of My Theory Problem 1

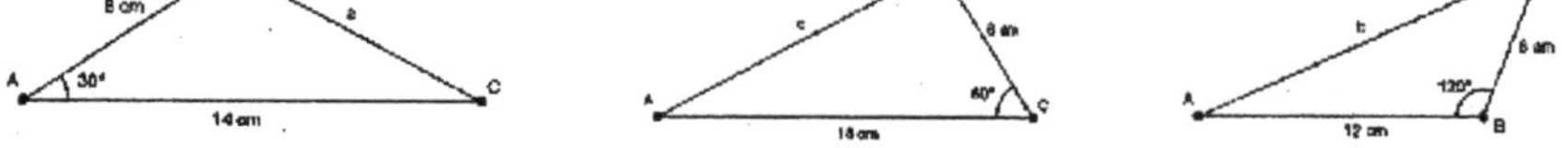

Figure 4. 2D Star Map

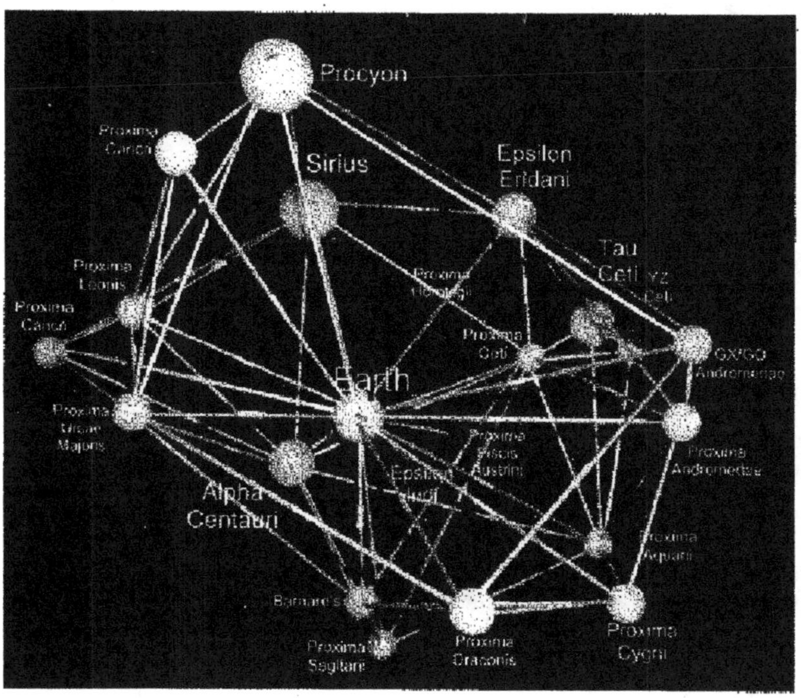

Procedures

During the first part of our lesson that focused on the implementation of the *My Theory* task, we went through the following instructional steps: 1) the introduction

Table 1: Distances of the star to the Earth

Name of stars	Distance from the Earth (light year)
Alpha Centauri	4.3
Barnard's	5.9
Sirius	8.7
Epsilon Eridani	10.7
Proxima Cygni	11.1
Epsilon Indi	11.3
Procyon	11.4

Figure 5. A snapshot from the applet

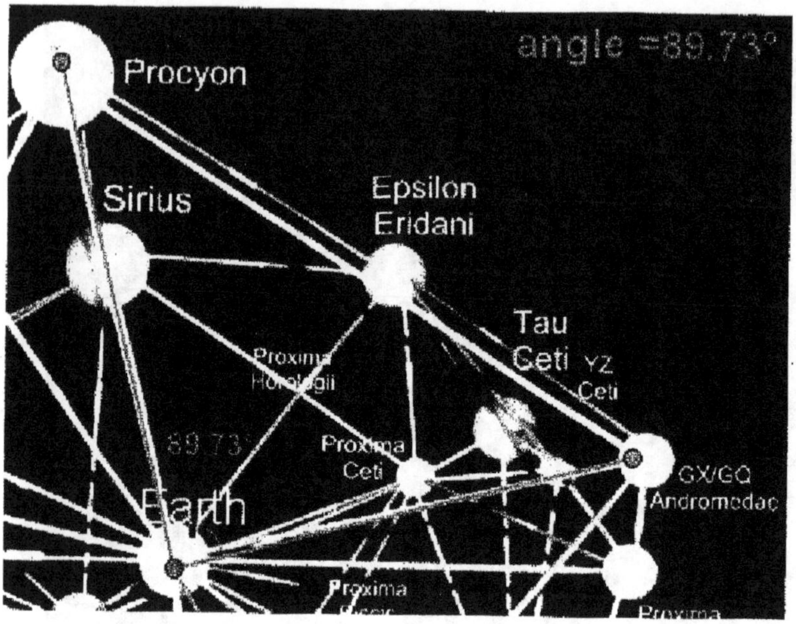

of the topic, 2) small group-work for the *My Theory* Problem 1, 3) whole class discussion about *My Theory* Problem 1, 4) small group-work for the *My Theory* Problem 2, and 5) whole class discussion about *My Theory* Problem 2 and the conclusion. During the introduction part, we first informed students about our goals and learning objectives for the lesson. Next, we refreshed students' memory about algebraic identities, trigonometric ratios, and the Pythagorean Theorem. For example, we covered the following mathematical principles during the introduction by orchestrating a whole class discussion (see Figure 6).

After the introduction, the students worked in their small groups to solve the *My Theory* Problem 1. During the whole class discussion, small groups who used algebraic identities or trigonometric ratios with some errors were initially asked to participate so that other students could contribute to the discussion for corrections. Due to the complexity of the *My Theory* Problem 2, we intentionally scaffolded students more during their small group work. In the conclusion part, we distributed the *My Theory* Solution Sheet and briefly described the Star Map task for the for next class hour.

We organized the second part of our lesson by 1) reviewing the law of cosines, 2) describing the Star Map task, and 3) wrapping up. Before students were assigned to work on the Student Worksheet Part 2 (i.e., the Star Map task), the first author

Figure 6. Major Mathematical Principles covered during the introduction of our lesson

$$(a+b)^2 = a^2 + 2ab + b^2$$
$$(a-b)^2 = a^2 - 2ab + b^2$$

$$x = z \cdot sin[\alpha]$$
$$y = z \cdot cos[\alpha]$$
$$(sin[\alpha])^2 + (cos[\alpha])^2 = 1$$

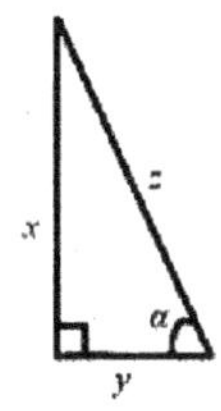

helped the students recall the law of cosines from the first part of the lesson. After the students completed their small group-work, we collected their Student Worksheets to inform the cooperative teacher about the students' learning gains from our intervention.

To wrap up our lesson, we presented the history of the law of cosines and its relationship with Earth and space sciences. More specifically, we described Euclid's way of proving the law of cosines and his contributions to spherical astronomy and Al Battani's conception of *angular distance* to measure the distances between stars in their times. Finally, to pique the students' interest in Earth and space sciences, we shared a YouTube Video (BrainStuff - HowStuffWorks, 2013) about the methods astronomers use to measure distances in space.

Extensions

We developed an extension of the Star Map task for those who finished the actual task earlier than the rest of the class. In the extension, we asked students to find the angle between two stars. To find the answer for the extension, students needed to use the inverse trigonometric function feature on their calculators. Students were then asked to check the same angle approximately with the applet:

If you know that the distance between Proxima Cygni and Barnard's stars is 9 light years, what should be the angular distance (angle) between these stars (look at the table of distance)?

CHALLENGES BASED ON STUDENT REFLECTIONS

At the end of our lesson, we collected all of the student worksheets and additionally asked students to fill out a two-question reflection form. The student reflections were one of the ultimate products of our lesson. The reflections provided student feedback on their views about instructional choices and their interest in learning about Earth and space sciences while covering the law of cosines. In other words, we administered student reflections to improve the quality of our lesson for revisions and future implementations. The following two questions were included in the student reflections form:

1. Did you see a connection between *My Theory* Problem 1 and Problem 2 in *My Theory* task? Do you think this pattern (connection between Problem 1 and 2) helped your construction of the law of cosines? How? If it did not, why?
2. Did the Star Map task and the use of Earth and space sciences as a theme pique your interest and motivate you in mathematics? Why or why not? Please explain.

Six out of seventeen students algebraically constructed the law of cosines. Whereas all students answered the *My Theory* Problem 1 question correctly, approximately two thirds of the class failed to algebraically construct the law of cosines. In order to facilitate the students' law construction, we applied a pattern from a concrete example to an abstract example in the *My Theory* task. It seems like the pattern of the task (i.e., from concrete to abstract), the explicit use of the mathematics language in the task, and the use of their previous knowledge helped some students construct the law of cosines. Thirteen out of seventeen students claimed that they saw the connection, and some students reasoned that the pattern helped them understand the law of cosines.

Even though most students claimed that they saw the connection between the concrete and abstract problems, as we mentioned before, eleven students could not construct the law of cosines by themselves. We speculate that ninety minutes was not a sufficient amount of time for students to develop a geometrical principle on their own. Many students discussed within their reflections that they needed to have more time to construct the law of cosines. Moreover, the student worksheets revealed that many students could not execute the algebraic manipulations appropriately. Students additionally indicated that they were not accustomed to producing proofs in their geometry class. Since our lesson was one of their early experiences with developing a proof for a geometrical principle, students needed to spend more time with the *My Theory* task.

Thirteen out of seventeen students liked the Star Map task and astronomy theme for the lesson due to its relationship to real life and the opportunity to represent mathematics untraditionally. Five students did not like the Star Map task due to their lack of interest in Earth and space sciences, unclarity of the relationship between the theory and practice, and overall difficulty of the task.

We may conclude from this experience that representing a pattern, through the use of a concrete example followed by an abstract example, can be helpful in developing students' understanding of mathematical proofs. However, the students' knowledge and experience with trigonometric ratios, algebraic manipulations, and mathematical proofs might be influential factors in determining the effectiveness of instruction. Furthermore, the majority of the students liked the context of astronomy for mathematics. The Star Map activity provided students with an opportunity to recognize the use of mathematics in real life. Regarding students' reflections, mathematics teachers should find similar ways to integrate science and technology content into their mathematics instruction. By observing the practical applications, students feel more engaged and motivated to learn advanced mathematical concepts.

RECOMMENDATIONS: HOW TO STEAM MORE

In this chapter, we presented a STEAM Education case for which mathematical content was the focus. As our lesson was at the Content Specific STEAM Education level, further efforts to make the disciplines more integrative are needed. To begin with, we consider the collaborative design and development of lessons, by teachers who are highly qualified in various subject areas, to promote the integration of the STEAM fields. Our lesson was only developed by mathematics teachers, including the authors of this chapter and the cooperating teacher; no science or computer science teacher was involved during the lesson development process. The gains from a collaboration among teachers having expertise in different STEAM fields would be twofold. First, secondary science and computer science teachers would rigorously contribute to the development of the students' scientific and computational thinking skills. Second, such a collaboration among multiple subject-matter teachers would enable students to perceive the lesson from a multidisciplinary standpoint instead of perceiving the lesson as an isolated mathematical principal.

Our lesson provided students with a ready-made applet so students would not need to spend their time applying computational thinking to develop an applet that would help them solve the given problem independently. An alternative lesson design in which students are guided to openly use dynamic geometry software during the problem solution process could also benefit students by allowing them to practice their computational thinking and technology skills.

The emphasis of engineering and art content in our lesson was limited. To amplify these two disciplines, we consider the use of data collection, analysis, and modelling techniques as the main instructional components. The "NASA's Open Data Portal" (n.d.) could enable teachers and students to get access to any data set of interest free of charge. After identifying the data and conducting some analyses on it, students could be asked to develop a 3D model to demonstrate their unique Star Map. That way, all letters within the STEAM acronym could be touched upon to some extent. We still anticipate possible revisions of our lesson to further integrate the disciplines of art and engineering, and to promote students' computational and problem-solving skills around a focused content.

Future efforts and detailed conceptions of STEAM Education should aim to elevate Content Specific implementations to a Universal STEAM curriculum, in which lessons are independent from any focused content/discipline. Developing a Universal STEAM curriculum should be the priority of teachers, teacher educators, and curriculum developers.

REFERENCES

BrainStuff – HowStuffWorks. (2013, October 4). *How Do Astronomers Measure Distance?* [Video File]. Retrieved from https://www.youtube.com/watch?v=_Lsj-Hz-NS4&t=161s

Bequette, J. W., & Bequette, M. B. (2012). A place for art and design education in the STEM conversation. *Art Education*, *65*(2), 40–47. doi:10.1080/00043125.2012.11519167

Bybee, R. W. (2010). Advancing STEM education: A 2020 vision. *Technology and Engineering Teacher*, *70*(1), 30.

Common Core State Standards Initiative. (2010). *Common Core State Standards for Mathematics (CCSSM)*. Washington, DC: National Governors Association Center for Best Practices and the Council of Chief State School Officers.

Gess, A. H. (2017). Steam education: Separating fact from fiction. *Technology and Engineering Teacher*, *77*(3), 39–41.

Hammer, D. (1997). Discovery learning and discovery teaching. *Cognition and Instruction*, *15*(4), 485–529. doi:10.12071532690xci1504_2

International Society for Technology in Education. (2016). *ISTE standards for students: A practical guide for learning with technology*. Eugene, OR: Author.

Marginson, S., Tytler, R., Freeman, B., & Roberts, K. (2013). *STEM: Country comparisons*. Melbourne, Australia: Australian Council of Learned Academies.

McMullin, L. (2003). How to solve any triangle: First, forget the law of sines and the law of cosines. *Mathematics Teacher*, *96*(6), 448–449.

Morrison, J., & McDuffie, A. R. (2009). Connecting science and mathematics: Using inquiry investigations to learn about data collection, analysis, and display. *School Science and Mathematics*, *109*(1), 31–44. doi:10.1111/j.1949-8594.2009.tb17860.x

Nasa's Open Data Portal. (n.d.). Retrieved from https://data.nasa.gov/

Next Generation Science Standards Lead States. (2013). *Next generation science standards: For states, by states*. Washington, DC: National Academies Press.

Tuovinen, J. E., & Sweller, J. (1999). A comparison of cognitive load associated with discovery learning and worked examples. *Journal of Educational Psychology*, *91*(2), 334–341. doi:10.1037/0022-0663.91.2.334

Venville, G., Rennie, L., & Wallace, J. (2004). Decision making and sources of knowledge: How students tackle integrated tasks in science, technology and mathematics. *Research in Science Education*, *34*(2), 115–135. doi:10.1023/B:RISE.0000033762.75329.9b

Yakman, G. (2010). *What is the point of STE@M? A brief overview*. Retrieved from https://annegilesclelland.typepad.com/files/whatissteam.pdf

KEY TERMS AND DEFINITIONS

Algebraic Identities: Set of equalities that are held valid for any group of variables. The standard algebraic product identity can be established using the Binomial Theorem. A subset of these equalities can also be used to find the factors for given formulas.

Angular Distance: The measurement of the angle between two objects based on the use of optical instruments by an observer. While angle on a 2D plane can be measured using protractor, identifying the angular distance between two objects in 3D space requires the use of angular distance by an observer.

Dynamic Geometry Software: Instructional technology commonly used for the geometry content. The affordances of this technology enables users to construct various geometrical figures, develop conjectures and assess the validity of these conjectures. GeoGebra is an example of the dynamic geometry software.

Information Communication Technologies: This is the broader term used for infrastructure and devices used for communication, application, collection and analysis of data.

Law of Cosines: This is the equation used to find one unknown side of any triangle when the lengths of the other two vertices and the angle between these two vertices are given.

Pythagorean Theorem: This theorem describes the numerical relationship among three vertices of any right triangle.

Trigonometric Ratios: Trigonometric relationships in any right triangle. There are 3 basic trigonometric ratios: sine, cosine, and tangent. This list can be expanded with the following additional trigonometric ratios: cotangent, secant, and cosecant.

ENDNOTE

1 Solution sheet was distributed at the end of the class meeting.

Chapter 14
Using STEAM in Marine Science:
Incorporating Graphic Design Into an Existing STEM Lesson

Callie (Van Koughnett) Dollahon
Wando High School, USA

EXECUTIVE SUMMARY

This case summarizes two perspectives on inclusion of Arts in STEM/STEAM education and how they influence the modification of an existing STEM lesson. Teachers are encouraged to use many instruction models throughout their careers, and inclusion of new methods can seem daunting. This case hopes to illustrate how STEAM education can be included in a classroom through intentional use of graphic design in an everyday lesson or a longer unit. Students in the case are asked to design and build a robotic arm that is capable of accomplishing a task such as move or grasp an object. The specific context is Marine Science in nature but can be adapted to many other content areas.

LITERATURE REVIEW

PBL, PLTW, DBQ, Inquiry, 5E, Backward by Design, STEM, STEAM. Educators are exposed to so many methods, models, and (hopefully) best practices throughout their careers that it is no wonder they sometimes question why and how they will incorporate them all. One challenge that faces teachers today is covering the breadth of district and state-tested content each year while incorporating methods that seem

DOI: 10.4018/978-1-5225-9631-8.ch014

"bigger" than the time allotted in the course. Specifically, in terms of Science, Technology, Engineering, Art, and Mathematics (STEAM) education, teachers are faced with the thought that in order to incorporate content and lessons from each of the five included disciplines, we must create an elaborate project-based lesson that might require weeks of instruction.

For example, the Dream Factory at Elizabeth Forward Middle School integrates all of these disciplines by having students set up their very own chocolate business. Students work with a local candy factory, print 3D molds for the chocolate, create artistic labels, analyze cost and sales, and look at the chemistry of the chocolate itself (Keruskin, 2015). Projects like this are inspirational. They are experiential in nature, provide links between disciplines, allow for connections to be made outside the classroom, and promote rigor in education in an engaging and fun way for students. However, some educators may find challenges such as time, technology, and support to recreate something similar in their own school.

STEAM education is interdisciplinary by nature, but it does not have to be an elaborate project like the Dream Factory. Some teachers are finding ways to inspire the creativity, collaboration, problem solving, and reflective learning that STEAM proponents encourage by simple tweaks to their everyday lessons. Take, for example, Henriksen's (2014) description of a teacher who integrates art into science lessons by having "students create visual advertisements to describe a science idea, or a concept, organism, and soon." The classroom described features daily use of STEAM-inspired lesson content, rather than a project or problem-based approach, and is still successful in developing the skills sought in quality STEAM education programs. Students are afforded opportunities to use creativity, visual communication, critical thinking (through comparison, explanation, and modeling), and collaboration,-just as in the project-based approach of the Dream Factory.

Why STEAM? (STEM + A)

My experience in the academic field as an educator indicates that it is possible to incorporate quality STEAM learning as the in-depth, time consuming project and as the everyday lesson that allows for coverage of the many standards. According to Beers (2011), quality Science, Technology, Engineering, and Math (STEM) education "exemplifies the cross-curricular learning that is the foundation of a twenty-first century curriculum." Beers (2011) writes that STEM education also promotes the four C's: Critical thinking and problem solving, Communication, Collaboration, and Creativity (p. 5). STEAM Education, then, should promote these skills while additionally providing greater opportunity for students to discover new ways of seeing, thinking about, and learning each discipline through the additions of arts education.

Take a look at any science textbook, engineer's notebook, NOVA video publication, or architectural installation, and it is clear that art is already a fundamental component of STEM. The engineering process mimics the artistic process, especially in a commercial setting. The purposeful practice that artists use to improve and enhance their craft is akin to the repeated iterations used by engineers to refine and test solutions. Each reproduction provides new information, from data to muscle memory, whether the learner is an artist, mathematician, scientist, engineer, or otherwise. In a graphic design course, students often keep a portfolio of ideas and variations of a concept, and it is through the process of reflection, evaluation, and sometimes collaboration, that these ideas are narrowed down into one product, whether art or engineering.

Personal experience in graphic design gave me an intuitive window into deeper math concepts in college. After working on a 4-H project involving embroidery design and print graphics in eighth grade, I opened a small business that I operated throughout high school. The business used graphic digital programs that exposed me to translations, rotations, 3Dimensional structure, vectors, and color theory that would later be the tangible mental foundation I used to explore and understand the same math, physics, and applied technology concepts in STEM courses. As Vaidyanathan (2012) states, "Digital design is a blend of art and technology, [built] on a foundation of math, science, and engineering" (p. 27). Art contributes deeply to my experience as a scientist, teacher, and STEM coordinator. STEM concepts also contribute to my work as a quilter and graphic designer as I determine scale, take measurements, find working angles, and use order of operations (particularly when building a complex quilt design).

Many artists work alone because the creative process and copyright are rather insular in nature. Guyotte, Sochacka, Costantino, Kellam, and Walther (2015) suggest that STEM disciplines can contribute to the arts themselves by allowing for artists to experience collaboration, especially a collaborative approach to creativity. Through the development of a STEAM course, the Transdisciplinary Design Studio, reinforces the process of collaborative creativity by purposefully crafting an authentic, real-world, and multidisciplinary assignment that forces students from multiple backgrounds to explore and define problems rather than create solutions. The assignment specifies that students create an art exhibit that illustrates the problem itself, not the solution, which is how problem development rather than solution finding was achieved. Additionally, students in the course kept "visual journals—a hybrid between the writer's reflective journal, artist's sketchbook, or engineer's design notebook" (Guyotte et al., 2015), that offered insight for students and instructors about the learning that took place during the course. Students found that the group exploration and reflection during the course allowed them to develop new understandings of what collaboration, empathy, creativity, and the learning process mean to them as students and as employees in their future careers.

Sketching Holds a Place in STEM

Directions to assemble a piece of furniture or use an appliance are made more effective with a visual reference in the form of a picture, sketch, or diagram. Similar to the Transdisciplinary Design Studio students' visual journal, Koester (in press) has developed a professional development program, Project Draw for Science, that applies basic theories of graphic design and sketching to science concepts and inquiry. Teachers and students create explanatory drawn models called "know"tations that represent the "4 main languages of science, 1) science terminology (the WORDS); 2) explanatory drawings (the IMAGES); 3) the equipment and procedures used during an investigation (the ACTIONS); 4) and mathematical equations, icons, graphs, and tables (the SYMBOLS)" (Koester, 2016). Part of the goal of creating a "know"tation is to provide authentic context to the content, which is a recommendation from quality STEM educators. Therefore, using visual art forms creates a strong STEM lesson, but including drawing as a meaningful contributing element creates a quality STEAM lesson. Ghanbari (2015) suggests that STEM needs arts integration because an integrated arts curriculum promotes collaboration and fun, engaged learning for students. Similarly, Koester and Simmons (n.d.) asserts "aesthetic inquiry can move students away from alienation and towards affiliation." Student engagement in any lesson is key, but it is a driving force behind the entire STEAM movement. The hope for educators is that engagement through STEAM lessons will help improve retention in the disciplines during rigorous future coursework and improve achievement gaps by creating entry points to rigorous learning for all learners.

Art in This Case

There are two ideas that drive the inclusion of arts education in this case: diversity of student engagement and art as an integral element in STEM. One idea promoting STEAM education is that the arts can be an entry point for learners, especially reluctant STEM learners (Koester, in press). It is imperative that educators recognize this information and use it, especially given the reports on STEM learning and careers (ACT, 2015). Results from the ACT 2015 Report on the Condition of STEM indicate that student in STEM is high, yet students expressing an interest in STEM are not prepared for the further rigorous study required for entry into STEM careers. Further, students who expressed an interest in STEM outperformed their peers. The nature of science and art are often considered quite similar, or even interrelated. Ghanbari compares arts coursework to the process of scientific inquiry stating that "A large facet of arts coursework is inquiry-based, which means it revolves around questioning and understanding concepts versus finding the answer to a given problem" (2015, p. 3). Art can be clearly linked to several interdisciplinary concepts found in the Next Generation Science Standards including patterns, scale, proportion, quantity, structure and function, and systems models (Koester & Simmons, n.d.). Alternatively, through the inclusion of meaningful aesthetic inquiry, all of the cross-cutting concepts can be explored in a deeper way. This is the true nature of a quality STEAM lesson: integrating and transcending multiple educational disciplines to provide students with authentic, universally meaningful context.

Creativity is an integral part of art, but it is also a component of great STEM practice. Visual media is by no means the only art form that can be incorporated into a STEAM lesson. However, by giving learners the opportunity to explicitly incorporate elements of graphic design, spatial reasoning, and idea sketching, educators can expand student access to reflective learning, explanations, and communication throughout the learning experience. There are indications that students who develop spatial thinking are more likely to be interested and successful in STEM careers and research (Newcombe, 2010).

For those hoping to incorporate STEAM by modifying existing lessons, STEM or otherwise, the question to ask is: How is this addition meaningful to the learner, with respect to the goals of STEAM education? Specifically, is the modification simply to add an assessment that is artistic in nature, or is it incorporating the arts as an entry point for students to the STEM content? Will including basic arithmetic enhance students' spatial reasoning, logic, and ability to think critically about how to solve the problem? By examining the goals of STEAM education and the objectives selected for the lesson, teachers should be able to avoid the addition of "art for art's sake" or "math for math's sake," respectively.

CASE DESCRIPTION

In the following case, students are asked to plan, design, build, and evaluate a robotic arm that can complete a given task. The lesson is complementary to a story in which a submarine is lost at sea and must be found in order to rescue the scientists on board. The lab is adapted from Lesson 9: Invent a Robot of Volume 2: How Do We Explore of The NOAA Ship Okeanos Explorer Education Materials Collection (National Oceanic and Atmospheric Association, 2014). The context is marine science, but can be adapted for virtually any discipline, given the pervasive status of simple machines and robotics in the modern world. Several resources included in the case provide examples of the lesson in other content areas. Content applications outside of marine science include medicine, factory automation, automotive, construction, and many more.

The examples, perspectives, and results described here are from a high school marine science course from a suburban large school setting, 25-30 students per class. Students take marine science as an elective, with Honors and College Prep options. Many students are interested in learning marine biology and do not anticipate engineering content. Students in marine science have access to 1:1 Chromebook devices during class. Google Apps for Education is used extensively throughout the entire course, including Google Drive, Classroom, Docs, Slides, Sheets, and Drawings. The lesson falls within an Exploration and Technology Unit that covers content addressing the following essential questions:

1. What's in the deep ocean?
2. How do we use technology to explore the ocean?
3. How do we locate and study phenomena found in the deep ocean?
4. What are challenges to studying the deep sea?
5. How is life supported in the deep sea?

Standards are adapted from the National Ocean Literacy Principles, but the lesson would address Next Generation Science Standards as well as other Common Core standards. Because the content is adaptable to various disciplines, only a brief list of relevant standards is included here.

Next Generation Science Standards

- **MS-ETS1-4 Engineering Design:** Develop a model to generate data for iterative testing and modification of a proposed object, tool, or process such that an optimal design can be achieved. (Grades 6 - 8)
- **MS-ETS1-2 Engineering Design:** Evaluate competing design solutions using a systematic process to determine how well they meet the criteria and constraints of the problem. (Grades 6 - 8)
- **HS-ETS1-3 Engineering Design:** Evaluate a solution to a complex real-world problem based on prioritized criteria and trade-offs that account for a range of constraints including cost, safety, reliability, and aesthetics as well as possible social, cultural, and environmental impacts. Performance Expectation, Grade: High School. (9-12)
    - **Ocean Literacy Principle #7:** The ocean is largely unexplored (Ocean Literacy Network, 2013).

- B. Understanding the ocean is more than a matter of curiosity. Exploration, inquiry and study are required to better understand ocean systems and processes.
- F. New technologies, sensors, and tools are expanding our ability to explore the ocean. Ocean scientists are relying more and more on satellites, drifters, buoys, subsea observatories and unmanned submersibles.

Authentic Context: The Student Challenge

In this unit of study, students progress through a sequence of lessons that are designed to promote understanding of ocean exploration and technology in both two and three dimensions. The lessons in the unit involve: 1) spatial reasoning topics such as mapping, scale, and elevation above or below sea level, 2) bathymetry topics such as seafloor topography, 2D and 3D mapping, and sonar, and 3) technology topics such as history, submarine, ROV/AUV, and marine sampling equipment. These lessons emphasize content that help students formulate answers to the Essential Questions from the unit, especially: "How do we use technology to explore the ocean?" and "How do we locate and study phenomena found in the deep ocean?"

Students first learn how to find objects on the surface of Earth using latitude and longitude coordinates. They then learn about technology used to find things beneath the surface, such as sides can and multibeam sonar. Students are told a story about a submarine, the *Johnson Sea Link*, that has been lost at sea and are asked to "find" it using simulated sonar in the classroom. The story is inspired by actual events involving an incident in 1973 that claimed the lives of two of the four passengers on board the *Sea Link*. In the simulated sonar lab, students search through boxed "models" of the sea floor that represent search quadrants that an underwater mapping device would map, and one of them contains a small replica of the *Sea Link* for students to find. Once the location of the submarine is found, students are asked to design a robotic arm that would be able to assist in the rescue of the trapped *Sea Link* and its passengers. The arm would hypothetically help move the debris or fishing gear that immobilized the submarine. This is only one example of the possible scenarios that could be used to capture and engage students' attention. ROV and robotic arm use for the 2010 Deepwater Horizon oil spill solution, general sampling in various marine environments, lost at sea airplanes or ships like the *Titanic* would all provide realistic context for the lesson.

Students are given the following challenge description when they meet in their team groups:

Your group is one of several teams working to design a robotic arm that is able to complete an underwater mission: to rescue the Johnson Sea Link Submarine that is trapped in 200 feet of water with 4 passengers on board. Today, you will test your ideas by building a robotic arm that will perform the task of MOVING an object. You will submit your final design in the form of a digital model in Google Drawings and an Engineering Report that discusses your design process. Your report should demonstrate why your team's arm is the best choice for the mission.

When students hear that they must build a working robotic arm, there is usually one of two responses: 1) That's really cool (or some variation thereof) OR 2) I can't do that; it's too hard; I'm not an engineer. Too often, students don't realize that the technology used in robotics and what they perceive as "difficult" engineering can be found right in front of them on a daily basis.

To begin the lesson, I ask them to draw a robotic arm. Usually I assign this as a bellwork, exit ticket, or small homework assignment. I don't give them any examples or content, just ask them to draw an arm that can move using parts like nuts, bolts, cardboard, or whatever else they feel might be needed. The purpose for limiting the amount of direction I give here is so that I can really see what they know already about the topic. Based on this pre-assessment, I can make specific student groupings, or determine how difficult I should make the challenge part for the arm. Some students may already have knowledge about hydraulics or joint movements in devices like the robotic arm, so it helps to see where they might be best placed in a group to share that prior knowledge.

The challenge portion of the build has two parts: 1) Moving an object and 2) Grasping an object. All groups must pass the first challenge of moving a large piece of coral (weight approximately 4 kilograms) a distance of 10 centimeters (cm). In context, the moving coral rock represents moving the obstacle that entangled the original *Johnson Sea Link* in 1973. Students will need to put effort into creating an arm sturdy enough to hold up the mass of the coral rock when applying force. Further, since the syringe plungers are only 8 cm in length, moving the coral 10 cm presents an additional challenge, discussed in detail in a later section. For the "Moving an object" challenge, students test their prototypes on a 10 cm track that is marked on a central lab table. The rock must move from one taped line and cross the second to pass the "move" test. The second challenge, "Grasping an object," is for groups who finish the first challenge quickly. When a group finishes well ahead of the others, it is great to have a transition or extension for them to complete while I continue to work with the rest of the class. In this case, I have them build a "grabber" at the end of the arm. This part of the lab can for be extra credit, classroom points, or just part of the overall grade for the lab.

The Lab

According to the NOAA lesson plan used, the materials needed for the device include (NOAA, 2014, p. 133):

- Cardboard or heavy poster board, approximately 12" x 12" (the stiffer the better)
- Duct tape, approximately 2" x 60"
- Machine screw with nut, #8 x 1"
- 2 - Flat washers, #8 hole
- 2 - Oral syringes
- Plastic tubing approximately 3/16" inside diameter x 12" (should fit snugly over the end of the syringes)
- Water
- Heavy scissors to cut cardboard ("bandage scissors" are inexpensive and work well)
- Nail or Phillips screwdriver to punch holes in cardboard
- Screwdriver to fit machine screw.

In addition to these materials, I also keep on hand some box cutters and a power drill. Safety lab goggles are obviously a must, and we will break down the devices and reuse metal parts at the end of the lab to keep everything cost effective. Lab materials are kept in a large "sink" tub for each station—a roll of duct tape, scissors, nuts and bolts and washers, ruler, syringes and tubing.

The most common way to build the arm is to join two rectangular sections of cardboard (suggested dimensions: 10in by 3in) at one end by a nut and bolt joint. This joint creates a hinge that will allow one piece of cardboard (the lever arm) to swing out from the other. The arm is created by attaching the syringe body to one piece of cardboard, and the plunger of the syringe to the lever arm section. The plunger will push out and retract the lever arm away from and to the syringe section of cardboard when the actuator plunger is pressed or pulled on the other end. By altering the angle at which the syringe is attached or the length and shape of the cardboard arm, students can achieve many different design purpose possibilities.

Student Instructions

On the day of the actual lab, each lab group meets and shares their ideas. We discuss the time frame, expectations, and materials they have to use for the lab. One key expectation I give is that all students in the group should be able to answer any question I have for the group during the lab. This means that no one student should be solely responsible for the work and understanding what the group is doing. Further, it is my intent to create an inquiry setting, so I do not give them any step-by-step instructions for creating a device. I tell students that the challenges they will face during the lab are purposeful and meant to foster the creativity and collaboration that I want to see from them. We have an established class norm that deep thinking and understanding is hard work, but that I will never give them an assignment they cannot do. There are some scaffolds that I put in place at various points throughout the lab to help students achieve success. This varies by class, and I look for signs of frustration from students that indicate they are giving up before adding additional scaffolds.

Before each group begins, I show the class an "actuator" that will help them build the device. I also show students an example of an existing arm (one that I premade or from another class), but they are not allowed to touch, hold, or look too closely at the example arm. They are allowed to touch, hold, and look closely at the actuator. As a class, we define the term "actuator" as a device that will convert their mechanical energy into motion on the arm by pushing a button or controller (NOAA, 2014). The actuator consists of two plastic syringes connected by plastic tubing. (The plastic tubing is similar to that used in a bubbler for an aquarium and can be purchased in the plumbing section at any hardware store). The example actuator chamber that I show the class is intentionally left empty without water so that only air fills the chamber. I tell students to begin by building an actuator and trying out different motions.

From there, it is a matter of observing them work, asking questions, and intervening with additional scaffolds as needed. Some students immediately create an elaborate cardboard extension at the end of the actuator syringe plunger. The key question that I would ask this group is, "Have you tested how far you can move an object with that design?" They will test, discover that the plunger only extends 8cm, and regroup to find another solution. At this point I would review the concept of displacement with this group because they will often attempt to add another "extension" to their initial extension design, which will not change the distance that the arm can actually travel. Depending on the group and their frustration level or the amount of time left in the class, I might suggest they look for ways to extend the range of motion of the arm by adding a lever or a hinged part. We look at the path a door or their own elbow travels, and they go back to try to incorporate those ideas. There are some groups who become so frustrated that I feel it is necessary to allow them to look carefully at an existing arm, which is the original prototype I showed the entire class. I stress to them that this is not the ONLY solution to the problem, but it is one possibility. Once they have an understanding of how the moving joint is created, I take away the arm, and they begin building again. It is important for them to know that not every engineering or technological feat comes from a novel idea; sometimes it is from refining an existing concept. The remaining refinements that they will need to make to this now "known" design which may include adding structural stability to the cardboard, adjusting the angle and distance the lever arm can travel, and creating the joint. These adjustments will still add to students' knowledge and understanding of the engineering design process and the science behind lever and robotic arms.

Some groups will design an arm that increases the distance that the arm can move by using two actuators in conjunction. This method is successful, but the instructor would have to decide if students would be allowed to use multiple actuators. Personally, I feel that the purpose of the lab in this case is more about problem solving and engineering design process skills rather than teaching lever concepts, so I encouraged using more than one solution. These unique designs make for much more interesting and developed class discussion and reflection at the end of the lab.

Once students have created an arm that moves far enough, they begin to tackle the problem of structural stability. Cardboard is not always the most stable building choice, so students may struggle with creating an arm that will hold up to the weight of a large piece of coral "rock." There are not specific scaffolds with which I respond to that challenge, but I instead send them to check out other groups' solutions around the room. This is sometimes accomplished with a lab meet and greet of sorts, in which one person is elected to stay at their lab station as a spokesperson who explains their design to the other groups' members who rotate through each station. Students will usually find many different ways of adding stability to the arm including multiple layers, folded cardboard tubes, and blocks of cardboard attached where the syringe plunger will apply pressure.

The last key scaffold that I implement is encouraging students to consider how well the actuator responds when it is only filled with air. They usually say it feels "weak" or "soft" when they apply pressure to a larger object. Often, they will begin thinking about what else could go in the cylinder and will ask to try water. I make a point to show them how to securely fasten the syringe to the hoses in an effort to avoid water accidents in the lab when the hoses disconnect under increased, now hydraulic, pressure. I often find students who switch to hydraulic actuators early have discovered, and solved, this problem with additional duct tape at the syringe/tube connection. At this point, the entire class will review prior lessons on gases, density, and compression, and discuss the applications of Pascal's Law, hydraulic, and pneumatic technology in detail.

Students work their way through many problems and challenges with their design throughout the lab. Each time they try their design at the testing station they record areas of the design that need modification. Each change is documented for their reports, and each change puts them closer to creating a working arm. Once the arm is complete, they move on to completing their engineering reports, or with enough time, the second grasping arm challenge.

STEAM Integration Summary

In general, the breakdown of the integrated content from each discipline is as follows.

- **Science:** ROV's, hydraulics, levers, compression, pressure.
- **Technology:** ROV's, syringes, joints, Google Apps, models.
- **Engineering:** Design process, iterations, modeling/build.
- **Art:** Google Drawings or SketchUp, creativity, design, advertisement.
- **Math:** Pressure calculations, measure, angles, displacement.

A more detailed description of how the disciplines are integrated and studied is found in the following sections. It is clear that separation of the disciplines is difficult, which is why STEAM integration is a natural and to this lesson. The math section discusses science content, the technology section illustrates science applications, the art section highlights the use of technology, the engineering section shows the reflective elements of the scientific and art process, and so on.

Science Content

Students will learn about many kinds of physical concepts during the design and build of a robotic arm, the extent of which can be tailored based on the age, level, time allowed, content area and interest. Generally, students would be expected to learn about hydraulic power and the applications of actuators. The content is extended to touch on the topics of displacement, levers, hydraulic versus pneumatic devices, pressure, compression, and marine technology such as submarines and ROV's.

The use of hydraulic and pneumatic power in technology is quite common. Many public exterior doors operate on a pneumatic cylinder that forces the door to close slowly. Car manufacturers utilize hydraulic cylinders to keep the trunk or hood open without the use of a manual prop. I remember the hydraulic fluid (some form of oil) leaking out of the trunk cylinder of my family's old Ford station wagon as a little girl. I always had to be mindful not to stand under it without careful attention, as, without a functional hydraulic cylinder, the trunk door would come crashing down under the force of its own weight. Robotic arms seen in factories and new medical facilities also use hydraulic and pneumatic devices. Large construction vehicles such as dump trucks, tractors, or front-end loaders (think Tonka truck) use hydraulic power to move and lift materials.

For this case, it is essential that students ultimately differentiate between pneumatics and hydraulics, especially in terms of compression and effectiveness in design. This differentiation will help to connect other technology design concepts such as the use of gasoline as a buoyancy device in the 1960 submarine *Trieste*'s voyage to the *Challenger Deep*, 11,000 kilometers down in the Pacific Ocean.

To complete the challenge described, moving a coral rock 10 cm, students must build a lever arm. This is key because the distance that the syringe plunger moves is limited to 8 cm, so in order to extend that distance, it is essential that students use a lever to extend the range of motion of the arm. Thus, it may be useful to review the concept of simple machines, especially the lever, before beginning the lesson. Additionally, depending on how far the teacher would like the mathematics and science content to extend, knowledge of Pascal's Law, concerning the pressure exerted in a fluid, would be appropriate.

I purposely limit the background information provided in the lab in order to facilitate inquiry, creativity, and collaboration during the design process. One of the content details that I do not provide students is that hydraulic powered devices are more efficient for the type of work that is needed to move the coral rock. Because water is less compressible than air, filling the syringes and tubing with water is an excellent way to achieve better transfer of energy from the actuator to the coral rock. This is where Pascal's Law applies: as pressure is applied to a liquid such as water, the force is evenly dispersed and without compression, that force is applied directly to the opposite end of the actuator. Students will discover that using the syringe with air allows for some spring or give in the actuator line, and it doesn't seem to move as fast or far as a syringe filled with water. As a facilitator, I ask them to explain WHY the water is a better choice and be sure to include that in their final report. I also note the responses that are initially made, which often include a focus on the fact that water is denser than air, rather than the concept of compression. This is significant to discuss and break down on a group basis and often as an entire class, especially because it relates to other units and lessons such as submarine technology and SCUBA diving.

Technology Content

There are many ways to include technology and improve student understanding of what it means to use technology in addition to electronic devices. This lesson explicitly involves the use of modern electronic technology such as digital design software and Chromebooks, but there are other less explicit technologies that play a role as well. The arm, materials, and process used to build the arm can be considered technology. The actuator is a form of technology used in all the designs. For very

sturdy cardboard, it is sometimes helpful to have students use power tools, such as a drill to punch holes for the nut/bolt hinges, as another form of technology in the classroom. The applications that a hydraulic arm can be applied to outside of the context of the student challenge provide numerous other examples of technology.

Additionally, students use cell phones to take pictures of their devices, Chromebooks to record their design processes, and SketchUp or Google Drawings to create diagrams of their initial build plan and ideas.

The technical design that combines art (whether media or graphic) with the engineering process is one of the most important parts of this lesson. For many labs, I use a sketch of the lab setup as a pre-assessment for student understanding of the lab before we begin. It is eye-opening sometimes to see how much they understand of the lab directions and can provide me with feedback on how well the instructions are written!

Engineering Content

One of the most important elements of engineering education, particularly for the goals of STEAM education, is the learner's participation in the Engineering Design Process. It is this reflective learning process, geared towards data collection, analysis, and modification that allows students to develop communication, collaboration, creativity, and problem-solving skills. This process can be enhanced through the use of digital technology in the classroom, which made more engaging for learners through art. The process also allows for tangible application of math and science concepts. There are several "models" of the engineering design process. For this case a combination of a district model and the NOAA lesson's model are summarized below.

1. Identify Problem,
2. Gather Relevant Background Information,
3. Generate Ideas,
4. Build a model,
5. Analyze possible solutions, make modifications as needed, and
6. Communicate Results: Document test results and reflect on modifications' effectiveness.

The main difference between the model presented in the NOAA curriculum and here is the addition of Step 4: Build a model and Step 6: Communicate Results. Step 4 is included to emphasize the build as a hands-on, real-world experience. Step 6 emphasizes explicit reflection in the student learning process. Students' final reports will allow them to reflect by assessing their initial design and detailing the changes they made in order to create a working model by the lab's end. The following text is an excerpt from a student reflection on the design process. It specifically reflects on modifications that were made to the arm in order to make it functional.

Student Reflection

Along the way there were modifications made to the design. One important change was the decision to change from pneumatic air-filled actuator to hydraulic water filled actuator at the beginning of the construction process. This change was due to the realization that water is not compressible like air and allows for force to be evenly distributed within the tubes, unlike a pneumatic actuator that would have been less effective given the easily compressible air in the tubes. This change made an improvement in the overall structure of the robotic arm and ensured that the lever would be able to be moved to its fullest potential and distance. Throughout the process adjustments were made as the trials went on. This mostly consisted of strengthening the cardboard by adding additional pieces as reinforcement and ensuring that the board did not bend or cave in with the pressure. The only way to figure out how thick the arm needed to be was to run trial after trial and see how much the arm moved on its own and then the difference when resistance, by way of a heavy rock was applied.

At this point in the lab this student would learn the term Pascal's Law using inquiry practices and make connections between the science content and the intuitive understanding of that content described in this reflection.

Art Content

Vaidyanathan (2012) states that the [creation of digital artwork] is an intuitive way to learn STEM subjects (pg. 27). Further, it is a way to increase student participation as most digital editing programs have an "undo" button which allows for freedom to try new ideas without repercussions (Vaidyanathan, 2012). This aligns with my own personal experience in graphic design, but it is exciting to see in the classroom setting where students are trying completely new things. In order to promote spatial reasoning skills that can contribute to STEAM learning, Newcombe (2010) recommends that students create sketches to explain their understanding of topics. Such a sketch must be completed in order to complete this lab in marine science. This can be done by hand, or if technology is available the use of an app or program such as Google Drawings or SketchUp would be a great way to integrate technology into the lesson. (Keep in mind that the design and model itself are a form of technology, and the introduction of electronic devices is just an additional element of technology use in the classroom.) Because I include the pre-design at the beginning, ALL students must use some type of creativity or applied knowledge to create a visual representation of their ideas for a robotic arm. When they first meet in their teams, they will share and discuss each individual design and come up with a team plan and design. Each lab group must present a diagram, completed in either SketchUp or Google Drawings, of their team plan for the arm after collaborating at the lab's beginning. Finally, they must turn in the design of the final working model. This may be in the form of a photo, improved digital design, or edited (notes on changes) initial team design. This last form is similar to "know"tations (Koester, in press), where arrows and text would explain changes in the diagram/sketch of the idea that were made to create the final working model.

Included are several examples of student work that illustrate their progression of design through a first idea sketch to initial team design to final 3Dimensional build. Student reflections show the thought processes that students used when modifying their designs during the build.

For students who have no idea how to start in Google Drawings, I share a template with them to help scaffold the digital design process (Figure 1). The template combines elements of digital design (using the program to create a design), science, and engineering (labels, scale, notations).

The context of the challenge is ultimately to design an arm that will be fitted to an ROV that can accomplish a variety of tasks underwater. Students learn about ROV and submarine technology during the unit and can apply that to their arm

Figure 1.

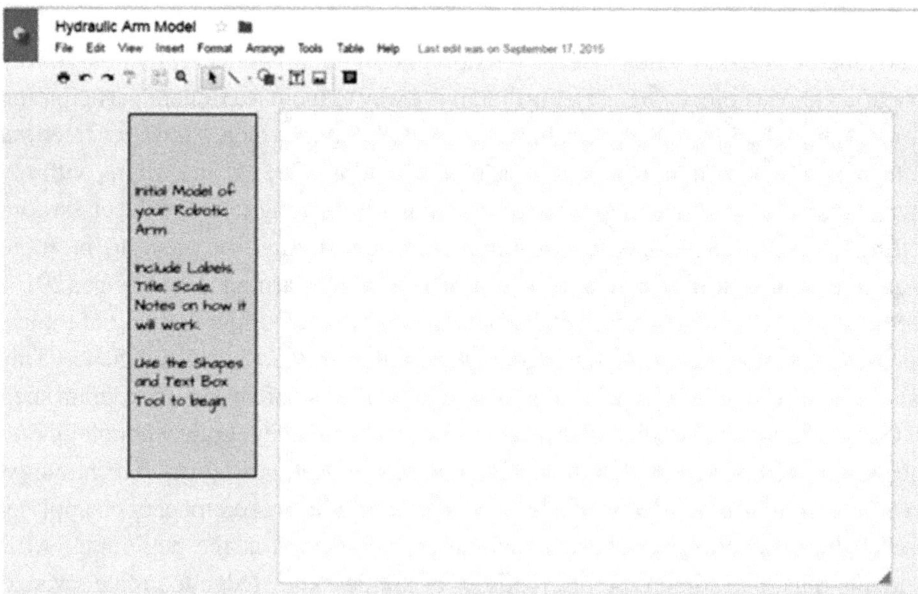

design in context. This student used that context to create a sketch of an ROV that would fit the team's actual model robotic arm (Figure 2). Students would reflect upon design functions and the modifications that formed the functioning arms in their reports.

The transition from digital diagram to working arm is a distinct one for many students. Some students work from the 3D working model backwards to create a final digital diagram for the team report, and others use the 3D model to explain changes that were made (Figure 3). As a teacher, I find this illustrative of the creative diversity found in each classroom.

Student designs between each team are highly varied, but the way each student represents the design digitally is also distinct (Figure 4). Even in an engineering setting, where the art product would be considered technical and almost scripted, variations in color, theme, symbol, and design are abundant.

The digital design coupled with graphic design concepts like labeling, text, arrows, and symbols is an excellent way for teachers to assess what students learned from a lesson like the hydraulic arm lab. However, the written reflection that accompanies this diagram in the report illustrates key misconceptions that are important to address before concluding the lesson (Figure 5). Art integrated with the other STEM content makes for a powerful assessment tool in this case.

Originally, my thought for this lesson was simply to have students create digital mock-ups of their ideas and creations, but with the use of other integrated technology such as camera phones or digital cameras, they surprised me again with their innovation and creativity. This student felt like it would be more effective to import and edit an image of the 3D model in order to explain the design and modifications for their report (Figure 6).

Figure 2.

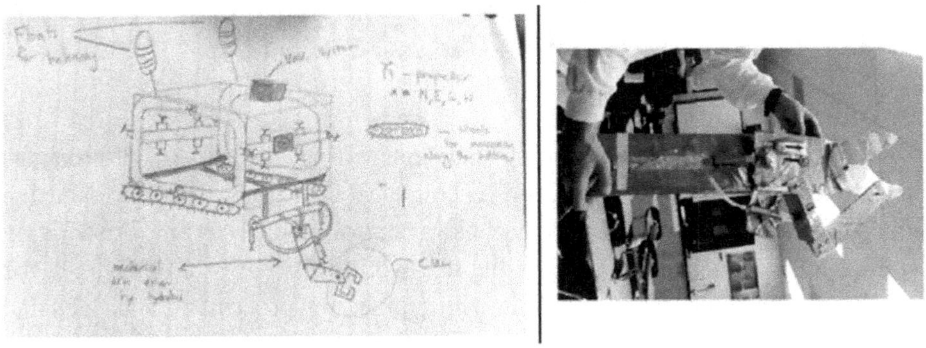

Figure 3.

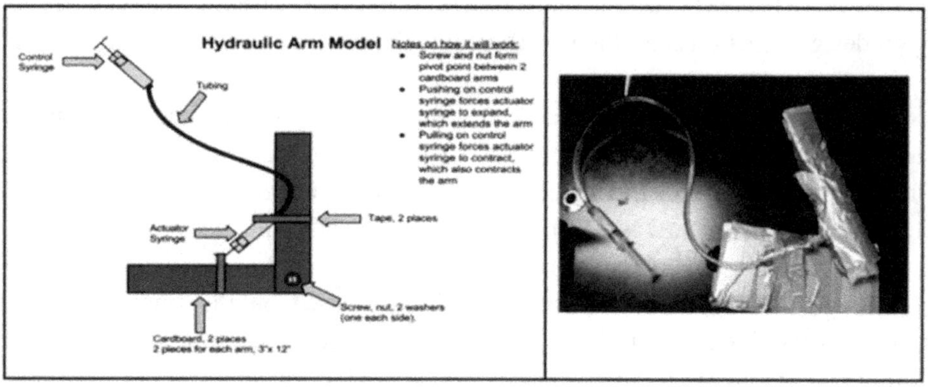

Figure 4.

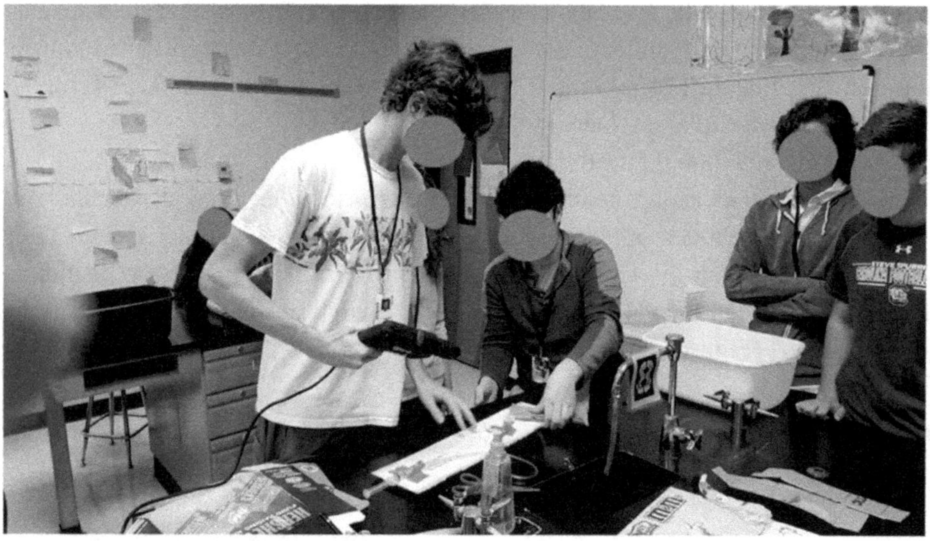

Even without the digital technology, students will create a variety of products. Once students have a working concept, I encourage them to make the exterior design more complex and incorporate elements like a control panel. They enjoy decorating the arms and naming them, just like the real-world ROVs we study. Overall, the variety of design and ways of explaining the engineering process, both visually and verbally, provide for an engaging lesson for both student and teacher. Each year we complete the lab, I learn new ways of tackling the design challenge, highlighting

Figure 5.

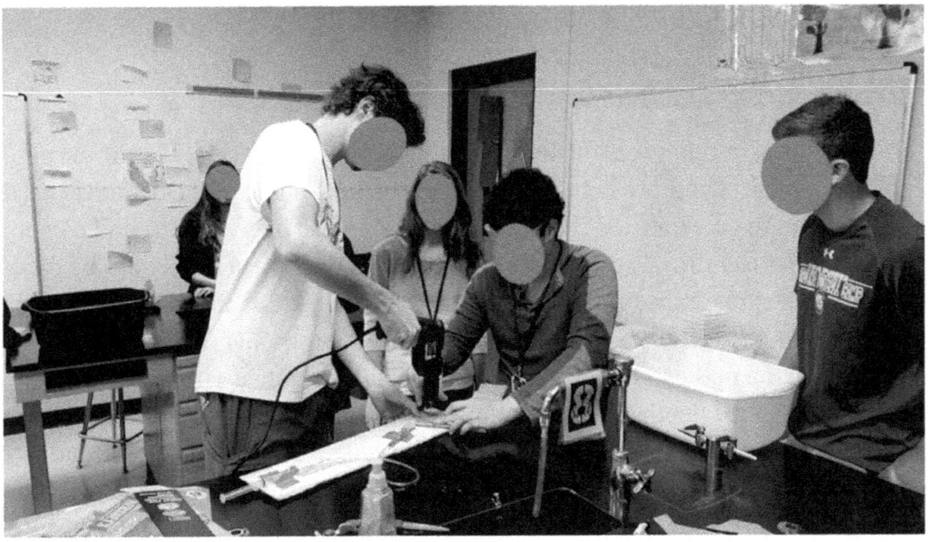

Figure 6.

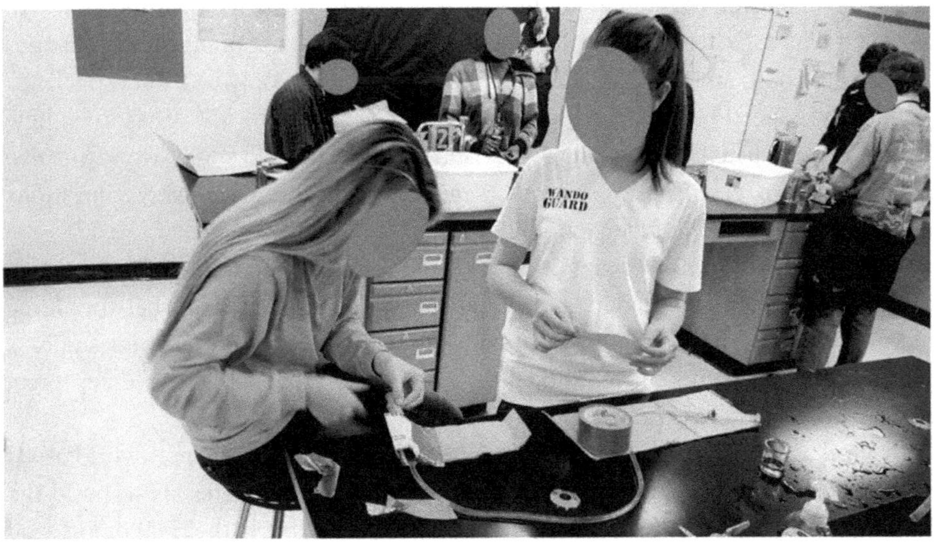

the skills of collaboration and innovation that I hope students carry with them to future classes and their careers.

Math Content

Math is the most implicit content area taught in this particular case. For the purpose of this example students are not asked to measure the volume of the cylinders used in the actuator to calculate the pressure applied to the arm, given time and content restraints. That is a direct application of applied math content that could be used, however. One explicit use of mathematics that is included is the concept of scale since student must include scale in their final design proposals. This idea is tied to a previous lesson in the unit sequence in which students create a scale model of the earth by plotting the depth and height of features around the earth in the classroom.

Additionally, in order to build the arm and create an accurate diagram, students measure the length of each part and assess the angle at which to place the syringe plunger which will apply force to the moving lever arm. Students test and adjust the angle until they successfully achieve the full range of motion required to move the coral rock 10 cm. The concept of displacement is cemented when students measure the distance the syringe plunger moves, a total of 8 cm, and then test the idea of adding an extension to the plunger head to increase the distance that the coral will move. Of course, in terms of displacement, adding bulk to the plunger only changes the starting point, and the displacement of the plunger can only be 8 cm. This leads them to the "speed bump" in the lesson that forces creativity and reliance on new ideas. Students usually try many angles and types of levers as they work to solve the problem. As each group proceeds, I move around the room asking questions and looking for student understanding of topics.

The Fluid Power Lab (TeacherGeek, 2008) has a great set of diagrams and practice problems to help students understand Pascal's Law and the relationship between force and pressure. The lab handouts have many student practice problems for students to use basic math skills of multiplication and division to calculate volume, force, energy, work, and pressure.

A great way to include additional math content that provides real-world skills and context is to have students budget for and "purchase" materials to build the arm. The cost effectiveness of their build would be included in the final report, and groups can compare total cost of each build to their effectiveness as a method of class reflection.

CURRENT CHALLENGES FACING TEACHERS

Teacher Concerns About STEAM Education

Do Lessons Have to be "Big" Lessons?

No. This lesson can be accomplished in one or two block periods, depending on the depth of assessment tasks assigned. In the case of the TeachEngineering (Chilton, Carpenito, & Milto, 2013) or Basketbots (Bigelow Middle School, n.d.) resource examples provided, you could also create a much longer unit. Just remember that as practitioners we should be reflective and explicit about what and why we are teaching. Reflecting upon these questions is just as essential to transdisciplinary learning that students reflect and make explicit the "what" and "whys" of what they did during the lesson.

Isn't it Hard to Grade Something Like This?

I use Google Sheets and an add-on, Doctopus (with Goobric) to grade the reports based on a rubric for student design. There are many ways to assess, but it depends on what your goals for learning are to determine the right assessment. This lesson seeks development of a working model with verbal and visual explanations of how the model was created, and demonstration of content and technology design skills. These elements were present in the rubric provided to students for their report.

I Don't Know Anything About Building Robots.

Sometimes, neither do your students. Go ahead, have fun LEARNING WITH THEM. It will help to create the caring teacher-student relationship that so much literature supports as a best practice!

I Don't Have Access to Technology.

A thousand years ago, technology was a formula, or a new way of gardening. Mechanization and computers are NOT the only forms of technology. My students sometimes put forth that written language is the most influential form of technology created. Without getting into a debate about whether language counts, it is true that technology in quality education does not just have to include 3D printers, expensive 1:1 device programs, robots, or laboratory equipment. Use whiteboards to promote collaboration and sketching or large Post-It posters and markers. Create a gallery walk. Start a BYOD (bring your own device) program. Maybe those little devices in the pockets or purses of your students could be put to use!

It can be helpful to ask students to define the technology they used in a lesson. In this lesson, the model and diagram are forms of technology. The media used count, too, (Guyotte, Sochacka, Costantino, Walther, & Kellam, 2014). Student reflections include that for the real device they would need to use something other than cardboard, or a waterproofing agent. By answering that question, they will begin to reflect on the use of tools and media (in all the disciplines) to define technology, not just the use of a digital device.

It Seems Hard to Include All Five Disciplines Equally.

I have found that not every lesson I teach can possibly encompass all five elements. As a curriculum team, we try to diversify the ways we use and connect each discipline. Include art as a performance assessment—have students create a video or advertisement at the end of the lesson that ties together the multiple content areas. Try creating a "know"tation or visual journal on an integrated topic. Start small, but think big. Start with a daily lesson and consider the topic. Guyotte et al. (2015) suggest that using a complex social issue as a topic is a good way to encourage collaboration, deeper understanding, and expanded perspective for students, particularly when coupled with artistic reflection. There are so many topics to choose from, especially in the environmental and sustainable design fields. Try finding a problem-based lesson and look for ways to incorporate or integrate the arts from there. Keep in mind Anne Jolly's caveat: "The purpose of STEAM is not to teach art, but to apply it in real situations. Just like the purpose of STEAM is not to just teach math, but to apply math in real situations" (2014). There are several recommendations for great STEM/STEAM programs in literature, but as a teacher, my best advice is to just try. Take a risk. It might be fun learning something new.

SOLUTIONS AND RECOMMENDATIONS

REFLECT. REFLECT. REFLECT. Be explicit about this. Student outcomes vary over the course of this lesson. One student enjoyed the creative process and challenge so much she changed her major to engineering. Others determine that engineering isn't for them, but become familiar with the reflective process that can be helpful for any discipline. STEAM is a meta discipline. High quality STEAM programs do not treat the disciplines separately. It follows that art should not be just a separate discipline in quality STEAM instruction, that all elements should be a relevant part of the learning process. However, the power of the programs described here isn't in the fact that all the disciplines are included equally; the power lies in the fact that students and teachers take the time to break down the process of what they did, how they did it, and what that tells them about the world. That reflection piece can be a group discussion, a survey, a written report, or a journal, but it must be a part of the learning process. It's great to include all elements equally, but without explicit breakdown of the learning process itself, the power and meaning of the learning may go unnoticed by the student. This is why it is imperative to include the final steps of the engineering process, communicate results, as a reflective piece. The power of that reflection is evident in both Guyotte et al. (2014) and Koester's (2016) work through visual journals and "know"tations, respectively.

Using the work of Guyotte et al. (2014), perhaps a useful modification to the lesson is to eliminate the restrictions on the challenge and instead give students an opportunity to define the problem themselves. In their efforts to push students toward collaborative creativity by emphasizing a problem exploration, Guyotte et al. (2014) report that students become more aware of the significance of group dynamics and [possibilities]. Pay careful attention to the "speed bumps" that can be set in place in a carefully constructed lesson to facilitate problem solving rather than solution finding. I find it similar to the twists included on modern cooking competition shows and game shows. Those carefully placed hurdles keep the learner engaged and force (the student) to rely on others, new ideas, or develop emotional intelligence and empathy.

In terms of reaching a diverse group of learners, I see students that struggle in core areas such as language arts, math, and science have moments to shine during this particular lesson. It is always rewarding as a teacher to watch these interactions. For example, one student who is interested in cars and enrolled in an automotive course created a set of channels in the arm for the syringe plunger stem to rest, adding strength and stability to the moving joint. Other students in the class had previously overlooked his efforts, but in this moment, he had their acknowledgement and praise for expertise. The moments happen in some way in each class, whether a female discovering she enjoys "engineering," a quiet, low performing student being asked to help the "A" students make their Google Drawing better, or a high performing student discovering that their first idea didn't work so they must rely on others to complete the task. The experience creates a change in the classroom dynamic, as more students are willing to listen to others' ideas and hope for creative solutions that stemmed from the mere diversity of experience available in the room.

Through analysis of Texas's TSTEM initiative, Kennedy and Odell (2014) propose that in order to engage more students in STEM, teaching must be altered to incorporate truly interdisciplinary and real-world contextual lessons. Additionally, they state (Kennedy and Odell, 2014):

High quality STEM education programs provide teachers with opportunities to collaborate with one another ununified efforts aimed at integrating the four subjects into one cohesive means of teaching and learning. It is when this objective is achieved that students gain access to meaningful curricular opportunities promoting critical thinking skills that can be applied to their academic as well as everyday lives. (p. 256).

Providing these opportunities and *time* for teachers to develop, practice, and explore STEAM curriculum and content integration skills is imperative to achieve education reforms' goals. Hopefully, as teachers gain that experience, daily STEAM lessons will be easier to implement, and, ultimately, the ideal project-based lesson like the Dream Factory will seem less daunting and perhaps become a reality.

REFERENCES

ACT. (2015). *The condition of STEM*. National ACT, Inc. Retrieved from http://www.act.org/content/dam/act/unsecured/documents/National-STEM-Report-2015.pdf

Beers, S. Z. (2011). 21st century skills: Preparing students for THEIR future. *STEM*. Retrieved from http://cosee.umaine.edu/files/coseeos/21st_century_skills.pdf

Bigelow Middle School. (n.d.). *Engineering technology: Basketbots*. Retrieved June 30, 2016, from http://www.bu.edu/gk12/alex/bioengineering_teacher.pdf

Chilton, E., Carpenito, K., & Milto, E. (2013). Hands-on activity: Hydraulic arm challenge. *TeachEngineering*. Retrieved June 28, 2016, from https://www.teachengineering.org/activities/view/wpi_hydraulic_arm_challenge

Ghanbari, S. (2015). Learning across disciplines: A collective case study of two university programs that integrate the arts with STEM. *International Journal of Education & the Arts*, *16*(7). Retrieved from http://www.ijea.org/v16n7/

Guyotte, K. W., Sochacka, N. W., Costantino, J. W., Walther, J., & Kellam, N. N. (2014). Steam as social practice: Cultivating Creativity in Transdisciplinary Spaces. *Art Education*, *67*(6), 12–19. doi:10.1080/00043125.2014.11519293

Guyotte, K. W., Sochacka, N. W., Costantino, T. E., Kellam, N. N., & Walther, J. (2015). Collaborative creativity in STEAM: Narratives of art education students' experiences in transdisciplinary spaces. *International Journal of Education & the Arts*, *16*(15). Retrieved from http://www.ijea.org/v16n15/

Henriksen, D. (2014). Full STEAM ahead: Creativity in excellent STEM teaching practices. *The STEAM Journal*, *1*(2), 15. doi:10.5642team.20140102.15

Jolly, A. (2014). STEM vs. STEAM: Do the arts belong? *Education Week Teacher*. Retrieved June 30, 2016, from http://www.edweek.org/tm/articles/2014/11/18/ctq-jolly-stem-vs-steam.html

Kennedy, T. J., & Odell, M. R. L. (2014). Engaging students in STEM education. *Science Education International*, *25*(3), 246–258.

Keruskin, T. (2015). Finding the sweet spot: Creativity, candy, and commerce. *Edutopia*. Retrieved June 28, 2016, from http://www.edutopia.org/blog/sweet-creativity-candy-commerce-todd-keruskin

Koester, M. (in press). Getting to "know" STEAM. In M. Van Sickle & J. Bazler (Eds.), *Cases in STEAM education in practice*. Hershey, PA: IGI-Global. doi:10.4018/978-1-5225-2334-5.ch004

Koester, M., & Simmons, S. (n.d.). *The collaborative art of teaching science through drawing* [PowerPoint Slides]. NAEA. Retrieved from http://naeaworkspace.org/naea16/The%20Collaborative%20Art%20of%20Teaching%20Science%20through%20Drawing.pdf

National Oceanic and Atmospheric Association. (2014). Lesson 9: Invent a robot. In *The NOAA ship Okeanos explorer education materials collection, volume 2: How do we explore* (pp. 125-136). Retrieved from http://oceanexplorer.noaa.gov/okeanos/edu/collection/media/vol2_hdwe.pdf

Newcombe, N. (2010). Picture this: Increasing math and science learning by improving spatial thinking. *American Educator*.

NGSS Lead States. (2013). *Next generation science standards: For states, by states*. Washington, DC: The National Academies Press.

Ocean Literacy Network. (2013). *Ocean literacy principle #7*. Retrieved June 30, 2016, from http://oceanliteracy.wp2.coexploration.org/ocean-literacy-framework/principle-7-v2/

TeacherGeek. (2008). *Fluid power lab*. June 30, 2016, from http://teachergeek.org/fluid_power.pdf

Vaidyanathan, S. (2012). Fostering creativity and innovation through technology. *Learning and Leading with Technology*, *39*(6), 24–27.

KEY TERMS AND DEFINITIONS

Actuator: Mechanical device that converts energy into motion, in this case converts muscle energy into motion of the syringe.

Graphic Design: Digital art that is organized using text, shape, and image that communicates a message, idea, or other topic.

Hydraulic: Liquid powered, in this case water.

Lever Arm: Arm that moves around a fulcrum, in this case a hinge or joint.

Pascal's Law: Pressure applied on a fluid is dispersed (evenly distributed) throughout the entire fluid.

Pneumatic: Air powered.

Pressure: Force per unit area.

ROV: An underwater remotely operated vehicle that is tethered to a ship or submarine and operated by a technician, used for various marine expeditions such as sampling and filming.

Chapter 15
"Imagioneering" a New Mission:
Space

Kyle Seiverd
Toms River Regional School District, USA

EXECUTIVE SUMMARY

STEAM education is a comprehensive approach to addressing content in the classroom. By using STEAM, educators present material utilizing multiple-intelligences. This chapter is geared towards high school and uses students' familiarity with Disney as a hook to address STEAM. Critical analysis is applied to the exterior and line-queue design of a famous attraction at Disney parks. Ride-layout is critiqued and improved upon via student collaboration. Students use their ability to analyze design to engineer a 2-D scale model that fits a particular purpose.

LITERATURE REVIEW

Walt Disney World: The New Standard

Even though most envision the famous, magical Walt Disney World theme parks at the mention of Walt Disney's name, the global powerhouse of a company that is Disney got its start in animation. The most iconic and famous creation of the Walt Disney Company is Mickey Mouse (Figure 1). By using Mickey Mouse as the mascot for his company, Walt Disney was able to convey messages and illicit emotions from audiences around the world. Disney's transition from animation to theme park was inspired by a family trip to a carnival. While his daughters rode

DOI: 10.4018/978-1-5225-9631-8.ch015

Figure 1.

the carousel, Disney reflected on the grounds that surrounded him. He noted the poor conditions and the lack of family-friendly entertainment. These observations sparked the ideas that would manifest as California's Disneyland.

As Disney's animators continued to pioneer music, film, and television, Walt Disney himself was exploring other avenues. He wanted to provide safe, wholesome family entertainment to all ages. Upon completing Disneyland, Disney continued to seek a better design for his "land of make believe." His restless pursuit of perfection gave rise to one of the most remarkable destinations on Earth. An article from Frommers.com (2016) explains that:

In 1964, Walt Disney began secretly buying millions of dollars' worth of Central Florida farmland. Some thought it was Howard Hughes; others, the space program. Speculation was rife almost to the very day, November 15, 1965, when Uncle Walt arrived in town and announced his plans to build the world's most spectacular theme park. In a 2-year construction effort, Disney employed 9,000 people. The total cost of the project by its October 1971 opening was $400 million. Mickey Mouse escorted the first visitor into the Magic Kingdom, and numerous celebrities, from Bob Hope to Julie Andrews, took part in the opening ceremonies. In Walt Disney World's first 2 years, the attraction drew 20 million visitors and employed 13,000 people. The sleepy citrus-growing town of Orlando had become the "Action Center of Florida," and the fastest-growing city in the state. (Para. 1).

Imagineers: The STEAM Experts

When Walt Disney began planning Disneyland and later Disney World, he relied on his most creative staff to assist in novel project designs. He termed these members 'Imagineers.' Imagineers are the creative force behind iconic Disney attractions and experiences," (WDI, 2016). Walt Disney knew that greatness could not be achieved alone and insisted all attractions utilize a team of Imagineers. Each Imagineer brought ideas that contributed to the development of iconic rides such as the Haunted Mansion and Pirates of the Caribbean.

When developing an attraction for Disney Parks, Imagineers had to know their limitations. Each attraction at Disney Parks has a clear objective. An attraction's purpose can range from fantasy to science-based. For example, when designing a space-themed ride it is important that concepts presented utilize factual scientific data. Meaning, the idea of finding a singing elephant on the surface of the sun is such a far-fetched idea that the attractions objective is lost.

Attraction design begins with a pitch. The pitch presents the needs for the attraction. If the pitch is approved, research into the limitation of design such as space, cost, and timeline are pursued. One of the first steps in attraction developing is to contract a model. Models come in a variety of formats. A 2D drawing, a 3D structure, and acting can all be considered different means of modeling. Constructing an accurately designed scale-model requires both artistic and mathematical skills. If a real-life moon was needed for a ride, a scaled down model would first be made. The size of the moon's distinctive features such as craters and valleys are scaled and designed using a medium that is believable and within engineering constraints.

Recently, Imagineers have been using cutting-edge technology to build their models virtually before sending them for construction. Computer programs allow 3D models to be designed and modified instantly before being approved. This saves money and allows Imagineers to catch flaws that may easily be missed. If the design is approved, 3D printers can make a scale-model of the attraction. This methodology was used in the 2015 expansion project to Fantasyland at the Magic Kingdom in Disney World. The new area was designed virtually allowing creative and technological collaboration before construction.

EPCOT

The gemstone of Disney Parks in Orlando, Florida is the Walt Disney Magic Kingdom. When designing the Magic Kingdom, Disney reflected on design flaws from Disneyland. He had the Imagineers build the entire park one story higher to prevent park guests from seeing trash disposal and cast member movement. Although the Magic Kingdom attracts the greatest number of guests per year, Walt Disney conceived an idea to build another theme park showcasing cutting-edge research and cultures from around the world. According to wdwinfo.com (2016), "Epcot, the Experimental Prototype Community Of Tomorrow (Figure 2), opened in 1982. EPCOT was conceived by Walt Disney to take its cue from the new ideas and new technologies that are now emerging from the creative centers of American industry." EPCOT provides guests an opportunity to explore new frontiers in the fields of agriculture, sustainability, and more.

Ride Design and Scale

All attractions at Disney Parks began as a simple idea before blossoming into the innovative rides they are today. When asked about how one should begin designing a ride, the response is to be a renaissance person in multiple areas. Brigante (2010) stated:

One should not be limited to just one or two specific tools. The ability to be creative across any medium is more important than expertise in one particular area. Universal designer Rich West emphasized the importance of the simplest design tool: a number 2 pencil. (Para. 11).

All designs must first be translated from mind to drawing. This happens when a 2D or 3D picture is drawn to illustrate an Imagineer's concept. The illustration will

Figure 2.

often be accompanied by a storyline for the ride. From there the idea is pitched, collaborated on, and a scale model is built.

Scale models allow developers to see a ride's design in a physical 3D way. Models are generally small-scale replicas of an actual object. Math skills are imperative to constructing an object to scale. By using models, Imagineers are able to reflect on and modify the design. Scale models can be made by hand using a variety of materials; however, 3D printing and computer software is becoming more common. Models come in a variety of scales, pending the stage of the design processes. Often, preliminary models are done to a higher degree of scale. Building preliminary models on a high degree of scale allows developers to get a feel for how the ride will work, and it provides a foundation for developers to build upon for later models. As the design progresses, models become lower in scale. Each succeeding model has greater detail and gives Imagineers an opportunity to refine the model further.

As the exterior of the ride is being constructed, ride vehicles and interactive guest experiences are 'play-tested.' Reviews of the play-tests are taken into consideration, and the ride is modified further according to feedback. Once the play-testing phase is complete, the ride goes into its final phase of construction. A ride is not solely the in-seat experience; the time waiting in the line-queue is an integral part of a park guest's ride experience as well. Imagineers use technology, art, and science to entertain guests while they wait. A line-queue can have noise, interactive displays, and more. By providing guests an engaging experience while waiting, excitement builds and the chance of the ride being vandalized decreases.

NASA: Mars Exploration

When Walt Disney World was being constructed, space exploration was headline news. Families gathered to watch Neil Armstrong set foot on the moon, and many traveled a great distance to watch NASA shuttle-launches from Fort Lauderdale, Florida. During the 1960s and 1970s, much of NASA's attention was focused on the moon. While the moon is still of interest, American culture's and NASA's attention has shifted to Mars. NASA's current goal is to send a manned mission to Mars by 2030 (NASA, 2014). The idea of sending humans to Mars is exciting, but it presents many scientific challenges. Current technology estimates that the trip to Mars would take over a year to complete. Research is being conducted to determine how humans will age in space; there is also a large focus on providing supplies to maintain a crew for over a year. Sending a manned mission to Mars will require ingenuity and new technologies.

Mars Voyage

In the early 2000s, construction began on a ride that would later be known as Mission: Space (Figure 3). According to wdwinfo.com (2016),

Mission: Space was created in collaboration with former NASA advisors, astronauts and scientists, along with the Walt Disney Imagineering team. The attraction uses new technology, which makes it the first ride system to take attraction visitors straight up in a simulated flight experience.

In Mission: Space guests are divided into teams of four, each being assigned a role on a shuttle to Mars. A simulation-based ride takes guests on a shuttle to Mars with a few complications along the way. The ride ends with the team on Mars, and guests exit the ride into an interactive space-themed area.

One unique aspect of Mission: Space is that guests are given an option when they enter the ride. This ride was a first for Disney because it gave riders choice in their experience. Imagineers created two experiences, one having greater G-forces than the other, but with the same plot. Providing this choice saves guests prone to motion-sickness from missing out on the ride's experience. A guest (Mumpower) described the ride as "(an) intense ride with a take on possible future missions to Mars offers a great experience. The ride spins really fast to produce the G forces you experience in the ride -Jacob F 2016."

LESSON PLAN

The Art and Math behind Ride Design

Mission: Space at Disney's EPCOT allows guests to view the solar system from a new perspective. The ride's entrance uses large-scale models of planets to entice guests to the ride. For this activity, students will re-design the entrance and line-queue to Mission: Space. The line-queue has interactive displays and showcases technologies that NASA uses. Due to the ever-changing pace of technology, space-themed rides can become outdated quickly. This activity gives students the chance to put themselves in an Imagineer's shoes and redesign one of the most iconic attractions at Walt Disney World.

Materials

- Markers

Figure 3.

- Colored Pencils
- Rulers
- Compasses
- Posterboard
- Pencils
- Cardstock
- Protractor
- Computer Paper
- Calculator
- Computers

Procedure:

1. Divide the students into groups not exceeding three. Smaller groups enable members to contribute more.
2. Explore: Tell the groups they are going to become Imagineers. Distribute a piece of cardstock and coloring materials to each group and have them develop a team name. They should design a team name/logo on the cardstock paper. If time permits, have students design their own name tags.

3. Engage: Play the following video on the Haunted Mansion Line-Queue:
    - youtube.com/watch?v=PUQxzdcL7ro
4. Lead a discussion about which aspects of the Haunted Mansion line-queue appealed to guests. Discussion prompts could be:
    - "How was art incorporated into the line-queue?"
    - "Have you been to this ride?"
    - "During which parts were the guests most/least engaged?"
    - "How was the line engineered to prevent bottlenecks from occurring?"
    - "What technology was used in the queue?"
    - "Is safety an issue with this layout?"
5. Play the following clip twice: Mission: SPACE - Green (Milder) Intensity: Epcot (Orlando, FL): youtube.com/watch?v=ldOPeg4KGqw
    - The first time, students should just watch the video clip.
    - The second time, students should reflect on the video. The reflection should include the pros and cons of the ride's line-queue and other design elements.
    - Following the video clip reflection, have students share aloud their thoughts on the ride.
6. Group task. Redesign the exterior and line-queue to Mission: Space.
    - The design must meet the following criteria:
        - Hold at least 100 people.
        - Expose guests to scientific data.
        - Have an entrance and exit, and flow in a logical fashion.

*Ride criteria is subject to revision by the instructor; however, it is essential that a minimum requirement is placed.

7. Give students time to collaborate and sketch out their ideas on paper. A time limit is recommended to keep groups progressing.
8. Teachers should rotate between groups and provide feedback about their designs.
9. Have groups select one representative to give a quick summary of their design to the class. Each group must ask the presenter one question.
10. After all groups present, give students time to reflect, sketch, and label their Mission: Space entrance/exit and line-queue. (The drawings need be to scale.)
11. Teachers must provide a written explanation for how to present space information.
12. Have groups draw their displays on space information using labels and cite resources.

Table 1. Mock rubric

	1	3	5
Science	**No** factual scientific data was used in the design of the line queue.	**Some** scientific data was used in the design of the line queue.	**Only** scientific data was used in the design of the line queue.
Technology	**No** current space technology was presented in the line queue.	**Some** space technology was presented in the line queue.	**Only** space technology was presented in the line queue.
Engineering	The line queue does **not** have a clear entrance and exit.	The line queue had **some** elements missing.	The line queue shows the flow and has a distinct entrance and exit.
Art	Line queue **lacks** original design and the pitch was not made.	Line queue is has **some** original design and the pitch was scripted.	Line queue was **mostly** original design and the pitch was unscripted.
Math	The line queue was **not** drawn to scale.	**Some** of the line queue was drawn to scale.	The **entire** line queue was drawn to scale.
Citations	**No** material was cited.	**Some** material was cited.	**All** material was cited.
Total Score:			

13. Each group will then be tasked with making a pitch for their project design. It is recommended that a guest judge comes in to be 'Walt Disney,' and that individual selects their favorite.
14. Grade the projects using a rubric.

CURRENT CHALLENGES FACING TEACHERS

Teachers face various hurdles regardless of the lessons they're teaching. This lesson's main obstacle is time. Presenting students to engineering, small-group discussion, planning, research, and drawing requires valuable class time that an educator may not have. When working in groups, students can become side-tracked and drift into off-topic conversation. This creates time delays by hindering progress. One way to troubleshoot this problem is to give check-points the groups need to hit every five or ten minutes.

Another hurdle is grading the projects. Each teacher has different expectations for their students, so a uniform rubric does not apply to all classes. Rubistar.org is a free online resource that allows teachers to design a rubric for various projects.

SOLUTIONS AND RECOMMENDATIONS

Disney's Imagineers stress (WDI, 2016) that a budget and time-frame is always provided when assigned a project. It is advisable that a time-limitation is placed to keep the project moving; however, this could cause quality of design to be compromised. One way to limit classroom time is to only have students design the outside of Mission: Space and not the line-queue or vice-versa. Parts of the project and scientific research can be assigned as homework, and each student is responsible for bringing a different component to class. For example, if a group is planning on doing their line-queue display on Mars, each member could bring in a list of facts about Mars. If a student fails to bring facts to class, participation points will be deducted from their rubric for the project. However, it is vital that students are given class time to discuss and collaborate on their project's design and layout.

If students are unfamiliar with Disney Parks, use a local amusement park or a common carnival ride to facilitate the idea of ride design. Teachers could also change the ride selection from Mission: Space to a ride such as Pirates of the Caribbean as long as it aligns with the curriculum. If only one ride pertains to content presented in class, play the ride video from an online source. This will give students perspective into the ride's theme and layout.

Pitching a project design can be daunting to some students. Public speaking is often cited as the number one fear people have (Ingraham, 2014). To eliminate this fear, groups can select one student to make the pitch instead of pitching the idea as a whole group. It is important to let students know that if they choose to have a spokesperson, that student's pitch will be reflective of all group members, and the grade will be uniform. If no members of a group are willing to make a pitch in front of their classmates, they can record a video which will be played in class the following day or posted online. The recorded video may be the best way to eliminate class time if the pitch is only for the teacher to view.

REFERENCES

Andrews. (2014). *6 early amusement parks*. Retrieved from http://www.history.com/news/history-lists/6-early-amusement-parks

Brigante, R. (2010). *Want to be a theme park designer? Disney Imagineer and Universal creative experts offer advice*. Retrieved from http://www.insidethemagic.net/2010/04/want-to-be-a-theme-park-designer-disney-Imagineer-and-universal-creative-experts-offer-advice/

Daines, G. (2017). *NASA's journey to Mars*. Retrieved from https://www.nasa.gov/content/nasas-journey-to-mars

Disney Careers. (2016). *Walt Disney Imagineering*. Retrieved from http://wdi.disneycareers.com/en/default/

EPCOT. (2016). *Theme parks. EPCOT Overview*. Retrieved from http://www.wdwinfo.com/wdwinfo/guides/epcot/ep-overview.htm

Frommers. (2016). *History of Walt Disney World*. Retrieved from: http://www.frommers.com/destinations/walt-disney-world/683469

Ingraham, C. (2014). *America's top fears: Public speaking, heights and bugs*. Retrieved from https://www.washingtonpost.com/news/wonk/wp/2014/10/30/clowns-are-twice-as-scary-to-democrats-as-they-are-to-republicans/?noredirect=on&utm_term=.d7b20652a417

Mission: Space Overview. (2016). *EPCOT-Mission Space*. Retrieved from: http://www.wdwinfo.com/wdwinfo/guides/epcot/epfw-mission.htm

Mumpower, D. (2016). *It was the most controversial Disney ride ever built. Here's why*. Retrieved from https://www.themeparktourist.com/features/20160407/31867/sordid-history-misson-space-sickest-ride-epcot

KEY TERMS AND DEFINITIONS

Aesthetic: Beauty to the designer.

Attraction: Something that brings attention to itself.

Bottleneck: Narrowing of a region.

Imagineering: Term coined by Walt Disney that combines imagination and engineering.

Mars: Planet in the Milky Way Galaxy that is fourth closest to the Earth's sun.

Modeling: A scaled representation of an actual object.

Line-Queue: Space for people waiting to board.

Pitch: The act of presenting an idea.

Play-Testing: When individuals interactive with a prototype for the purpose of error spotting.

Reflect: Looking back at something from a different viewpoint.

Theme Park: Area designed to replicate a specific place and/or time.

Section 4
Differentiating Instruction

Chapter 16
Differentiating Instruction in the Forensics Classroom

Tracy L. Mulvaney
Monmouth University, USA

Kathryn Lubniewski
Monmouth University, USA

EXECUTIVE SUMMARY

This chapter will discuss differentiated instruction within the STEAM classroom. An example of a differentiated instruction case that was used in a Forensics Science class will be referenced. The case study focuses on fingerprint recovery, identification, and classification. After a series of lessons about the science of fingerprinting, a mock crime scene is set up to allow students the opportunity to become forensic scientists. Students use the forensic tools to recover them, and then identify and classify them using the process taught through direct and supplemental instruction. Some issues with differentiating instruction that arise are professional development around differentiated instruction, the time it takes to differentiate (amount of planning), lack of classroom time to complete projects, and lack of support or collaboration with key stakeholders are discussed.

LITERATURE REVIEW

Students differ from one another socially, academically, physically, cognitively, and emotionally. They also learn differently (Gregory & Chapman, 2007). Teachers can no longer just "teach the lesson" and hope that everyone understands. It important to first take into consideration who each learner is, their readiness levels, interests,

DOI: 10.4018/978-1-5225-9631-8.ch016

preferred learning styles, and strengths and needs (Tomlinson, 2005). This is the first step in differentiated instruction.

Differentiated instruction allows all students to access the same classroom curriculum by providing accommodations tailored to meet their individual needs (Tomlinson, 2005; Watts-Taffe, Laster, Broach, Marinak, Connor & Walker-Dalhouse, 2012). After teachers know, understand, and assess each learner, it is critical to use that information to design instruction. This can be done through curriculum compacting, grouping, and adjustments to students' learning.

Curriculum compacting is a strategy that can be used if a student has already mastered the content; the teacher then designs the instruction to "dive in deeper" or "compacts it" and moves on to the next skill (Gregory & Chapman, 2007; Logan, 2011). If, for example, a first-grade lesson's is to use observations of the sun, moon, and stars to describe patterns, the student should be able to identify and discuss patterns like how the sun and moon rise in one part of the sky and move across the sky. To compact the curriculum, the teacher would design "deeper instruction" where the student collects data on specific times that the sun rises or data on the location of the sunrise and sunset. The other option for curriculum compacting would be for the teacher to move the student on to the next skill within the curriculum. For example, if they mastered that topic, then the teacher would move the student onto making observations at different times of year to relate the amount of daylight to the time of year (since this is the next-generation science standard) or the next standard in the curriculum.

An important feature in differentiated instruction is the use of flexible grouping and different grouping strategies. The best instruction is when students have a balance of working alone, with a partner, and in small flexible groups. Flexible grouping means students are mixed with other students based on their ability and interest level. By incorporating flexible grouping, students "maximize their learning time based on their performance levels" (Gregory & Chapman, 2007). The critical components in flexible grouping are that students work with a variety of their peers, and the groups are not always the same. There are a variety of grouping strategies (see Figure 1: Differentiated Grouping Strategies), and based on content the teacher can determine if another grouping strategy would better meet the needs of the students.

Figure 1.

$$f(x) = xw(x) - \sqrt{x^2 w^2(x) + f_0^2}.$$

The final step in differentiated instruction is making adjustments to students' learning. Teachers can accomplish this by making adjustments to the content, process, and product. Differentiating the content means identifying "what we teach or what we want students to learn" (Tomlinson, 2005). There are two ways to differentiate the content: one is adapting "what" is taught. For example, I'm teaching a sixth-grade science class about the solar system. I can differentiate the content because I may have some students first reviewing and relearning the characteristics about the earth while others are working on all of the planets together, in turn, differentiating "what" they learned. The second way to differentiate content is to adapt "how we give students access to what we want them to learn" (Tomlinson, 2005). For example, in the same science lesson, some students may be reading about the solar system on a sixth grade–level text; some are watching a movie; and others are reading on a fifth-grade level, and the access is adapted to how they are acquiring the information.

The second step in making adjustments to students' learning is through differentiating the process. When the *process* is differentiated, the teacher is adapting *how* the students are learning the content. For example, I'm teaching a second-grade class about the life cycle of a butterfly. I typically teach it through worksheets where the students read about it, then cut out the different steps, and put them in order. I'm going to adapt the process so that instead of using the worksheet, the students are going to visually see the different models of each step in the process. They are going to act out in small groups each step in the process. Then they are going to play a game with a partner to match the different steps in the process. Finally, students are going to write a story about a butterfly that includes the steps of the life cycle. In this example, a "good" differentiated activity is something students do or make that has a variety of support (or scaffolding), various modes of learning, uses essential skill(s) and essential information, and uses the essential idea to answer an essential question (Tomlinson, 2005).

Differentiating the *product* is the final step in making adjustments for students. This can be done when the teacher makes adaptations to the culminating assignment. Differentiating the product should help students rethink, use, and extend what they have already learned. When students differentiate the product, it should be done completed independently, and students should take ownership of what they have learned. There are tips in creating a powerful product (see Figure 2: Tips to Creaking a Powerful Product).

An example of adapting the product is a unit completed with ninth graders where they are designing an iPad case. Typically, at the end of the unit the students make a model in groups and share it with the class. To adapt the product, I instead could give the students a choice to prepare a report of all of the steps, rationale, logistics, and final design of the iPad case; compose a rap song or rhyme that describes their iPad case with specific details; have them describe using a video the sequence or

Figure 2.

Step	Be sure to include
Identify the goals/ objectives	Facts, concepts, and skills. A list of specific standards in content area A list of specific standards in other STEAM content areas Key academic language terminology
Brainstorm a list of multiple formats for the product	Multiple formats based on differing learning styles or interests Requirements that need to be consistent between all of the products Ideas that build on students' strengths A list of exploratory and investigative assignments for critical thinking
Identify the specific product you will use & the particular part of the assessment that matches the goals/objectives.	A chart to identify the specific goal/objective and the exact part of the assessment when that goal/objective is being measured. Direct connections between the two and double check that in fact that is an accurate way to measure that goal/objective.
Determine your expectations	The content by reflecting on your expectations of the information, materials, ideas, and concepts. The process by reflecting on your expectations of their participation during the lesson. The product by reflecting on your expectations of the construction, size, and parts of the assessment.
Scaffold the assignment to provide success!	A criteria for success Breaking down a large assignment into smaller "steps" or assignments A timeline Support to help them in planning Individual conferencing if appropriate Peer-to-peer feedback to help support revising and editing
Clearly present the assignment	A reflection of the content to be sure that it is things that you have taught and they have practiced A review of your steps and process of the assignment A criteria for how the students will reach the specific level of quality
Adapt assignment based on individual students	Student readiness Student interest ELL/ESL learners Learners with disabilities (including gifted)
Facilitate for Achievement!	The best way to contact you Specific feedback and growth throughout the process

Adapted from Tomlinson, C. A. (2005). How to differentiate instruction in mixed-ability classrooms (2nd ed.). Upper Saddle River, NJ: Pearson.

process of how they got to the point and final design of the iPad case; or the original idea of having them create a model and share it. This way, the students are in charge of their learning because they choose how they will show what they learned based on their learning preferences, and the objective of the lesson has not been altered.

To complete this process, teachers should start small after creating learning profiles and/or gathering information about students. They should then try one component (i.e., content, process, product) with one lesson, and build upon knowledge. It is critical in differentiation to reflect. Differentiation should not end with just planning but be a continual process (Parsons, Dodman, & Burrowbridge, 2013). Some ways to do this are to consistently assess student progress, stay up to date on effective pedagogy and how students learn. Finally, reflect "on action" and "in action." When you reflect "on action," you are looking back and critiquing what happened after the instruction has been completed. Reflecting "in action" occurs in the midst of teaching, where you can make immediate changes or interventions based on the students. Upon reflection, the key is to make changes based on what was learned.

Through differentiated instruction teachers push students' individual limits to maximize their academic success. This empowers teachers to be proactive rather than reactive to unique abilities, personalities, and backgrounds in the classroom (Anderson, 2007). This can be completed by knowing the learner and adjusting the curriculum through alterations in the content, process, and product of the lesson.

Content Information

To gain the most out of this case study, teachers will first need to think of their students and complete the initial process of differentiation. Understanding what is and is not differentiated instruction is crucial; consider using the ASCD infographic found in the additional resources below (ASCD, 2015). The following section of this case study will have ideas on differentiating the content, process, and product. Integration of science, technology, engineering, art, and mathematics will be demonstrated throughout the lessons.

CASE DESCRIPTION

This case is a unit of instruction delivered in a high school Forensics Science class focused on fingerprint recovery, identification and classification. After a series of lessons about the science of fingerprinting, a mock crime scene is set up to allow students the opportunity to become forensic scientists. Students use the forensic tools to recover them, and then identify and classify them using the process taught through direct and supplemental instruction. The Next Generation Science Standard included in this case is **HS-LS1 from Molecules to Organisms: Structures and Processes.**

Fingerprint Recovery

Students receive direct instruction on the Anthropometry and the Bertillon methods which investigators employ to keep track of criminals based on the measurement of various parts. Activities are also provided to teach students how to lift latent fingerprints using physical and chemical methods. Direct instruction, along with textbook readings, are the sources of instruction for the lesson. This leaves the teacher with many creative ways in which to differentiate instruction. The *content* is differentiated by providing videos, readings and web-based activities. There is an abundance of information available to teachers to differentiate in order to meet the needs of students of all levels. The key is to check the readability of the content to ensure that students above or below reading level are reading based on their ability. Depending on their grade level and ability, another way to differentiate the *content* is to provide search terms, well-designed instructional games or Web links. This way the students are accessing the most appropriate and relevant materials and ensuring that it is not too overwhelming based on their individual needs.

Some strategies to differentiate the *process* of this lesson are to (1) provide a checklist that outlines how to access relevant information, (2) use videos, (3) review roles and expectations in group work, and (4) use technology. The lesson requires students to research and identify relevant information about fingerprint recovery. To differentiate this *process* the teacher could also provide a checklist of how to recover fingerprints. Depending on ability level, this can be a poster on the wall, a piece of paper on students' desks, or on an iPad or personal device. The checklist can even be projected on a smartboard. It can also include pictures to represent the steps along with the instructional narrative. This can be reviewed with specific students if necessary. It can be given as an option for students to use if they deem it necessary. As a part of this lesson, students are required to read about fingerprint recovery from the textbook. To differentiate the *process* in this section, the teacher could also provide a video as an option if students wanted to view the content in that format. The third requirement, but not in the exact order, is the use of cooperative grouping. This is a critical component where differentiating the *process* could really support students. One way to do this is to review the roles of students and the expectations during group work. Another idea is to have a list of roles and contracts with which students can identify roles and responsibilities, and then sign off on the contracts. This holds all students accountable. The last option in differentiating the *process* would be to use technology. This lesson requires the use of a textbook; some students may prefer to use technology to collaborate (e.g., GoogleDocs, GoogleHangout, GoToMeeting), complete their forms (e.g., Weebly, GoogleForms), or simply just to type instead of writing by hand. They can watch video examples of lifting latent prints or review websites from forensic scientists or the FBI, showing best practices in the process.

The final component in differentiation is the *product*. For this, the teacher should challenge their learning by requiring students to choose different ways to reflect on what they have learned. Each student in this case study is required to "dust for fingerprints" and recover at least two latent prints from the mock crime scene. Students are then asked to write a summary of their process. Some students may want to create a rap or song to share their findings. Others may want to complete a chart about their results. Another option could be to write a few paragraphs on what they learned. Students could also create a short video segment in the style of any popular television investigation show. These are just a few ways to show what they learned, expanding on the original lesson of a discussion about fingerprints and how to recover latent prints.

Classification of Fingerprints

In this lesson, students are taught how to classify fingerprints (loops, arches, whorls). They practice identifying using their own fingerprints and that of their classmates. The students then are given a written test identifying multiple fingerprint samples. One way to differentiate the *content* of this lesson is to provide a list of Web sites and videos for students so that the reading materials are more appropriate for their level. Large posters can be created by the teacher and strategically placed around the room. Another way to differentiate the *content* is to show students of all levels how the classification of fingerprints is relevant to them. For example, many students have phones that use fingerprint identification instead of a passcode to unlock their phone. Some high school students may volunteer and have to have their fingerprints for certain volunteer positions. It would be helpful to connect the purpose to more than identifying criminals.

There are four ways that differentiating the *process* would be beneficial: (1) sequential task mastery, (2) assigned groups and roles, (3) electronic version of the materials, and (4) assigning specific questions for students to complete. The first way to differentiate is to task analyze the *process* and ensure students complete the tasks in deliberate, sequential order. For example, the steps outlined could come with a sequential chart or checklist. Students must follow each step deliberately prior to completing the next. They can either peer or self-evaluate mastery using a chart or checklist. They then go on to the next step. The second way to differentiate the *process* is by intentional cooperative grouping. During this unit the teacher assigns roles to the group members. One way to support this would be through the use of *Task Bookmarks (see Figure 3: Task Bookmark Example)*. Task Bookmarks can help team members stay focused on their duties that facilitate team success. Each section or "role" is given to a specific student to complete. It would be important to consider strengths and needs of the students, as well as mixing up the groups. Additionally, the teacher may want to think about assigning the students group in order to challenge students while holding them accountable. The third way to differentiate the *process* would be providing websites and videos online that may help students understand the steps and best practices for the procedures. Additionally, forensic apps that include fingerprinting can be downloaded on class iPads or Chromebooks (e.g., MobileDetect, CrimePad, bMOREsafe). The final way to differentiate the *process* would be to assign specific questions and/or have students create an additional question on their own. This would help students think about the material differently way and match them to their strengths. (In some cases, this may purposely match them based on their needs so they can continue improving.)

In this lesson students are required to write a one-page report connecting theory to the practice of lifting (or recovering) latent fingerprints as their final product.

To differentiate the *product,* the teacher could give students the option of creating an informational brochure, a *CSI*-like video (after the popular show), or a website (e.g., Wix, Weebly) that includes all of the information in a report but allows them to choose the format based on their strengths.

Figure 3.

Title	Role
Production Manager	Oversee the project Assign final roles after discussion Ensure that everyone is doing their job Keep track of the project Keep the team on track Give reminders to different students based on their role Manage deadlines Communicate with the teacher
Technology Manager	Provide resources from the Internet Assist in computer and technology aspects Design the presentation (if using a technology format)
Information Manager	Ensure that materials are accurate Listen and clarify ideas Follow written directions Communicate updates to the group based on findings
Resource Manager	Gather materials Store all materials for projects Make sure that materials are available Communicate needs of materials with production manager

Adapted from Gregory, G. H., & Chapman, C. (2007). Differentiated Instructional Strategies: One Size Doesn't Fit All (2nd ed.). Thousand Oaks, CA: Corwin Press.

Identification of the Criminal

The final assessment challenges students to locate, recover, classify, and identify fingerprints. Students use knowledge from the previous two lessons on recovering and classifying fingerprints to complete the final process of identifying prints given multiple sets of prints of possible "suspects." The crime scene is set up with five sets of fingerprints belonging to fictional characters. There is one set of fingerprints found on the weapon used to commit the crime that is shared with the students. Students must match the fingerprints found at the crime scene with those on the weapon. The whole process is interactive, inquiry-based, and completed in teams of investigators (i.e., cooperative groups).

Figure 4.

Name	Time	Task	Observed Behavior

Since this unit relies heavily on group work, a way to assess students during that time would be to collect anecdotal observations. One way to do this is by doing a "clipboard cruise" (*see Figure 4: Clipboard Cruise Example*). Teachers record their findings on a sheet attached to their clipboard. The information that is typically included is the name of the student, the date, time, center or task, and the behavior that is observed. It is essential to evaluate and provide feedback to students while they are working in groups. This is one way to have more concrete data to share with students and families.

To differentiate the *process* of the students' final project, teachers can create graphic organizers to help students who may lack organization or other executive functioning skills keep their data organized. The graphic organizer will help as a visual aid for students to write out common characteristics of the fingerprint sets. Another way to differentiate the *process* is to create a rubric that measures students' skills throughout the project. Therefore, the students are focused on the progression of skills, and will not fail if they make an error on the final step. This will help them see where in the process they were not successful, and the group can discuss how they can remedy their errors in subsequent activities.

In the final assessment of the unit, students are asked to write a full report with their groups to state who they believe the guilty suspect is. They must include the process they used and discuss evidence to support their findings. To differentiate the *product*, the students could create a short crime-show video or news report about how they solved the crime. They could also use web-based software (e.g., Prezi, Powerpoint, Weebly) to create a presentation report. Students are able to build on their strengths and gain a deeper understanding of the unit due to the various available options to demonstrate their knowledge.

These are just some ideas to differentiate the content, process, and product of different lessons within a forensic unit. Continual reflection on research-based practices and pedagogical practices that affect student learning are critical to the success of differentiating instruction.

CURRENT CHALLENGES FACING TEACHERS

Teachers face two main challenges in the differentiated STEAM classroom: time and lack of support. It is important to explore these challenges as well as potential solutions to make the STEAM classroom as successful as possible.

Overall, differentiating instruction in the STEAM classroom takes time. Educators need to take time to learn about methods of differentiating within the classroom. They should also attend professional development to learn more about differentiation. Another way teachers are limited by time is the additional planning required to successfully differentiate in the classroom. Getting to know students and how they learn the best, then planning to reflect that is time consuming as well. Finally, teachers may struggle to find adequate class time to complete STEAM projects.

The other challenge teachers face is the possible lack of support, particularly from administration and special education teachers. Administration may not understand the components of a successful differentiated STEAM classroom. They also may not know how to support teachers in this setting. Special education teachers also might not offer adequate support for STEAM teachers because of other students on their caseload, common planning time, or instruction time.

It is crucial that teachers keep in mind time and support as they begin to implement STEAM in the classroom. Being able to know and recognize these potential challenges are critical so that teachers can plan ahead for what may occur.

SOLUTIONS AND RECOMMENDATIONS

To address the first challenge of the time necessary to learn the ways to differentiate in the STEAM classroom, use websites that offer professional development online, such as the Association for Supervision and Curriculum Development (ASCD) or the Council for Exceptional Children (CEC). These professional development sessions are frequently recorded so they are accessible at any time. It's also helpful to search to see what professional development conferences a state offers so that distant travel is not necessary. Often, principals are willing to offer financial support to attend conferences that require travel as well.

One way for educators to improve time management is to require themselves to plan their time and find uninterrupted time. This is very hard to do as a teacher—especially during the day—but it does help to stay regimented. When I was a teacher, I had to close my door during planning to try to minimize the distractions. Having an accountability partner is helpful as well. This means that educators tell co-workers or other professionals about their work and plans. This will help keep the work a priority, and it may benefit the partner's teaching as well!

There never seems to be enough time in the classroom to cover everything that that educators want to cover. It's important to think about how to lay out the assignments to make class time as smooth and time efficient as possible. Creating and delivering clear directions is crucial to ensuring class time is being used efficiently. Often, unclear or lengthy directions can overwhelm students, so considering how directions will be provided orally and on paper is paramount. As stated earlier, it is also important to assign students to groups and seating areas; listing them on the board helps create smoother transitions to which students can refer. (This is more efficient than calling students' names in order and sending them to different areas.) Having a plan for students to get help when the teacher is busy with a group or another student can be very helpful in creating an effective classroom. Some classrooms utilize an "expert for the day" where one or more students who are skilled in an area respond to questions before the teacher. Another useful idea is for educators to have a question board, where students go up and write their questions on chart paper and the teacher answers them as time allows. It's also important to think about the students or groups that will get finished first. Teachers should be sure to have a plan for them so as to promote on-task behavior and discourage rushed assignments. The final strategy in managing the classroom time is to break larger assignments into "chunks." This means each day the teacher helps students manage their time because only have access to a portion of the assignment. This helps students learn time management, and eventually the educator can scaffold instruction so that they build up where they have longer tasks and greater amounts of time.

Collaborating with the administration is another critical factor. Educators should invite principals into their classrooms to observe (even when it's not a scheduled observation). Stopping into their offices to share student work and data that they have collected that show the impacts of the STEAM classroom. Ensuring that the administration understands the content and depth of lessons is very beneficial. This way, teachers can approach administration to ask for common planning time with other content area teachers, special education teachers, or English as a second language (ESL) teachers. Educators unable to build this type of relationship prior to meeting should be sure to go with existing data that support the reason behind any requests they are making. Administrators want to know that what you are doing is supported by data and that it can be successful.

The ability to work with special education or support teachers for students in the STEAM classroom can be challenging as well. It's important to ask support teachers about their caseload and when are the best times to meet. Meeting in person may not be necessary, but identifying preferred methods of communication is important so that collaboration happens to enhance lessons. GoogleDocs is a great way to share lessons and materials with special education teachers who may not have the time to physically meet at convenient times. Sharing a curriculum map of upcoming topics for the special education teacher to provide potential materials of lesson differentiation may be helpful as well. It's not mandatory that the special education teacher be involved in differentiation, but their expertise is extremely invaluable for students who either are above level or are struggling.

Differentiating in the STEAM classroom is possible! Starting small to build confidence will help educators see the positive impacts on the achievement of their students. Teachers should remember to first get to know them and then make adjustments to their curriculums. This way, students are able to access the content, have multiple ways to process the information, and have a variety of assessments to show what they learned. These key ingredients are the recipe for success!

REFERENCES

Anderson, K. (2007). Differentiating instruction to include ALL students. *Preventing School Failure*, *51*(3), 49–54. doi:10.3200/PSFL.51.3.49-54

Autodesk. (2016). *Tinkercad.* Retrieved from https://www.tinkercad.com/about/features

Dixon, F. A., Yssel, N., McConnell, J. M., & Hardin, T. (2014). Differentiated instruction, professional development, and teacher efficacy. *Journal for the Education of the Gifted*, *37*(2), 111–127. doi:10.1177/0162353214529042

Gregory, G. H., & Chapman, C. (2007). *Differentiated Instructional Strategies: One Size Doesn't Fit All* (2nd ed.). Thousand Oaks, CA: Corwin Press.

Logan, B. (2011). Examining differentiated instruction: Teachers respond. *Research in Higher Education*, 13.

National Center for Educational Statistics. (n.d.). *Create a graph.* Retrieved from https://nces.ed.gov/nceskids/createagraph/

Office Applications. (n.d.). *Writer.* Retrieved from https://writer.bighugelabs.com/login

Parsons, S. A., Dodman, S. L., & Burrowbridge, S. C. (2013). Broadening the view of differentiated instruction. *Phi Delta Kappan*, *95*(1), 38–42. doi:10.1177/003172171309500107

Rock, M. L., Gregg, M. G., Ellis, E., & Gable, R. A. (2008, Winter). REACH: A framework for differentiating classroom instruction. *Preventing School Failure*, *52*(2), 31–47. doi:10.3200/PSFL.52.2.31-47

Tomlinson, C. A. (2005). *How to differentiate instruction in mixed-ability classrooms* (2nd ed.). Upper Saddle River, NJ: Pearson.

Watts-Taffe, S., Laster, B. P., Broach, L., Marinak, B., Connor, C. M., & Walker-Dalhouse, D. (2012). Differentiated instruction: Making informed teacher decisions. *The Reading Teacher, 66*(4), 303–314. doi:10.1002/TRTR.01126

Zygomatic. (n.d.). *Chart Tool*. Retrieved from http://www.onlinecharttool.com/

ADDITIONAL READING

ASCD. (2008). *Nonnegotiables of differentiated instruction: Respectful tasks*. Retrieved from http://www.ascd.org/ascd-express/vol4/425-video.aspx

ASCD. (2015). *What is (and is not) differentiated instruction?* Retrieved from http://www.ascd.org/ASCD/pdf/siteASCD/publications/Differentiation_Is-IsNot_infographic.pdf

BER Staff Development. (2011, March 17). *Differentiating instruction, 6-12, Part 1: Student choice and multiple modes of learning*. Retrieved from https://www.youtube.com/watch?v=akvDT9KFZPw

BER Staff Development. (2011, March 17). *Differentiating instruction, 6-12, Part 11: Tiered assignments and grouping*. Retrieved from https://www.youtube.com/watch?v=1ob4eGz04G4

Castaneda, R. (2012, July 10). *Getting started on differentiated instruction*. Retrieved from https://www.youtube.com/watch?v=LGYa6ZacUTM&NR=1&feature=endscreen

Cox, J. (n.d.). *Differentiated instruction strategies: Learning stations*. Retrieved from http://www.teachhub.com/differentiated-instruction-strategies-using-learning-stations

Edutopia. (2012, July 11). *Differentiating instruction through interactive games*. Retrieved from https://www.youtube.com/watch?v=XZ0BGXMf83U

Edutopia (2015, September 9). *Differentiated instruction: Resource roundup*. Retrieved from http://www.edutopia.org/article/differentiated-instruction-resources

Forest Lake Elementary School. (2010, May 3). *Tech-fueled differentiated instruction engages elementary school students*. Retrieved from http://www.edutopia.org/stw-differentiated-instruction-technology-elementary-video

Grant, J. (2014). *Staff development for educators: Differentiated instruction.* Retrieved from http://www.sde.com/teacher-resources.asp

Hocket, J. (2015, February 19). *How I learned differentiation.* Retrieved from http://www.teachthought.com/learning/personalized-learning/how-i-learned-differentiation/?utm_content=556b54cf04d3014c38000001

Huck, C., & Amaral, B. (2014, October). *Scaffolding for success: Best practices for secondary ELLs.* Retrieved from http://newsmanager.commpartners.com/tesolc/issues/2014-10-01/3.html

Kluth, P. (2015). *20 ways to adapt the science lab.* Retrieved from http://www.paulakluth.com/readings/differentiating-instruction/20-ways-to-adapt-the-science-lab/

Kluth, P. (2016). *Rewriting history, and nine other ways to adapt textbooks.* Retrieved from http://www.paulakluth.com/readings/differentiating-instruction/rewriting-history-and-nine-other-ways-to-adapt-textbooks/

Roberts, E. J. (2014, October 22). *Differentiating instruction: In the classroom and at home.* Retrieved from http://home.edweb.net/differentiated-instruction-classroom-home/

Sousa, D. A. (n.d.). *How can differentiation be achieved-without putting too much burden on teachers?* Retrieved from http://www.centerforlearning.org/t-free-resources.aspx

Teaching Channel. (2011). *Content differentiation in 3rd grade s science.* Retrieved from https://www.teachingchannel.org/videos/content-differentiation-science

Teaching Channel. (2011). *New teacher survival guide: Differentiating instruction.* Retrieved from https://www.teachingchannel.org/videos/differentiating-instruction?fd=1

Teaching Channel. (2012). *Differentiating with learning menus.* Retrieved from https://www.teachingchannel.org/videos/differentiating-instruction-strategy

The IRIS Center. (2015). *Differentiated instruction: Maximizing the learning of all students.* Retrieved from http://iris.peabody.vanderbilt.edu/module/di/

Tomlinson, C. A. (2014, May 15). *Revisiting the differentiated classroom: Looking back and ahead.* Retrieved from http://www.ascd.org/professional-development/webinars/revisiting-the-differentiated-classroom-webinar.aspx

Tomlinson, C. A., Brimijoin, K., & Narvaez, L. (2008). *Setting the stage for change toward differentiation*. Retrieved from http://www.ascd.org/publications/books/105005/chapters/Setting-the-Stage-for-Change-Toward-Differentiation.aspx

Tomlinson, C. A., & Britt, S. (2012, June 6). *Common Core State Standards: Where does differentiation fit?* Retrieved from http://www.ascd.org/professional-development/webinars/tomlinson-and-britt-webinar.aspx

University of Texas. (2013). Differentiating instruction: Key to student success. Retrieved from http://buildingrti.utexas.org/online-modules/differentiated-instruction-key-student-success

KEY TERMS AND DEFINITIONS

Accommodation: A change to the environment or curriculum without changing the objective.

Adaptation: A change in the environment or curriculum that allows the learner more appropriate access to learning based on his or her abilities and needs. An adaptation can be an accommodation or modification.

Anthropometry Method: Designed by Alphonse Bertillon, began in 1890 and lasted approximately 20 years before being replaced by fingerprint identification.

Arches: Ridges in a fingerprint that creates a wave-like pattern and runs from one side to the other, making no backward turn.

Bertillon Method: An anthropometric system of physical measurements of body parts, especially components of the head and face, to produce a detailed description of an individual.

Collaboration: When at least two individuals work together towards a common goal.

Curriculum Compacting: A strategy that can sometimes be used if a student has already mastered the content, the teacher then designs the instruction to "dive in deeper" or "compacts it" and moves on to the next skill.

Differentiated Instruction: A proactive approach to meeting individual's abilities and needs.

Loops: Part of a fingerprint that recurves back on themselves to form a loop shape.

Modification: A change to the curriculum that alters the objective so that the student can be in an inclusive setting with more appropriate expectations.

Scaffolding Instruction: A process where teachers model or demonstrate how to solve a problem, then step back, and offer support as needed.

Whorls: A friction ridge pattern of the fingerprint when some of the ridges make a turn through at least one circuit.

Conclusion

Throughout this book, the authors have conceptualized STEAM teaching as a form of communal and improvisational *making*—one that unites the head, heart, and hands in teaching and learning performances, whose outcomes cannot be prescribed. In each chapter, the case presented has the possibility of resulting in a diverse array of learning artifacts. Students personally construct knowledge and design solutions to problems as they engage in three-dimensional science learning that features aesthetic practices. Our authors support the "Yes, and" message of improvisational theatre and fully recognize that some cases may evoke responses that range from full acceptance to full rejection. Our hope is that at least a few of these cases of STEAM education in practice inspire you try STEAM in your own classrooms!

Throughout this text, our authors have regarded the arts as an equal partner to the STEM disciplines—an essential and necessary component to creative teaching and learning. In all cases, STEAM teaching and learning are regarded as co-created, produced, and/or performed as a way of artfully crafting something that was not there before. In all cases, STEAM serves as a pathway toward inclusive education because it is a non-prescriptive, multimodal approach to discipline integration and creative problem solving. In many cases, STEAM education is portrayed as an active metaphor: as musicianship and an ethical act of care (Van Sickle & Koester); as a way of bringing a problem to light (Dollahan; Belling); as a design process (Bazler; Dibiase, MacDonald, & Strong); as a way of developing mental models in the mind's eye through practice and visualization (Smith; Seiverd); as a system of signs and symbols employing a multimodal form of literacy semiosis (Koester; Romagnoli); and as embodied simulation through movement and creative drama (Lindquist, James-Hassan, & Linquist; Rizzuto, Henning, & Ducket; Jablon). Groff brings many of these metaphors together in her description of teaching STEM through the creation of graphic novels. Mulvany and Lubniewski offer specific examples of ways STEAM is powerful pedagogy for special needs students. Zambak and Mulyono develop a STEAM PBL case in which both mathematical content and reasoning are applied to build a personalized understanding of the law of cosines through a *Star Map* task. Creative options for exploration include the use of dynamic geometry software.

Valdiviezo and Graybill call for STEAM teaching that helps students practice twenty-first century skills, urging the creation of partnerships and purposeful co-teaching with special educators, general educators, and a science educator. Similarly, Koester advocates for a kind of STEAM that features research and practice partnerships characterized by a constellation of C's: *Caring Collaboration* among *Content* specialists, *Cultural relevance*, *Cross-training, Curating*, *Creative process*, *Community*, and the achievement of a *Collective* impact. By employing these "C'ing" practices, participants can collectively build and share lessons with which they have found success, as well as reflect upon obstacles they may have encountered in the process. This book is one such sharing.

We have set forth many ways of experimenting with STEAM as an inclusive teaching practice. We propose the letting go of prescribed outcomes. Instead, we advocate for the design of lessons that can lead to a construction of deep, personal meaning by students, afforded the freedom associated with making their own ideas, performances, and/or artifacts. Our theoretical framework is inherently progressive and *relational*, seeking not to privilege any one discipline over another as to its "worthitness". Instead, the proposed STEAM approach situates synergy and synthesis of diverse ideas as best practice and argues against the contrived, often institutionalized separation of the disciplines. Employing artistic habits of mind and emergent mindsets in the process of science curriculum-making and enacting can add energy into the teaching/learning system, making possible all kinds of exciting and meaningful work.

Merrie Koester
University of South Carolina, USA

Compilation of References

Abadzis, N. (2007). *Laika*. New York: First Second.

ACT. (2015). *The condition of STEM*. National ACT, Inc. Retrieved from http://www.act.org/content/dam/act/unsecured/documents/National-STEM-Report-2015.pdf

Adams, A. E., & Pegg, J. (2012). Teachers enactment of content literacy strategies in secondary science and mathematics classes. *Journal of Adolescent & Adult Literacy*, *2*(56), 151–161. doi:10.1002/JAAL.00116

Adjibolosoo, S. (1995). *The human factor in developing Africa*. Westport, CT: Praeger Publishers.

Ainsworth, S. E. (2006). DeFT: A conceptual framework for learning with multiple representations. *Learning and Instruction*, *16*(3), 183–198. doi:10.1016/j.learninstruc.2006.03.001

Alenljung, B., Persson, A., & Wohlin, A. A. (2003). The fundamental nature of requirements engineering activities as a decision-making process. *Information and Software Technology*, *45*(14), 945–954. doi:10.1016/S0950-5849(03)00096-X

Anderson, C. (2004). Learning in "as-if" worlds: Cognition in drama in education. *Theory into Practice*, *43*(4), 281–286. doi:10.120715430421tip4304_6

Anderson, K. (2007). Differentiating instruction to include ALL students. *Preventing School Failure*, *51*(3), 49–54. doi:10.3200/PSFL.51.3.49-54

Andrews. (2014). *6 early amusement parks*. Retrieved from http://www.history.com/news/history-lists/6-early-amusement-parks

Archambault, I., Janosza, M., Fallu, J., & Pagania, L. S. (2008). Student engagement and its relationship with early high school dropout. *Journal of Adolescence*, *32*(3), 651–670. doi:10.1016/j.adolescence.2008.06.007 PMID:18708246

Arts Council. (2003). *Arts council guidance on drama education: Drama in schools*. London: The Arts Council of Great Britain.

Ashley, S. (2001). When the twin towers fell. *Scientific American*. Retrieved from https://www.scientificamerican.com/article/when-the-twin-towers-fell/

Ashton-Hay, S. (2005). Drama: Engaging all learning styles. In *Proceedings 9th International INGED (Turkish English Education Association) Conference*. Economics and Technical University.

Aubusson, P. J., & Fogwill, S. (2006). Role-play as analogical modelling in science. In P. J. Aubusson, A. G. Harrison, & S. M. Ritchie (Eds.), *Metaphor and analogy in science education* (pp. 93–104). Springer. doi:10.1007/1-4020-3830-5_8

Aubusson, P., Fogwill, S., Barr, R., & Perkovic, L. (1997). What happens when students do simulation-role-play in science? *Research in Science Education*, *27*(4), 565–579. doi:10.1007/BF02461481

Autodesk. (2016). *Tinkercad.* Retrieved from https://www.tinkercad.com/about/features

Bachelis, G. F., Maxim, B. R., James, D. A., & Stout, Q. F. (1994, April). *Bringing algorithms to life: Cooperative computing activities using students as processors*. Academic Press.

Bailey, S. D. (1993). *Wings to fly: Bringing theatre arts to students with special needs*. Rockville, MD: Woodbine House.

Baird, S. (2015). *How Disposable Coffee Cup Design Became High Art*. Retrieved September 30, 2016, from http://www.eater.com/drinks/2015/8/14/9153849/how-disposable-coffee-cup-design-became-high-art

Barton, K., & Smith, L. (2000). Themes or motifs? Aiming for coherence through interdisciplinary outlines. *The Reading Teacher*, *54*(1), 54–63.

Bedetti, G. (2015). Collaborative college playwriting and performance: A core course "Trespassing" onto the dramatic arts. *International Journal of Education & the Arts*, *16*(19).

Beers, S. Z. (2011). 21st century skills: Preparing students for THEIR future. *STEM*. Retrieved from http://cosee.umaine.edu/files/coseeos/21st_century_skills.pdf

Bennett, J., & Hogarth, S. (2009). *Would YOU want to talk to a scientist at a party? Students' attitudes to school science and science*. University of York, Department of Educational Studies Research Paper 2005/08.

Bequette, J. W., & Bequette, M. B. (2012). A place for art and design education in the STEM conversation. *Art Education*, *65*(2), 40–47. doi:10.1080/00043125.2012.11519167

Berk, R., & Trieber, R. (2009). Whose classroom is it anyway? Improvisation as a teaching tool. *Journal on Excellence in College Teaching*, *20*(3), 29–60.

Bianchini, J., & Colburn, A. (2000). Teaching the nature of science through inquiry to prospective elementary teachers: A tale of two researchers. *Journal of Research in Science Teaching*, *37*(2), 177–209. doi:10.1002/(SICI)1098-2736(200002)37:2<177::AID-TEA6>3.0.CO;2-Y

Bigelow Middle School. (n.d.). *Engineering technology: Basketbots*. Retrieved June 30, 2016, from http://www.bu.edu/gk12/alex/bioengineering_teacher.pdf

Bigelow, B. (2009). Once upon a genocide: Columbus in children's literature. In W. Au (Ed.), *Rethinking multicultural education* (pp. 73–85). Milwaukee, WI: Rethinking Schools.

Blinder, S. M. (2007). *The Coffee Cooling Problem.* Retrieved June 07, 2016, from http://demonstrations.wolfram.com/TheCoffeeCoolingProblem/

Bloom, B. S., Engelhart, M. D., Furst, E. J., Hill, W. H., & Krathwohl, D. R. (1956). *Taxonomy of educational objectives: The classification of educational goals.* New York: David McKay Company.

Boal, A. (2002). *Games for actors and non-actors.* New York: Routledge.

Bordieu, P. (2003). The forms of capital. In A. H. Halsey, H. Lauder, P. Brown, & A. S. Wells (Eds.), *Education: Culture, economy, society* (pp. 46–58). Oxford, UK: University Press.

Bracha, B. A. (2007). *The integration of creative drama into science teaching* (Doctoral dissertation). Kansas State University. Retrieved from https://krex.k-state.edu/dspace/bitstream/handle/2097/531/BrachaArieli2007.pdf?sequence=1

BrainStuff – HowStuffWorks. (2013, October 4). *How Do Astronomers Measure Distance?* [Video File]. Retrieved from https://www.youtube.com/watch?v=_Lsj-Hz-NS4&t=161s

Bransford, J. (2000). *How people learn: Brain, mind, experience, and school.* Washington, DC: National Academies Press.

Braund, M. (2015). Drama and learning science: An empty space? *British Educational Research Journal, 41*(1), 102–121. doi:10.1002/berj.3130

Braund, M., Lubben, F., Scholtz, Z., Sadeck, M., & Hodges, M. (2007). Comparing the effect of scientific and socio-scientific argumentation tasks: Lessons from South African. *The School Science Review, 88*(324), 67–76.

Bredekamp, K., & Copple, S. (2009). *Effective practices in early childhood education: Building a foundation* (3rd ed.). Upper Saddle River, NJ: Pearson.

Bresler, L. (2005). What musicianship can teach educational research. *Music Education Research, 7*(2), 169–183. doi:10.1080/14613800500169399

Bresler, L. (2006). Toward connectness: Aesthetically based research. *Studies in Art Education, 48*(1), 52–69. doi:10.1080/00393541.2006.11650499

Brigante, R. (2010). *Want to be a theme park designer? Disney Imagineer and Universal creative experts offer advice.* Retrieved from http://www.insidethemagic.net/2010/04/want-to-be-a-theme-park-designer-disney-Imagineer-and-universal-creative-experts-offer-advice/

Brown, J. S., Collins, A., & Duguid, P. (1989). Situated cognition and the culture of learning. *Educational Researcher, 18*(1), 32–42. doi:10.3102/0013189X018001032

Brozo, W. G., Moorman, G., & Meyer, C. K. (2014). *Wham! Teaching with graphic novels across the curriculum*. New York: Teachers College Press.

Bryant, J. (2008). *Newton's Law of Cooling Demonstration*. Retrieved June 07, 2016, from http://demonstrations.wolfram.com/NewtonsLawOfCooling/

BSCS - Biological Science Curriculum Study. (2005). *BSCS Science and technology: Investigating physical systems*. Dubuque, IA: Kendall Hunt.

Buchanan, M. (2016). Why teach drama? *Child Drama*. Accessed June 3, 2016 from http://www.childdrama.com/the-essay.html

Burj Khalifa. (2011). *Civil Engg. Seminar*. Retrieved from http://civilenggseminar.blogspot.com/2011/09/burj-khalifa.html

Burke, R. J., & Mattis, M. C. (2007). *Women and minorities in science, technology, engineering and mathematics: Upping the numbers*. Northhampton, NJ: Edward Elgar Publishing, Inc. doi:10.4337/9781847206879

Butler, J. (1989). Science learning and drama processes. *Science Education*, *73*(5), 569–579. doi:10.1002ce.3730730505

Bybee, R. W. (1989). *Science and technology education for the elementary years: Frameworks for curriculum and instruction*. Washington, DC: The National Center for Improving Instruction.

Bybee, R. W. (2010). Advancing STEM education: A 2020 vision. *Technology and Engineering Teacher*, *70*(1), 30.

Caldwell, K. (2011). Drama as a teaching tool. *Teaching Through the Arts*. Accessed June 3, 2016 from http://teachingthroughthearts.blogspot.com/2011/07/drama-as-teaching-tool.html

Carnine, L., & Carnine, D. (2004). The interaction of reading skills and science content knowledge when teaching struggling secondary students. *Reading & Writing Quarterly*, *20*(2), 203–218. doi:10.1080/10573560490264134

Carter, J. B. (Ed.). (2007). *Building literacy connections with graphic novels: Page by page, panel by panel*. Urbana, IL: National Council of Teachers of English.

Case study: The Leaning tower of Pisa. (n.d.). *Madrid Engineering*. Retrieved from http://madridengineering.com/case-study-the-leaning-tower-of-pisa/

Catchen, R. (2013). Reflections: How STEM becomes STEAM. *The STEAM Journal*, *1*(1), 22. doi:10.5642team.201301.22

Cawthon, S., Dawson, K., & Ihorn, S. (2011). Activating student engagement through drama-based instruction. *Journal for Learning through the Arts*, *7*(1). Retrieved from http://www.escholarship.org/uc/item/6qc4b7pt

Cazden, C., Cope, B., Fairclough, N., & Gee, J. (1996). A pedagogy of multiliteracies: Designing social futures. *Harvard Educational Review, 66*(1), 60–92. doi:10.17763/haer.66.1.17370n67v22j160u

Ceylan, E., Kerim Polat, R., Akpınar, M., & Kalender, S. (2015). Designing Science Instruction based on Creative Drama: The Effect on 6th Grade Students' Understanding and Elimination of Misconceptions. *International Online Journal of Educational Sciences*, *7*(4), 1–10.

Chaille, C., & Britain, L. (2002). The young child as scientist (3rd ed.). Boston: Allyn & Bacon.

Chautauqua Lake Central School District. (n.d.). *History of project lead the way*. Retrieved from http://www.clake.org/District/Class/28-Engineering-and-Technol- ogy-Department/1447-HistoryOfPLTW.html

Chilton, E., Carpenito, K., & Milto, E. (2013). Hands-on activity: Hydraulic arm challenge. *TeachEngineering*. Retrieved June 28, 2016, from https://www.teachen-gineering.org/activities/view/wpi_hydraulic_arm_challenge

Christa McAuliffe Quotes. (n.d.). Retrieved from https://www.brainyquote.com/quotes/christa_mcauliffe_134582

Close, H., & Scherr, R. (2015). Enacting Conceptual Metaphor through Blending: Learning activities embodying the substance metaphor for energy. *International Journal of Science Education, 37*(5-6), 5–6, 839–866. doi:10.1080/09500693.2015.1025307

Cobern, W. W. (1993). Contextual constructivism: The impact of culture on the learning and teaching of science. In K. Tobin (Ed.), *The Practice of constructivism in science education*. Washington, DC: American Association for the Advancement of Science.

Coffey, H. (n.d.). *Interdisciplinary teaching*. Retrieved October 3, 2016 from http://www.learnnc.org/lp/pages/5196

Coleman, J. S. (2003). Social capital in the creation of human capital. In A. H. Halsey, H. Lauder, P. Brown, & A. S. Wells (Eds.), *Education: Culture, economy, society* (pp. 80–95). Oxford, UK: University Press.

Collins, J. M. (2013). *The new world champion paper Airplane book*. New York: Crown Publishing.

Columbus, C. *(1492). Journal of the first voyage (Diario del primer viaje) 1492* (B. W. Ife, Trans. & Ed.). Warminster, UK: Aris and Phillips Ltd.

Common Core State Standards Initiative. (2010). *Common Core State Standards for Mathematics (CCSSM)*. Washington, DC: National Governors Association Center for Best Practices and the Council of Chief State School Officers.

Conezio, K., & French, L. (2002). Science in the preschool classroom. *Young Children, 42*(3), 1–5.

Conquergood, D. (1998). Beyond the text: Toward a performative cultural politics. In S. Dailey (Ed.), *The future of performance studies: Visions and revisions* (pp. 25–36). Annandale, VA: National Communication Association.

Cornett, C. E., & Smithrim, K. E. (2001). *The arts as meaning makers: Integrating literature and the arts throughout the curriculum*. Don Mills, Canada: Pearson.

Cottrell, J. (1987). Creative drama in the classroom grades 4-6. Lincolnwood, IL: National Textbook Company.

Covell, K., & Howe, R. B. (2001). Moral education through the 3 Rs: Rights, respect and responsibility. *Journal of Moral Education*, *30*(1), 29–41. doi:10.1080/03057240120033794

Covell, K., McNeil, J., & Howe, R. B. (2009). Reducing teacher burnout by increasing student engagement: A children's rights approach. *School Psychology International*, *30*(3), 282–290. doi:10.1177/0143034309106496

Craven, J. (n.d.). The World Trade Center twin towers, 1973-2001. *About.com*. Retrieved from http://architecture.about.com/od/worldtradecenter/ss/worldtrade.htm#step1

Cristofi, C., & Davis, M. (1991, January). Science through drama. *Education in Science*, (141): 28–29.

Csikszentmihalyi, M. (1997). *Finding flow: The psychology of engagement with everyday life*. New York, NY: Basic Books.

Daines, G. (2017). *NASA's journey to Mars*. Retrieved from https://www.nasa.gov/content/nasas-journey-to-mars

Darwin, C. (1909). *The Origin of Species*. New York: P.F. Collier & Son Company.

Davidson, C. D., & Simms, W. (2017). Science theater as STEAM: A case study of "Save It Now". *e STEAM Journal, 3*(1), 14. Available at http://scholarship.claremont.edu/steam/vol3/iss1/14

Deans, J. (2016). Thinking, feeling and relating: Young children learning through dance. *Australian Journal of Early Childhood*, *41*(3), 46–57. doi:10.1177/183693911604100307

Decoene, J., & DeCock, R. (1996). The children's rights project in the primary school "Devrijdagmarkt" in Bruges. In E. Verhellen (Ed.), *Monitoring children's rights*. The Hague: Martinus Nijhoff.

DeLoache, J. S. (2000). Dual representation and young children's use of scale models. *Child Development*, *71*(2), 329–338. doi:10.1111/1467-8624.00148 PMID:10834468

Dewar, A. (2014). *One minute paper airplanes kit*. Rutland, VT: Tuttle Publishing Company.

Dewey, J. (1938). *Experience and education*. New York, NY: Kappa Delta Pi.

Diaz, D., & King, P. (2007). *Adapting a post-secondary STEM instructional model to K-5 mathematics instruction*. Clemson, SC: Clemson University.

Disney Careers. (2016). *Walt Disney Imagineering*. Retrieved from http://wdi.disneycareers.com/en/default/

Dixon, F. A., Yssel, N., McConnell, J. M., & Hardin, T. (2014). Differentiated instruction, professional development, and teacher efficacy. *Journal for the Education of the Gifted*, *37*(2), 111–127. doi:10.1177/0162353214529042

Dockett, S. (1995). *You make me alive! Developing understanding through play*. Deakin, ACT: Australian Early Childhood Association.

Dorion, K. R. (2009). Science through drama: A multiple case exploration of the characteristics of drama activities used in secondary science lessons. *International Journal of Science Education*, *31*(16), 2247–2270. doi:10.1080/09500690802712699

Dow, C. B. (2010). Young children and movement: The power of creative dance. *Young Children*, *65*(2), 30–35.

Drake, S. (1973). Galileo Gleanings XXII: Galileo's experimental confirmation of horizontal inertia: Unpublished manuscripts. *Isis*, *64*, 291–305. doi:10.1086/351124

Dubson, M., & Loeblein, T. (2010). *Curve Fitting*. Retrieved June 07, 2016, from http://phet.colorado.edu/en/simulation/legacy/curve-fitting

Duckert, C. L., & DeStasio, E. A. (2016). Setting the stage for science communication: Improvisation in an undergraduate life science curriculum. *The Journal of American Drama and Theatre*, *28*(2), 1–18. Retrieved from https://jadtjournal.org/2016/05/27/setting-the-stage-for-science-communication-improvisation-in-an-undergraduate-life-science-curriculum/

Duckworth, A. L., Peterson, C., Matthews, M. D., & Kelly, D. R. (2007). Grit: Perseverance and passion for long-term goals. *Journal of Personality and Social Psychology*, *92*(6), 1087–1101. doi:10.1037/0022-3514.92.6.1087 PMID:17547490

Duckworth, E. (2006). *The having of wonderful ideas: And other essays on teaching and learning* (3rd ed.). New York, NY: Teachers College Press.

Durso, C. (2014). Coffee Cups of The World, A Photo Series of Beautiful Takeaway Cups. Retrieved September 30, 2016, from http://www.foodiggity.com/coffee-cups-of-the-world-a-photo-series-of-beautiful-takeaway-cups/

Duveen, J., & Solomon, J. (1994). The great evolution trial: Use of role-play in the classroom. *Journal of Research in Science Teaching*, *31*(5), 575–582. doi:10.1002/tea.3660310510

Dweck, C. (2006). *Mindset: The new psychology of success*. New York, NY: Ballantine Books.

Eckard, S. (Ed.). (2015). *Yin and Yang in the English Classroom: Teaching with Popular Culture Texts*. Lanham, MD: Rowman & Littlefield.

Educational Theater Association. (2017). *2017 EdTA National Conference – The rise of STEAM. Nashville, TN September 14-20, 2017*. Retrieved from https://www.americansforthearts.org/events/artsmeet/2017-edta-national-conference-the-rise-of-steam-real-world-theatre-education-for-the-21st-century

Ehrenfeld, T. (2011). Reflections on mirror neurons. *Observer, 24*(3). Retrieved from http://www.psychologicalscience.org/index.php/publications/observer/2011/march-11/reflections-on-mirror-neurons.html

Eiffel Tower. (2006). *Engineering.com*. Retrieved from http://www.engineering.com/Library/ArticlesPage/tabid/85/ArticleID/62/Eiffel-tower.aspx

Eisner, E. (1991). *The enlightened eye: Qualitative inquiry and the enhancement of educational practice*. New York: Macmillan.

Eisner, E. (2002). *The arts and the creation of mind*. New Haven, CT: Yale University Press.

Eisner, W. (1978). *A contract with God*. New York: Baronet Books.

Eisner, W. (1985). *Comics & Sequential Art: Principles & Practice of the World's Most Popular Art Form*. Poorhouse Press.

Elert, G. (2018). *The physics hypertextbook*. Retrieved from https://physics.info/projectiles/

EPCOT. (2016). *Theme parks. EPCOT Overview*. Retrieved from http://www.wdwinfo.com/wdwinfo/guides/epcot/ep-overview.htm

Farthest Flight by a Paper Aircraft. (n.d.). Retrieved from http://www.guinnessworldrecords.com/world-records/farthest-flight-by-a-paper-aircraft

Fels, L., & Meyer, K. (1997). On the edge of chaos: Co-evolving world(s) of drama and science. *Teaching Education, 9*(1), 75–81. doi:10.1080/1047621970090113

Finson, K., & Pederson, J. (2011). What are visual data and what utility do they have for science education? *Journal of Visual Literacy, 30*(1), 66–85. doi:10.1080/23796529.2011.11674685

Fisher, D., & Frey, N. (2015). Selecting texts and tasks for content area reading and learning. *The Reading Teacher, 68*(7), 524–529. doi:10.1002/trtr.1344

Fitzgerald, A., & Schneider, K. (2013). What teachers want: Supporting primary school teachers in teaching science. *Teaching Science, 59*(2), 7–14.

Fortus, D., Krajcikb, J., Dershimerb, R., Marxc, R., & Mamlok-Naamand, R. (2005). Design-based science and real-world problem-solving. *International Journal of Science Education, 27*(3), 855–879. doi:10.1080/09500690500038165

Freire, P. (2000). *Pedagogy of the oppressed* (M. B. Ramos, Trans.). New York: Continuum Publishing.

Frommers. (2016). *History of Walt Disney World.* Retrieved from: http://www.frommers.com/destinations/walt-disney-world/683469

Gallant, J. (2002). The shape of the Eiffel tower. *American Journal of Physics*, *70*(2), 160–162. doi:10.1119/1.1417530

Gallese, V. (2005). Embodied simulation: From neurons to phenomenal experience. *Phenomenology and the Cognitive Sciences*, *4*(23), 23–48. doi:10.100711097-005-4737-z

Gardner, H. (2011). *The unschooled mind: How children think and how schools should teach* (2nd ed.). New York, NY: Basic Books.

Gazzaninga, M. S. (1989). Organisation of the human brain. *Science*, *245*(4921), 947–952. doi:10.1126cience.2672334 PMID:2672334

Gerde, H., Schachter, R., & Wasik, B. (2013). Using the scientific method to guide learning: An integrated approach to early childhood education. *Early Childhood Education Journal*, *41*(5), 315–323. doi:10.100710643-013-0579-4

Gess, A. H. (2017). Steam education: Separating fact from fiction. *Technology and Engineering Teacher*, *77*(3), 39–41.

Ghanbari, S. (2015). Learning across disciplines: A collective case study of two university programs that integrate the arts with STEM. *International Journal of Education & the Arts*, *16*(7). Retrieved from http://www.ijea.org/v16n7/

Gomes, L. F. A. M. (2016). *Teaching decision making analytical skills to engineers: A new paradigm.* Retrieved from http://citeseerx.ist.psu.edu/viewdoc/download?doi=10.1.1.201.3517&rep=rep1&type=pdf

Gonick, L., & Criddle, C. (2005). *The Cartoon Guide to Chemistry.* New York: Collins Reference.

Goodnough, K. (2001). Multiple intelligences theory: A framework for personalizing science curricula. *School Science and Mathematics*, *101*(4), 180–193. doi:10.1111/j.1949-8594.2001.tb18021.x

Gopnik, A. (2012). Scientific thinking in young children: Theoretical advances, empirical research, and policy implications. *Science*, *337*(28), 1623–1627. doi:10.1126cience.1223416 PMID:23019643

Grady, S. (2000). *Drama and diversity: A pluralistic perspective for educational drama.* Portsmouth, NH: Heinemann.

Grant, R. G. (2007). *The complete history.* London, UK: Dorling Kindersley New Jersey Department of Education Standards for all curriculum area. Retrieved May 4, 2016 from http://www.state.nj.us/education/educators/standards/

Greene, M. (2001). *Variations on a blue guitar: The Lincoln Center Institute lectures on aesthetic education*. New York, NY: Teachers College Press, Columbia University.

Gregory, G. H., & Chapman, C. (2007). *Differentiated Instructional Strategies: One Size Doesn't Fit All* (2nd ed.). Thousand Oaks, CA: Corwin Press.

Grennon-Brooks, J. (2011). The courage to be constructivist. *Educational Leadership*, *57*(3), 18–24.

Grocholski, E. (2018). *Research reveals boys interests in STEM careers declining; Girls interest unchanged.* Junior Achivement USA News. Retrieved from https://www.juniorachievement.org/web/ja-usa/press-releases/-/asset_publisher/UmcVLQOLGie9/content/research-reveals-boys'-interest-in-stem-careers-declining-girls'-interest-unchanged

Groff, J. S. (2013). Expanding our frames of mind for education and the arts. *Harvard Educational Review*, *83*(1), 15–39. doi:10.17763/haer.83.1.kk34802147665819

Guteral, F., & … . (2010). How we are not learning science. *Scientific American*, 23–30.

Guyotte, K. W., Sochacka, N. W., Costantino, J. W., Walther, J., & Kellam, N. N. (2014). Steam as social practice: Cultivating Creativity in Transdisciplinary Spaces. *Art Education*, *67*(6), 12–19. doi:10.1080/00043125.2014.11519293

Guyotte, K. W., Sochacka, N. W., Costantino, T. E., Kellam, N. N., & Walther, J. (2015). Collaborative creativity in STEAM: Narratives of art education students' experiences in transdisciplinary spaces. *International Journal of Education & the Arts*, *16*(15). Retrieved from http://www.ijea.org/v16n15/

Halverson, E. (2016). *The art of improvisation in teaching and learning*. Paper presented at the American Educational Research Association Conference, Washington, DC.

Hammer, D. (1997). Discovery learning and discovery teaching. *Cognition and Instruction*, *15*(4), 485–529. doi:10.12071532690xci1504_2

Hargreaves, H. (2016). *Coffee Cups of the World (@coffeecupsoftheworld) • Instagram photos and videos*. Retrieved September 30, 2016, from https://www.instagram.com/coffeecupsoftheworld/

Haring, D., & Kelner, T. (2015). Why we got serious about interdisciplinary teaching. *Educational Leadership*, *73*(4), 68.

Hashweh, M. Z. (1987). Effects of subject-matter knowledge in the teaching of biology and physics. *Teaching and Teacher Education*, *3*(2), 109–120. doi:10.1016/0742-051X(87)90012-6

Hassett, D. D., & Scheible, M. B. (2010). Finding space and time for the visual in K-12 literacy instruction. In M. Cappello & B. Moss (Eds.), *Contemporary Readings in Literacy Education* (pp. 325–331). Thousand Oaks, CA: SAGE Publications.

Heathcoate, D., & Bolton, G. (1995). *Drama for learning: Dorothy Heathcote's mantle of the expert approach to education*. Portsmouth: Heinemann.

Heathcote, D. (1991). *Collected writings on education and drama*. Evanston, IL: Northwestern University Press.

Heaverlo, C. (2011). *STEM development: A study of 6th-12th grade girls' interest and confidence in mathematics and science*. Graduate Theses and Dissertations. Paper 10056. Retrieved from http://lib.dr.iastate.edu/cgi/viewcontent.cgi?article=1035&context=etd

Hendrix, R., Charles, E., & Shannon, D. (2012). The Integration of Creative Drama in an Inquiry-Based Elementary Program: The Effect on Student Attitude and Conceptual Learning. *Journal of Science Teacher Education*, *23*(7), 823–846. doi:10.100710972-012-9292-1

Henriksen, D. (2014). Full STEAM ahead: Creativity in excellent STEM teaching practices. *The STEAM Journal*, *1*(2), 15. doi:10.5642team.20140102.15

Hetland, L., Winner, E., Veenema, S., & Sheridan, K. (2013). *Studio thinking 2: The real benefits of visual arts education.* Academic Press.

Hetland, L., Winner, E., Veenema, S., & Sheridan, K. (2013). *Studio thinking 2: The real benefits of visual arts education.* New York, NY: Teachers College Press.

History of Catapults - Physics of Catapults. (n.d.). Retrieved January 17, 2016 from https://sites.google.com/site/physicsofcatapults/home/history-of-catapults

HMI. (1989). Curriculum matters 17: Drama from 5-16. London: HMSO.

Hodson, D., & Reid, D. (1988). Changing priorities in science education: Part I. *School Science Review*, *70*, 101-108.

Hoffman, J. L., Collins, M. F., & Schickedanz, J. A. (2015). Instructional challenges in developing young childrens science concepts. *The Reading Teacher*, *68*(5), 363–372. doi:10.1002/trtr.1325

Hohenadel, K. (2014). *Traveling the World in Search of the Most Unique Disposable Coffee Cup Designs (Photos).* Retrieved September 30, 2016, from http://www.slate.com/blogs/the_eye/2014/07/03/photographer_henry_hargreaves_coffee_cups_of_the_world_highlights_the_best.html

Howcast. (2013). *How to Demonstrate Insulation Principles.* Retrieved June 07, 2016, from https://www.youtube.com/watch?v=8I96BzSNNqE

Howe, R. B., & Covell, K. (2007). *Empowering children: Children's rights education as a pathway to citizenship*. Toronto: University of Toronto Press.

Ingraham, C. (2014). *America's top fears: Public speaking, heights and bugs.* Retrieved from https://www.washingtonpost.com/news/wonk/wp/2014/10/30/clowns-are-twice-as-scary-to-democrats-as-they-are-to-republicans/?noredirect=on&utm_term=.d7b20652a417

Inoa, R., Weltzek, G., & Tabone, C. (2014). A Study on the Relationship Between Theater Arts and Student Literacy and Mathematics Achievement. *Journal for Learning Through the Arts*, *10*(1), 1–19. doi:10.21977/D910123495

Inspiring Breakthrough. (2005). *Visual spatial intelligence*. Retrieved from http:// www.inspiring-breakthrough.co.uk/learning-styles/visual-spatial-learning.html

Insulated Bottle. (2006). Retrieved June 09, 2016, from http://www.madehow.com/Volume-4/Insulated-Bottle.html

International Society for Technology in Education. (2007). *National educational technology standards for students*. Retrieved June 7, 2016, from http://www.iste.org/standards/iste-standards/standards-for-students

International Society for Technology in Education. (2016). *ISTE standards for students: A practical guide for learning with technology*. Eugene, OR: Author.

International Technology Education Association. (2007). *Standards for technological literacy: Content for the study of technology* (3rd ed.). Retrieved from https://www. iteea.org/File.aspx?id=67767&v=b26b7852

Ivanitskaya, L., Clark, D., & Montgomery, G. (2002). Article. *Innovative Higher Education*, *27*(95). doi:10.1023/A:1021105309984

Jablon, P. (1989). *A Documentation and impact analysis of a project-oriented, interdisciplinary, team-taught, peer-influence program for secondary schools and its effects upon science learning* (Doctoral Dissertation). New York University. (Dissertation Abstracts 9016410)

Jablon, P. (2010). *Analogous interrelationship role-plays: Engaging the previously disengaged in science*. Paper presented at The Association of Science Teacher Educators International Conference, Sacramento, CA.

Jablon, P. (2013). *Get real! Walking the walk to inform talking the talk. Full-time teaching in an urban high school. In Science teacher educators as K-12 science teachers: Practicing what we teach*. New York: Springer.

Jablon, P. (2014). *The synergy of inquiry: Engaging students in deep learning across the content areas*. Huntington Beach, CA: Shell Publications.

Jablon, P., & Born, T. (1988). STS works in the city. *SSTS Reporter.*, *4*(2), 5–8.

Jacobs, H. H. (1989). The growing need for interdisciplinary curriculum content. In H. H. Jacobs (Ed.), *Interdisciplinary Curriculum: Design and Implementation* (pp. 1–11). Alexandria, VA: ASCD.

Johnson, J. (1999). The forum on early childhood science, mathematics, and technology education. In Dialogue on Early Childhood Science, Mathematics, and Technology Education. Washington, DC: AAAS.

Jolly, A. (2014). STEM vs. STEAM: Do the arts belong? *Education Week Teacher*. Retrieved February 21, 2016, from http://www.edweek.org/tm/articles/2014/11/18/ctq-jolly-stem-vssteam.html

Jolly, A. (2014). STEM vs. STEAM: Do the arts belong? *Education Week Teacher*. Retrieved June 30, 2016, from http://www.edweek.org/tm/articles/2014/11/18/ctq-jolly-stem-vs-steam.html

Jolly, A. (2014). *STEM vs. STEAM: Do the Arts Belong?* Retrieved from http://www.edweek.org/tm/articles/2014/11/18/ctq-jolly-stem-vs-steam.html

Jones, G. (2004). *Men of tomorrow: Geeks, gangsters and the birth of the comic book*. New York: Basic Books.

Kafai, Y. B. (2006). Constructionism. In K. Sawyer (Ed.), *Cambridge handbook of the learning sciences* (pp. 35–46). Cambridge, MA: Cambridge University Press.

Kakalios, J. (2005). *The Physics of Superheroes*. New York: Gotham Books.

Kamen, M. (1991, August). *Creative drama and the enhancement of elementary school students' understanding of science concepts* (Doctoral Dissertation). The University of Texas at Austin.

Kamen, M. (1991). Use of creative drama to evaluate elementary school students. In G. Kulm & S. Malcom (Eds.), *Science assessment in the service of reform*. Washington, DC: AAAS.

Kase-Polisini, J., & Spector, B. (1992). Improvised drama: A tool for teaching science. *Youth Theatre Journal*, *7*(1), 15–19.

Keller, M., & Fuller, N. R. (2009). *Charles Darwin's On the Origin of Species: A Graphic Interpretation*. Rodale.

Kelley, J. A. (Producer & Director), & Wyanski, J. (Producer & Director). (1987). *Project lead the way - Where it started 1987 Dick Blais* [video]. Retrieved from http://youtu.be/iIdomsEhmYM

Kelly, G. J., Carlsen, W. S., & Cunningham, C. M. (1993). Science education in sociocultural context: Perspectives from the sociology of science. *Science Education*, *77*(2), 207–220. doi:10.1002ce.3730770208

Kelner, L. B. (1993). *The creative classroom: A guide for using creative drama in the classroom, PreK-6*. Portsmouth, NH: Heinemann.

Kennedy, T. J., & Odell, M. R. L. (2014). Engaging students in STEM education. *Science Education International*, *25*(3), 246–258.

Kentish, B. (1995). Hypotheticals: Deepening the understanding of environmental issues through ownership of learning. *Australian Science Teachers Journal*, *41*(1), 21–25.

Keruskin, T. (2015). Finding the sweet spot: Creativity, candy, and commerce. *Edutopia*. Retrieved June 28, 2016, from http://www.edutopia.org/blog/sweet-creativity-candy-commerce-todd-keruskin

Kesar, S. (2018). *Closing the STEM gap: Why STEM classes and careers still lack girls and what we can do about it*. Digital Skills/Microsoft. Retrieved from https://query.prod.cms.rt.microsoft.com/cms/api/am/binary/RE1UMWz

Khan, S. (2016). *Linear vs. exponential growth.* Retrieved June 07, 2016, from https://www.khanacademy.org/math/algebra/introduction-to-exponential-functions/comparing-exponential-and-polynomial-functions/v/linear-exponential-models

Kinsella, K. (1996). Designing group work that supports and enhances diverse classroom work styles. *TESOL Journal*, *6*(1), 24–31.

Koester, M. (1989). *Science and art together again* (Unpublished master's thesis). University of Hawaii Manoa.

Koester, M. (2011). What's your story? How can we bring students to science through reading and the arts? Keynote address delivered at the South Carolina Science Council meeting, Myrtle Beach, SC.

Koester, M. (2014*). Project Draw for Science.* Retrieved from http://merriekoester.wix.com/project-dfs

Koester, M. (2017). Deep learning in science: An argument for an aesthetic paradigm of science education. In F. V. Tochon & D. Busciglio (Eds.), Deep education in the disciplines. Blue Mounds, WI: Deep University Press.

Koester, M., & Simmons, S. (n.d.). *The collaborative art of teaching science through drawing* [PowerPoint Slides]. NAEA. Retrieved from http://naeaworkspace. org/naea16/The%20 Collaborative%20Art%20of%20Teaching%20Science%20 through%20Drawing.pdf

Koester, M. (2015). *Science teachers who draw: The red is always there.* Blue Mounds, WI: Deep University Press.

Koester, M. (2015b). *Agnes pflumm and the stonecreek science fair* (3rd ed.). Charleston, SC: Read for Science Publishing.

Koester, M. (2015c). *No place like periwinkle* (3rd ed.). Charleston, SC: Read for Science Publishing.

Koester, M. (2015d). *Pond scum and agnes pflumm* (3rd ed.). Charleston, SC: Read for Science Publishing.

Koester, M. (2015e). *Agnes pflumm and the secret of the seven* (2nd ed.). Charleston, SC: Read for Science Publishing.

Koester, M. (in press). Deep learning in science: An argument for an aesthetic paradigm of science education. In F. V. Tochon & D. Busciglio (Eds.), *Deep Education Across the Discipline sand Beyond: A 21st Century Transdisciplinary Breakthrough.* Blue Mounds, WI: Deep University Press.

Koester, M. (in press). Getting to "know" STEAM. In M. Van Sickle & J. Bazler (Eds.), *Cases in STEAM education in practice*. Hershey, PA: IGI-Global. doi:10.4018/978-1-5225-2334-5.ch004

Konie, R. (2011). *A brief overview of laban movement analysis*. Retrieved April 27, 2016 from http://www.movementhasmeaning.com/wp-content/uploads/2010/09/LMA-Workshop-Sheet.pdf

Lakoff, G., & Johnson, M. (2003). *Metaphors we live by*. Chicago, IL: University of Chicago Press. doi:10.7208/chicago/9780226470993.001.0001

Leaning tower of Pisa. (n.d.). In *Wikipedia*. Retrieved from https://en.wikipedia.org/wiki/Leaning_tower_of_Pisa

Lederman, N. (2007). Nature of science: Past, present, and future. In S. Abell & N. Lederman (Eds.), *Handbook of research on science education*. Abingdon, UK: Taylor and Francis.

Lederman, N. G. (1992). Students' and teachers' conceptions of the nature of science: A review of the research. *Journal of Research in Science Teaching*, *29*(4), 331–359. doi:10.1002/tea.3660290404

Lederman, N. G., & Lederman, J. S. (2015). What is a theoretical framework? A practical answer. *Journal of Science Teacher Education*, *26*(7), 593–597. doi:10.100710972-015-9443-2

Lee, O. (2003). Equity for culturally and linguistically diverse students in science education: Recommendations for a research agenda. *Teachers College Record*, *105*(3), 465–489. doi:10.1111/1467-9620.00247

Lee, O., Fradd, S. H., & Sutman, F. X. (1995). Science knowledge and cognitive strategy use among culturally and linguistically diverse students. *Journal of Research in Science Teaching*, *32*(8), 797–816. doi:10.1002/tea.3660320804

Lemke, J. (2002). Enseñar todos los lenguajes de la ciencia: palabras, símbolos, imágenes y acciones. In M. Benlloch (Ed.), *La educación en ciencias: ideas para mejorar su práctica* (pp. 159–186). Barcelona, Spain: Paidos.

Lemke, J. L. (1999). Typological and Topological Meaning in Diagnostic Discourse. *Discourse Processes*, *27*(2), 173–185. doi:10.1080/01638539909545057

Litts, B. (2015). *Making learning: Makerspaces as learning environments* (Doctoral dissertation). Retrieved from www.informalscience.org/sites/.../Litts_2015_Dissertation_Published.pdf

Llewellyn, D. (2002). *Inquire within* (2nd ed.). Thousand Oaks, CA: Corwin Press.

Llewellyn, D. (2004). *Teaching high school science through inquiry: A Case study approach*. Corwin Press.

Logan, B. (2011). Examining differentiated instruction: Teachers respond. *Research in Higher Education*, 13.

Loucks-Horsley, S. (1996). Principles of effective professional development for mathematics and science education: A synthesis of standards. *NISE Brief, 1*(1). Retrieved from http://eric.ed.gov/?id=ED409201

Maeda, J. (2013). Artists and scientists: More alike than different [Web log comment]. Retrieved from http://blogs.scientificamerican.com/guest-blog/artists- and-scientists-more-alike-than-different/

Mallett, M. (2010). Choosing and using fiction and non-fiction 3-11: A comprehensive guide for teachers and student teachers. New York, NY: Routledge.

Mandigo, J., Francis, N., Lodewyk, K., & Lopez, R. (2009). *Position paper: Physical literacy for educators.* St. Catherines, ON, Canada: Brock University. Retrieved May 10, 2016 from https://www.phecanada.ca/sites/default/files/pl_position_paper.pdf

Mardirosian, G. H., & Lewis, Y. P. (2009, March 6). How to use theater to teach at-risk students. *The Chronicle of Higher Education, 55*(26), 34–37.

Marginson, S., Tytler, R., Freeman, B., & Roberts, K. (2013). *STEM: Country comparisons.* Melbourne, Australia: Australian Council of Learned Academies.

Martin, D. J. (2011). *Elementary science methods: A constructivist approach* (6th ed.). Belmont, CA: Wadsworth Publishing.

Maslow, A. H. (1998). *Toward a psychology of being* (3rd ed.). New York, NY: Van Nostrand.

May, R. (1985). *The courage to create.* Toronto, Canada: Bantam.

McCaslin, N. (1981). *Children and drama* (2nd ed.). New York: University Press.

McCaslin, N. (1996). *Creative drama in the classroom and beyond.* London: Longman Publishers.

McCloud, S. (1993). *Understanding comics.* Tundra Publications.

McGregor, D., & Precious, W. (2015). Dramatic science: Inspired ideas for teaching science using drama ages 5–11. New York: Routledge.

McMullin, L. (2003). How to solve any triangle: First, forget the law of sines and the law of cosines. *Mathematics Teacher, 96*(6), 448–449.

Merriam-Webster. (n.d.). *Definition of text.* Retrieved Oct 4, 2016 from http://www. merriam-webster.com/dictionary/text

Metcalfe, R. J. A., Abbott, S., Bray, P., Exley, J., & Wisnia, D. (1984). Teaching science through drama: An empirical investigation. *Research in Science & Technological Education, 2*(1), 77–81. doi:10.1080/0263514840020109

Miller, C., & Saxton, J. (2004). *Into the story: Language in action through drama.* Portsmouth, NH: Heinemann.

Mirth, J., & Findley, A. (2015). Enduring design: Examining the relationship between art, engineering, and creativity. *Proceedings from ASEE '15: 2015 ASEE Zone III Conference.* Retrieved from https://www.asee.org/documents/zones/zone3/2015/Enduring-Design-Examining-the-Relationship-Between-Art-Engineering-and- Creativity.pdf

Mission: Space Overview. (2016). *EPCOT-Mission Space.* Retrieved from: http://www.wdwinfo.com/wdwinfo/guides/epcot/epfw-mission.htm

Morrison, G., & Quitely, F. (2011). *All Star Superman*. New York: DC Comics.

Morrison, J., & McDuffie, A. R. (2009). Connecting science and mathematics: Using inquiry investigations to learn about data collection, analysis, and display. *School Science and Mathematics*, *109*(1), 31–44. doi:10.1111/j.1949-8594.2009.tb17860.x

Mumpower, D. (2016). *It was the most controversial Disney ride ever built. Here's why*. Retrieved from https://www.themeparktourist.com/features/20160407/31867/sordid-history-misson-space-sickest-ride-epcot

Nasa's Open Data Portal. (n.d.). Retrieved from https://data.nasa.gov/

National Center for Educational Statistics. (n.d.). *Create a graph*. Retrieved from https://nces.ed.gov/nceskids/createagraph/

National Coalition for Core Arts. (2014). *National core arts standards*. Retrieved June 7, 2016, from http://www.nationalartsstandards.org/

National Governors Association Center for Best Practices & Council of Chief State School Officers. (2010). *Common Core State Standards for Mathematics*. Washington, DC: Author.

National Oceanic and Atmospheric Association. (2014). Lesson 9: Invent a robot. In *The NOAA ship Okeanos explorer education materials collection, volume 2: How do we explore* (pp. 125-136). Retrieved from http://oceanexplorer.noaa.gov/okeanos/edu/collection/media/vol2_hdwe.pdf

National Research Council (NRC). (2012). *A framework for K–12 science education: Practices, crosscutting concepts, and core ideas*. Washington, DC: National Academies Press.

National Research Council. (1996). *National science education standards*. National Academies Press.

National Research Council. (2000). Retrieved from http://www.nationalacademies.org/nrc/

National Research Council. (2012). *A framework for K-12 science education: Practices, crosscutting concepts, and core ideas. Committee on a Conceptual Framework for New K-12 Science Education Standards. Board on Science Education, Division of Behavioral and Social Sciences and Education*. Washington, DC: National Academies Press.

National Science Foundation. (2006). *The art of engineering* [press release]. Retrievedfromhttp://www.nsf.gov/news/news_summ.jsp?cntn_id=107990

Newcombe, N. (2010). Picture this: Increasing math and science learning by improving spatial thinking. *American Educator*.

Newmann, F. (1981). Reducing student alienation in high schools: Implications of theory. *Harvard Educational Review*, *51*(4), 546–564. doi:10.17763/haer.51.4.xj67887u87l5t66t

Newton's Law of Cooling. (2015). Retrieved June 07, 2016, from https://www.khanacademy.org/math/differential-equations/first-order-differential-equations/modeling-with-differential-equations/v/newtons-law-of-cooling

NGSS Lead States. (2013). *Next generation science standards: For states, by states*. Washington, DC: The National Academies Press.

NGSS Lead States. (2013). *Next Generation Science Standards: For States, By States*. Washington, DC: The National Academies Press.

NOAA Office of Ocean Exploration. (2016). *Ocean Explorer*. Retrieved from http://oceanexplorer.noaa.gov/

Noddings, N. (2003). *Caring: A feminine approach to ethics and moral education*. Berkeley, CA: University of California Press.

Noddings, N. (2005). *The challenge to care in schools: An alternative approach to education* (2nd ed.). New York, NY: Teachers College Press.

NSTA Position Statements. (2018). Retrieved from https://bit.ly/2TiXolx

O'Neill, C. (1995). *Drama worlds: A Framework for process drama*. Portsmouth: Heinemann.

Oakes, J. (2005). *Keeping track: How schools structure inequality*. New Haven, CT: Yale University.

Ocean Literacy Framework. (2005-20015). Retrieved from http://oceanliteracy.wp2.coexploration.org/

Ocean Literacy Network. (2013). *Ocean literacy principle #7*. Retrieved June 30, 2016, from http://oceanliteracy.wp2.coexploration.org/ocean-literacy-framework/principle-7-v2/

Ødegaard, M. (2003). Dramatic science. A critical review of drama in science education. *Studies in Science Education*, *39*(1), 75–101. doi:10.1080/03057260308560196

Office Applications. (n.d.). *Writer*. Retrieved from https://writer.bighugelabs.com/login

Olness, R. (2007). *Using literature to enhance content area instruction: A guide for K-5 teachers*. Newark, DE: International Reading Association.

Onwukwe, E., Akuezuilo, E., & Ijioma, B. (2014). Science drama: Clue to underachievement in secondary school chemistry *STEMplanet*, *11*. Retrieved from http://stemstates.org/journal/stemplanet-2014/2014-vol-11.html

Osborne, J., Simon, S., & Collins, S. (2003). Attitudes towards science: A review of the literature and its implications. *International Journal of Science Education*, *25*(9), 1049–1079. doi:10.1080/0950069032000032199

Ottaviani, J., Cannon, Z., & Cannon, K. (2009). *T-Minus: The Race to the Moon*. New York: Aladdin.

Oxford, R. L., Hollaway, M. E., & Horton-Murillo, D. (1992). Language learning styles: Research and practical considerations for teaching the multicultural tertiary ESL/EFL classroom. *System*, *20*(4), 439–456. doi:10.1016/0346-251X(92)90057-A

Pagnucci, G. S. (2004). *Living the narrative life: Stories as a tool for meaning making*. Portsmouth: Heinemann.

Papacosta, P. (2006). Humanities & Arts to the Rescue of Science. *Forum on Public Policy.* Retrieved from http://forumonpublicpolicy.com/archive07/papacosta.pdf

Papert, S., & Harel, I. (1991). Situating constructionism. In I. Harel & S. Papert (Eds.), *Constructionism* (pp. 1–11). Norwood, NJ: Ablex.

Park, G., Lubinski, D., & Benbow, C. (2010). Recognizing spatial intelligence. *Scientific American.* Retrieved from http://www.scientificamerican. com/article/recognizing-spatial-Intel/

Park, M. Y. (2014). *A Brief History of the Disposable Coffee Cup*. Retrieved June 09, 2016, from http://www.bonappetit.com/entertaining-style/trends-news/article/disposable-coffee-cup-history

Parsons, S. A., Dodman, S. L., & Burrowbridge, S. C. (2013). Broadening the view of differentiated instruction. *Phi Delta Kappan*, *95*(1), 38–42. doi:10.1177/003172171309500107

Patrick, H. (2014). *Engaging young children with informational books*. Thousand Oaks, CA: Corwin. doi:10.4135/9781452284903

Pedings-Behling, K. E. (2015). *How Do We Keep it Hot?* Hartsville, SC: South Carolina Governor's School for Science and Mathematics.

Peleg, R., & Baram-Tsabari, A. (2011). Atom Surprise: Using Theatre in Primary Science Education, Special Issue on Science Education in Preschools and Primary Schools: Classrooms, Teachers, and Children. *Journal of Science Education and Technology*, *20*(5), 508–524. doi:10.100710956-011-9299-y

Pestalozzi, J. H. (1898). How Gertrude teaches her children (2nd ed.). Syracuse, NY: Bardeen.

Pestalozzi, J. (1973). *How Gertrude teaches her children.* New York, NY: Gordon Press. (Original work published 1894)

Piaget, J. (1955). *The construction of the reality in the child.* Basic Books.

Pieczura, P. (2013). Decidedly dramatic! The power of creative drama in social studies. *Social Studies and the Young Learner*, *25*(3), 9–12.

Pliskow, T. (2008). *Reaching new heights in middle school science: Straw tower competition builds*. Academic Press.

Polanyi, M. (1966). *The tacit dimension*. New York, NY: Doubleday.

Post Illuminance. (n.d.). *Custom graph: Isometric paper*. Retrieved from http:// customgraph. com/piart.php?art=590

Pratap, T. (2015). What were the engineering challenges in building the Burj Kahalifa? [Msg 3]. Message posted to https://www.quora.com/What-were-the-engineering-challenges-in-building-the-Burj-Khalifa

Project Lead the Way. (2016). *Engaging students from New York to Honolulu*. Retrieved from https://www.pltw.org/about-us

Pulkkinen, L. (1982). Self-control and continuity from childhood to adolescence. In Life-span development and behavior (vol. 4). New York, NY: Academic Press.

Ratey, J. J., & Hagerman, E. (Collaborator). (2008). Spark: The revolutionary new science of exercise and the brain. New York, NY: Little, Brown and Co.

Reiff, J. (1992). *What research says to the teacher: Learning styles*. Washington, DC: National Education Association.

Reilly, E. (2015). Superheroes in Math Class: Using Comics to Teach Diversity Awareness. *Works & Days*, *32*(1 & 2), 61–74.

Rhode, M. (2013). *The Sketchnote handbook: The illustrated guide to visual notetaking*. Berkeley, CA: Peachpit Press.

Rinne, L., Gregory, E., Yarmolinskaya, J., & Hardiman, M. (2011). Why arts integration improves long-term retention of content. *Mind, Brain and Education: the Official Journal of the International Mind, Brain, and Education Society*, *5*(2), 89–96. doi:10.1111/j.1751-228X.2011.01114.x

Roberts, D. (2012). *Fatal invention: How science, politics, and big business re-create race in the twenty-first century*. New York: The New Press.

Rock, M. L., Gregg, M. G., Ellis, E., & Gable, R. A. (2008, Winter). REACH: A framework for differentiating classroom instruction. *Preventing School Failure*, *52*(2), 31–47. doi:10.3200/PSFL.52.2.31-47

Romagnoli, A. (2015). Up, Up, and Away: Superman in the Composition Classroom. In S. Eckard (Ed.), *Yin and Yang in the English Classroom: Teaching with Popular Culture Texts* (pp. 123–135). Lanham, MD: Rowman & Littlefield.

Romagnoli, A., & Pagnucci, G. (2013). *Enter the Superheroes: American Values, Culture, and the Canon of Superhero Literature*. Lanham, MD: Scarecrow Press.

Ross-Nazzal, J. (2004). *NASA Johnson space center oral history project*. Johnson Space Center History Collection. Retrieved from http://www.jsc.nasa.gov/history/oral_histories/BlackburnGC/BlackburnGC_5-4-09.htm

Salisbury, B. T. (1996). *Theatre arts in the elementary classroom* (2nd ed.). New Orleans, LA: Anchorage Press, Inc.

Sartre, J.-P. (1976). *Anti-Semite and Jew* (G. J. Becker, Trans.). New York, NY: Schocken. (Original work published 1948)

Satchwell, R., & Loepp, F. L. (2002). Designing and Implementing an Integrated Mathematics, Science, and Technology Curriculum for the Middle School. *Journal of Industrial Teacher Education*. Retrieved February 14, 2014 from http://scholar.lib.vt.edu/ejournals/JITE/v39n3/satchwell.html

Settlage, J., & Southerland, S. A. (2007). *Teaching science to every child: Using culture as a starting point*. Abingdon, OX: Taylor & Francis.

Shapell, B., & Willen, S. (2012). *The Wright brothers: Inventing the first successful airplane*. Retrieved from http://www.shapell.org/collectionfilters/the-wright-brothers

Sharapan, H. (2012). From STEM to STEAM: How early childhood educators can apply Fred Rogers' approach. *Young Children*, *67*(2), 36–40.

Shaywitz, S. (2003). *Overcoming dyslexia: A new and complete science-based program for reading problems at any level*. New York, NY: Vintage Books.

Shulman, L. S. (1987). Knowledge and teaching: Foundations of the new reform. *Harvard Educational Review*, *57*(1), 1–23. doi:10.17763/haer.57.1.j463w79r56455411

Siegesmund, R. (2010). Aesthetics as a curriculum of care and responsible choice. In T. Constantino & B. White (Eds.), *Essays on aesthetic education for the 21st century* (pp. 81–92). Rotterdam: Sense Publishers.

Smilansky, S. (1968). *The effects of sociodramatic play on disadvantaged preschool children*. John Wiley.

Smilansky, S. (1990). Sociodramatic play: Its relevance to behavior and achievement in school. In E. Klugman & S. Smilansky (Eds.), *Children's play and learning: Perspectives and policy implications* (pp. 18–42). Teacher College Press.

Snow, C. (2015). Rigor and realism: Doing educational science in the real world. 2014 Wallace Foundation distinguished lecture. *Educational Researcher*, *44*(9), 460–466. doi:10.3102/0013189X15619166

Snow, K., & Van Hamel, C. (2008). *Early childhood assessment: Why, what, and how*. The National Research Council.

Sousa, D. A., & Pilecki, T. (2013). *From STEM to STEAM: Using brain-compatible strategies to integrate the arts*. Thousand Oaks, CA: Corwin Publishing Company.

Spitzer, H. (2013). Introduction of interdisciplinary teaching: Two case studies: Commentary on Teaching Science, Technology, and Society to Engineering Students: A Sixteen Year Journey. *Science and Engineering Ethics*, *19*(4), 1451–1454. doi:10.100711948-013-9475-y PMID:24085355

Steinert, Y. (1993). Twelve tips for using role-playing in clinical teaching. *Medical Teacher, 15*(4), 283–291. doi:10.3109/01421599309006651 PMID:8139402

Stencel, J., & Barkoff, A. (1993). Protein Synthesis: Role-Playing in the Classroom. *The American Biology Teacher, 55*(2), 102–103. doi:10.2307/4449595

Strong, R., Silver, H., & Perini, M. (2001). *Teaching what matters most: Standards and strategies for raising student achievement.* Association for Supervision and Curriculum Development.

Stroupe, D., & Kramer, A. (2014, October). Students Modeling Molecule Movement Through Science Theater. *Science Scope, 38*(2), 70–77. doi:10.2505/4s14_038_02_70

Sturm, B. (2009). The drama of science. *Science in Schools, 13*, 29-33. Retrieved from http://www.scienceinschool.org/sites/default/files/teaserPdf/issue13_drama.pdf

Tatalovic, M. (2009). Science comics as tools for science education and communication: A brief, exploratory study. *JCOM: Journal of Science Communication, 8*(4), 1–17. doi:10.22323/2.08040202

TeacherGeek. (2008). *Fluid power lab.* June 30, 2016, from http://teachergeek.org/ fluid_power.pdf

The Common Core State Standards Initiative. (2018). Retrieved September 1, 2016 from http://www.corestandards.org/ELA-Literacy/introduction/key-design- consideration/

The Science of Insulation Explained. (2019). Retrieved January 05, 2019, from https://www.knaufinsulation.com.au/homeowners/energy-efficiency/science-of-insulation-explained

The White House. (2014). *Fact Sheet: The economic challenge posed by declining pollinator Populations.* Retrieved from: https://obamawhitehouse.archives.gov/the-press-office/2014/06/20/fact-sheet-economic-challenge-posed-declining-pollinator-populations

Thompson, J. (2012). *Applied theatre: Bewilderment and beyond* (4th ed.). Oxford, UK: Peter Lang; doi:10.3726/978-3-0353-0309-4

Tochon, F. V. (2013). *Educational semiotics: Signs and symbols in education.* Blue Mounds, WI: Deep University Press.

Tomlinson, C. A. (2005). *How to differentiate instruction in mixed-ability classrooms* (2nd ed.). Upper Saddle River, NJ: Pearson.

Trefil, J. (2008). *Why Science?* New York: Teachers College Press.

Tseng, T., & Yang, M. (2011). The role of spatial-visual skills in a project-based engineering design course. *Proceedings from ASEE 2011 Annual Conference & Exposition.* Retrieved from http://web.mit.edu/~mcyang/www/papers/2011- tsengYang-ASEE.pdf

Tu, T. (2006). Preschool science environments: What is available in an early childhood classroom. *Early Childhood Research Journal, 33*(4), 245 – 251.

Tuovinen, J. E., & Sweller, J. (1999). A comparison of cognitive load associated with discovery learning and worked examples. *Journal of Educational Psychology, 91*(2), 334–341. doi:10.1037/0022-0663.91.2.334

Uhrmacher, P. B. (2010). The power to transform: Implementation as aesthetic awakening. In T. Constantion & B. White (Eds.), *Essays on aesthetic education for the 21st century* (pp. 183–203). Rotterdam: Sense Publishers.

Uhrmacher, P. B., Conrad, B., & Lindquist, C. (2010). The power to transform: Implementation as aesthetic awakening. In T. Constantino & B. White (Eds.), *Essays on aesthetic education for the 21st century* (pp. 183–203). Rotterdam: Sense Publishers.

United Nations Educational, Scientific and Cultural Organization. (2003). *Literacy, a UNESCO perspective.* Retrieved May 10, 2016 from http://unesdoc.unesco. org/images/0013/001318/131817eo.pdf

United Nations. (2002). *56/116. United Nations Literacy Decade: education for all. Resolution adopted by the General Assembly, Fifty-Sixth Session.* Retrieved May 10, 2009 from http://www.undemocracy.com/A-RES-56-116.pdf

United States Department of Education. (2006). *The Condition of education.* Washington, DC: National Center for Education Statistics, U.S. Government Printing Office.

Vaidyanathan, S. (2012). Fostering creativity and innovation through technology. *Learning and Leading with Technology, 39*(6), 24–27.

Van Sickle, M., & Spector, B. (1996). Caring relationships in science classrooms: A symbolic interaction study. *Journal of Research in Science Teaching, 33*(4), 433–453. doi:10.1002/(SICI)1098-2736(199604)33:4<433::AID-TEA5>3.0.CO;2-T

Varelas, M., Pappas, C.C., Tucker-Raymond, E., Kane, J., Hankes, J., Ortiz, I., & Keblawe-Shama, N. (2010). Drama activities as ideational resources for primary-grade children in urban science classes. *Journal of Research in Science Teaching, 47*(3), 302–325.

Venkateswaran, T. V. (2008). *Communicating science through plays, games and simulations.* New Delhi: Scifun.

Venville, G., Rennie, L., & Wallace, J. (2004). Decision making and sources of knowledge: How students tackle integrated tasks in science, technology and mathematics. *Research in Science Education, 34*(2), 115–135. doi:10.1023/B:RISE.0000033762.75329.9b

Voyager. (n.d.). Retrieved from http://voyager.jpl.nasa.gov/mission/fastfacts.html

Wagner, B. J. (1979). *Dorothy Heathcote: Drama as a learning medium.* London: Hutchinson Education.

Watts-Taffe, S., Laster, B. P., Broach, L., Marinak, B., Connor, C. M., & Walker-Dalhouse, D. (2012). Differentiated instruction: Making informed teacher decisions. *The Reading Teacher, 66*(4), 303–314. doi:10.1002/TRTR.01126

Weld, J. D. (2000). Fly Guys. *Teacher Magazine, 11*(8), 56–58.

Westby, C. (2010). Multiliteracies: The changing world of communication. *Topics in Language Disorders, 30*(1), 64–71. doi:10.1097/TLD.0b013e3181d0a0ab

White, H. P. (2010). *STEAM not STEM whitepaper: An agreement on what drives the US economy in the future*. Retrieved from http://steam-notstem.com

Wicklein, R. C. (2006). Five good reasons for engineering as the focus for technology education. *Technology Teacher, 65*(7), 25–29.

Willis Tower. (n.d.). Retrieved from http://www.willistower.com

Wilson, M. (2016). *The ancient history of projectiles*. Retrieved from resources21.org/cl/files// project492.../Ancient%20History%20of%20Projectiles.pptx

Wright, P. (2016). *The unmixed question: the spatial music of Charles Ives*. American Mavericks. Retrieved from http://musicmavericks.publicradio.org/features/ feature_unmixed_question.html

Yager, R. E. (1993). The constructivist learning model. *Science Teacher (Normal, Ill.), 60*, 27–31.

Yakman, G. (2010). *What is the point of STE@M? A brief overview*. Retrieved from https:// annegilesclelland.typepad.com/files/whatissteam.pdf

Yopp, R. H., & Yopp, H. K. (2012). Young childrens limited and narrow exposure to informational text. *The Reading Teacher, 65*(7), 480–490. doi:10.1002/TRTR.01072

Zygomatic. (n.d.). *Chart Tool*. Retrieved from http://www.onlinecharttool.com/

About the Contributors

Judith Bazler earned her EdD with a focus on Science Education and Curriculum from University of Montana. Having taught in secondary, and university settings in Missoula, Montana and at Lehigh University in Bethlehem, Pennsylvania, where she founded a hands-on science museum, she is now a full professor at Monmouth University in New Jersey, mainly in charge of the elementary and secondary science education programs, Here major research interest includes equity in the science classrooms, informal science, technology in the classroom and global education. her work has been published in nation and international journal such as Journal of Research in Science Teaching, Journal of Science for Person with Disabilities, Journal of Science Teacher Education, Proteus, Journal of Career Development, Journal of General Education, The Clearing House, The American Biology Teacher, The Science Teacher, Technology and Teacher Education Annual, School Science and Mathematics, GASAT (Women and Science and Technology), Science Scope, Journal of Career Assessment, The Informal Sciences Review and World speaker. She also has written three books for Greenwood Press and authored and edited numerous reference books for Facts On File, Inc and co-authored two chapters in Cases on Instructional Technology in Gifted and Talented Education.

Meta Van Sickle, PhD, is a Full Professor at the College of Charleston. A teacher with over 20 years experience, Meta earned her Ph.D. in Science Education at the University of South Florida. One of her two recent IGI Global publications is about teacher education candidates who are gifted and talented. The first book chapter reviews the history of gifted and talented teaching through the science and mathematics education policies and practices. The second book chapter is a cross analysis of gifted and talented teacher education candidates. The most recent of her IGI publications is as an editor of Identifying, Describing and Developing Gifted and Talented Teachers.

Warren DiBiase is the middle grades/secondary school science educator at UNC Charlotte. He has 23 years working in higher education. In addition, Dr. DiBiase has 20 years experience as a high school science teacher.

Callie (Van Koughnett) Dollahon received her BS in Marine Science from the University of South Carolina Honors College in 2007 and MEd in Science and Math for Teachers from the College of Charleston in 2013. In 2010 she moved to Copenhagen, Denmark, as a Fulbright Scholar, studying the integration of national science standards in Danish teaching. She has taught Honors and CP high school courses in Anatomy, Biology, Biomedical Science, and Marine Science, and Earth Science. She is currently the Earth Science team leader at Wando High School, where she has taught Honors and CP high school courses in Anatomy, Biology, Biomedical Science, and Marine Science, and Earth Science.

Catherine Duckett, PhD, is a professor at Monmouth University.

Letitia Graybill taught science and mathematics in public schools for many years before coming to Monmouth University. She also supervised science and mathematics K-12 teachers in these schools. Her extensive experience at both the public school and university levels informs her ability to assess the effectiveness of teachers at all levels. Dr. Graybill was instrumental in developing the Partners in Science Program for gifted science students in New Jersey. This was part of the New Jersey Liberty Science Center outreach to gifted high school science students whereby these identified students were placed in research laboratories throughout New Jersey to observe and participate in real world science research.

John Henning is Dean of the School of Education at Monmouth University in West Long Branch, New Jersey. Previously, he served as the Associate Dean in The Patton College and the Chair of the Teacher Education Department at Ohio University. Dr. Henning has more than 50 publications; including three books, over thirty refereed journal articles, and seven book chapters. His current research interests include teacher development, clinically-based teacher education, and classroom discourse.

Paul Jablon, Ph.D., taught in the New York City public schools for 19 years where he and his colleagues created classrooms, programs, and school that utilized inquiry teaching and interdisciplinary, project-based learning to engage those students previously disengaged from school. As a faculty member at Brooklyn College, the University of Massachusetts, and Lesley University, he had the opportunity to work for over two decades with both preservice and practicing teachers. Leading systemic-change projects in urban and suburban school districts, he has collaborated with thousands of teachers to understand how inquiry teaching is best implemented with a diverse set of students from kindergarten through college.

Martha James-Hassan, EdD earned her undergraduate degree from the University of Wisconsin-Madison in Kinesiology, her Master's from Hamline University in St. Paul, Minnesota, and her Doctorate in Critical Pedagogy from the University of St. Thomas in Minneapolis, Minnesota. She is a graduate from the AASA/Howard University Urban Superintendents Academy and the Saint Paul Public Schools Leadership Academy. She is researcher, teacher, and leader with nearly 25 years of experience in urban public education. Dr. Hassan is currently an Assistant Professor at Morgan State University in Baltimore, Maryland. Her research agenda involves investigating in-service teacher professional development, cultural fluency, and interdisciplinary education. Dr. Hassan serves her community as a Commissioner on the Baltimore City Board of School Commissioners.

Merrie Koester, Ph.D., author of the nationally implemented Agnes Pflumm science education novels, is the director of Project Draw for Science action research initiative at the University of South Carolina Center for Science Education. Her book, Science Teachers Who Draw: The Red Is Always There, is the narrative of teacher researchers who learned to teach science through drawing and visualization in order to engage previously failing students-primarily struggling readers and/or special needs learners. As a Science and arts integration specialist, Merrie serves on the advisory council of the Innovation Collaborative.

Nathan Lindquist, PhD., received his M.S in Physics and his Ph.D. in Electrical Engineering from the University of Minnesota. In addition to teaching courses on computer programming, classical mechanics, electricity and magnetism, and materials physics, his research interests include interdisciplinary nanotechnology and super-resolution imaging. In his labs at Bethel University, he tinkers with lasers, microscopes and molecules. His research with undergraduate students has received major grant support from the National Science Foundation including a Faulty Early Career Development (CAREER) award.

William Lindquist, PhD, received his MA in Curriculum and Instruction: Learning Technology from the University of St. Thomas in Saint Paul, MN and PhD in Curriculum and Instruction: Elementary Science Education from the University of Minnesota. Dr. Lindquist came to Hamline University where he serves as an Associate Professor after a 25 year career in public elementary settings. He teaches "Teaching Science in the Elementary School" and is chair of the School of Education. His research interests include inquiry-based science, science and literacy, and science notebooks.

Kathryn Lubniewski, Ed.D., earned a doctorate from West Virginia University in special education. Her B.S. in multi-disciplinary studies and M.A. in elementary education were also from West Virginia University. She acquired additional certification in multi-categorical special education and early childhood education. She has taught inclusion for reading and mathematics in grades K, 1, and 2, and has taught in a self-contained setting for grades 9-12. Dr. Lubniewski worked as an Assistant/Associate Professor at Elmhurst College for 6 years. She specializes in collaboration, differentiated instruction, and teacher training. Dr. Lubniewski is a member of the Council for Exceptional Children. She served as an executive board member of Esperanza Community Services Center and also as an executive board of the Teacher Education Division for the Council for Exceptional Children. Her current scholarship focuses on The Collaborative Student Teacher Training Model, differentiated instruction, and how iPads impact student engagement.

Judith McDonald has been in science education all her professional career. She is an associate professor at Belmont Abbey College in the Department of Education. She taught in the middle grades and elementary classroom for many years prior to teaching pre-service teachers at the college level. Her research interests are inclusion in the STEM classroom.

Tracy Mulvaney, Ed.D., has over 20 years in K-12 and higher education. She obtained her BA and MS in Special Education from the University of Arizona and her Ed.D in Educational Leadership from Northern Arizona University. She spent her K-12 career as a cross categorical resource teacher, Principal of a day school for students with severe emotional disabilities, Director of Alternative Education, and an educational advocate for youth in foster care. Dr. Mulvaney was the Department Chair/Faculty Lead at Central Arizona College where she also served as a Faculty Senate Vice President and President. In 2016, Dr. Mulvaney was honored to receive the position of Assistant Dean of the School of Education at Monmouth University where she oversees accreditation, SOE advising, clinical placement, school partnerships and edTPA.

Kathryn Pedings-Behling has both bachelor's and master's degrees in mathematics with a secondary teaching certificate. During the first part of her career, Kathryn was in the high school classroom teaching and holding several leadership roles. Kathryn then moved in to a curriculum development position and was responsible for running summer camps across the state of South Carolina. She is now an adjunct faculty instructor at the College of Charleston in the mathematics department.

Kerry Carley Rizzuto has worked as a classroom teacher in general and special education, a Reading Specialist for grades K-12, a Reading Recovery teacher, a staff developer, and school administrator in both New York and New Jersey. In addition to her teaching position at Monmouth University, Dr. Rizzuto has completed the Learning Disabilities Teacher-Consultant program at Monmouth University and earned her doctorate from Rowan University in educational leadership, with a focus on early childhood. Dr. Rizzuto is the Program Director for the Early Childhood program at Monmouth University. In addition to developing science education in early childhood classrooms, her research interests include developing social justice dispositions in pre-service teachers and developing culturally responsive pedagogical practices.

Kyle Seiverd is a science educator in central New Jersey. His passions are wildlife conservation and elephants. He believes that traveling can serve as a gateway to life-long learning. His role models are his mother and Dr. Jane Goodall.

Kellan Strong assists the CTL staff with the student response system, Poll Everywhere, and with Canvas training and support. She is a doctoral student in the Curriculum & Instructional Literacy program at UNC Charlotte. She has 9 years of experience in K-12 teaching, literacy research, and instructional design. Kellan holds a Bachelor of Arts degree in English from Johnson C. Smith University, a Masters of Arts in African American Studies from University of Wisconsin-Madison, and a Masters of Arts in Education from Gardner-Webb University.

Jena Valdiviezo is the Science Director of Grades 6 through 12 at the Long Branch School District in New Jersey.

Vecihi S. Zambak completed his M.S. in Mathematics and Science Education at the University of Amsterdam and his Ph.D. in Curriculum and Instruction at Clemson University, both with a concentration in mathematics education. Both his Master and Ph.D. programs provided him several teaching and research opportunities in various school and cultural settings. Those experiences included university supervision and teaching instructional technology and mathematical content courses for elementary and secondary mathematics majors. Before his current position at Monmouth University, Dr. Zambak was a postdoctoral researcher in the Department of Mathematics, Statistics and Computer Science at Marquette University. In his postdoctoral appointment, he engaged in research focusing on the quality of K-8 preservice teachers' mathematical argumentation and professional noticing skills.

Index

A

B

C

D

N

O

P

R

S

T

www.ingramcontent.com/pod-product-compliance
Lightning Source LLC
LaVergne TN
LVHW080606300926
849162LV00005B/1062
* 9 7 8 1 5 2 2 5 9 6 3 1 8 *